Combinatorial Computational Biology of RNA

Combinatorial Computational Biology of RNA

Pseudoknots and Neutral Networks

Christian Reidys

Nankai University
Tianjin, China

 Springer

Christian Reidys
Research Center for Combinatorics
Nankai University
Tianjin 300071, China
reidys@nankai.edu.cn

ISBN 978-0-387-76730-7   e-ISBN 978-0-387-76731-4
DOI 10.1007/978-0-387-76731-4
Springer New York Dordrecht Heidelberg London

Library of Congress Control Number: 2010937101

Mathematics Subject Classification (2011): 05-02, 05E10, 05C80, 92-02, 05A15, 05A16

Printed on acid-free paper

Springer is part of Springer Science+Business Media (www.springer.com)

Preface

The lack of real contact between mathematics and biology is either a tragedy, a scandal or a challenge, it is hard to decide which.

Gian-Carlo Rota, *Discrete Thoughts.*

This book presents the discrete mathematics of RNA pseudoknot structures and their corresponding neutral networks. These structures generalize the extensively studied RNA secondary structures in a natural way by allowing for cross-serial bonds. RNA pseudoknot structures require a completely novel approach which is systematically developed here. After providing the necessary context and background, we give an in-depth combinatorial and probabilistic analysis of these structures, including their uniform generation. We furthermore touch their generation by present the ab initio folding algorithm, `cross`, freely available at `www.combinatorics.cn/cbpc/cross.html`. Finally, we analyze the properties of neutral networks of RNA pseudoknot structures.

We do not intend to give a complete picture about the state of the theory in RNA folding or computational biology in general. Three decades after the seminal work of Michael Waterman great advances have been made the representation of which is beyond the scope of this book. Instead, we focus on integrating a variety of rather new concepts and ideas, some – if not most – of which originated from pure mathematics and are spread over more than fifty research papers. This book gives graduate students and researchers alike the opportunity to understand in depth the theory of RNA pseudoknot structures and their neutral networks.

The book adopts the perspective that mathematical biology is both mathematics *and* biology in its own right and does not reduce mathematical biology to applying "mathematical" tools to biological problems. Point in case is the reflection principle – a cornerstone for computing the generating function of RNA pseudoknot structures. The reflection principle represents a method facilitating the enumeration of a non-inductive combinatorial class. Its very formulation requires basic understanding of group actions in general and the Weyl group, in particular, none of which are standard curriculum in mathematical biology graduate courses. In the following the reader will find all details on how to derive the generating function of pseudoknot RNA structures via k-noncrossing matchings from the reflection principle. We systematically develop the theoretical framework and prove our results via symbolic enumeration, which reflects the modularity of RNA molecules.

The book is written for researchers and graduate students who are interested in computational biology, RNA structures, and mathematics. The goal is to systematically develop a language facilitating the understanding of the basic mechanisms of evolutionary optimization and neutral evolution. This book establishes that genotype–phenotype maps into RNA pseudoknot structures exhibit a plethora of structures with vast neutral networks.

This book is centered around the work of my group at Nankai University from 2007 until 2009. The idea for the construction of k-noncrossing structures comes from the paper of Chen et al. [25], where a bijection between k-noncrossing partitions and lattice paths is presented. Our first results were Theorem 4.13 [76] and Problem 4.3 [77]. Shortly after, we studied canonical structures via cores [78] (Lemma 4.3) and derived a precursor of Theorem 4.9. A further milestone is the uniform generation of k-noncrossing structures presented in Chapter 5 [26] connecting combinatorics and probability theory. Only later we realized the modularity of RNA structures; see [108]. The central result on the structure of neutral networks is Theorem 7.11 due to [105].

I owe special thanks to Andreas Dress, Gian-Carlo Rota, and Michael Waterman. They influenced my perspectives and their research provided the basis for the material presented in this book.

Thanks belong to Peter Stadler, with whom I had the privilege of collaborating for many years. I also want to thank Victor Moll and Markus Nebel for their helpful comments. This book could not have been written without the help of my students. In particular I am grateful to Fenix W.D. Huang, Jing Qin, Rita R. Wang, and Yangyang Zhao. Finally, I wish to thank Vaishali Damle, Julie Park, and the Springer Verlag for all their help in preparing this book.

Tianjin, China, October 2010,   Christian Reidys

Contents

1

Introduction

Almost three decades ago Michael Waterman pioneered the combinatorics and prediction of the ribonucleic acid (RNA) secondary structures, a rather non-mainstream research field at the time. What is RNA? On the one hand, an RNA molecule is described by its primary sequence, a linear string composed of the nucleotides **A**, **G**, **U** and **C**. On the other hand, RNA, structurally less constrained than its chemical relative DNA, does fold into tertiary structures.

RNA plays a central role within living cells facilitating a whole variety of biochemical tasks, all of which are closely connected to its tertiary structure. As for the formation of this tertiary structure, it is believed that this is a hierarchical process [18, 133]. Certain structural elements fold on a microsecond timescale affecting the assembly of the global fold of the molecule. RNA acts as a messenger linking DNA and proteins and furthermore catalyzes reactions just as proteins. Consequently, RNA embodies both genotypic legislative and phenotypic executive.

The discovery that RNA combines features of proteins and DNA led to the "RNA world" hypothesis for the origin of life. It states that DNA and the much more versatile proteins took over RNA's functions in the transition from the "RNA world" to the present one.

Around 1990 Peter Schuster and his coworkers studied the RNA world in the context of evolutionary optimization and neutral evolution. This line of work identified the genotype–phenotype map from RNA sequences into RNA secondary structures and its role for the evolution of populations of erroneously replicating RNA strings.

Recent discoveries suggest that RNA might not just be a stepping stone toward a DNA–protein world exhibiting "just" a few catalytic functions. Large numbers of very small RNAs of about 22 nucleotides in length, called microRNAs (miRNAs), were identified. They were found in organisms as diverse as the worm *Caenorhabditis elegans* and *Homo sapiens* exhibiting important regulatory functions.

These novel RNA functionalities motivated to have a closer look at RNA structures. An increasing number of experimental findings as well as results

C. Reidys, *Combinatorial Computational Biology of RNA*,
DOI 10.1007/978-0-387-76731-4_1,
© Springer Science+Business Media, LLC 2011

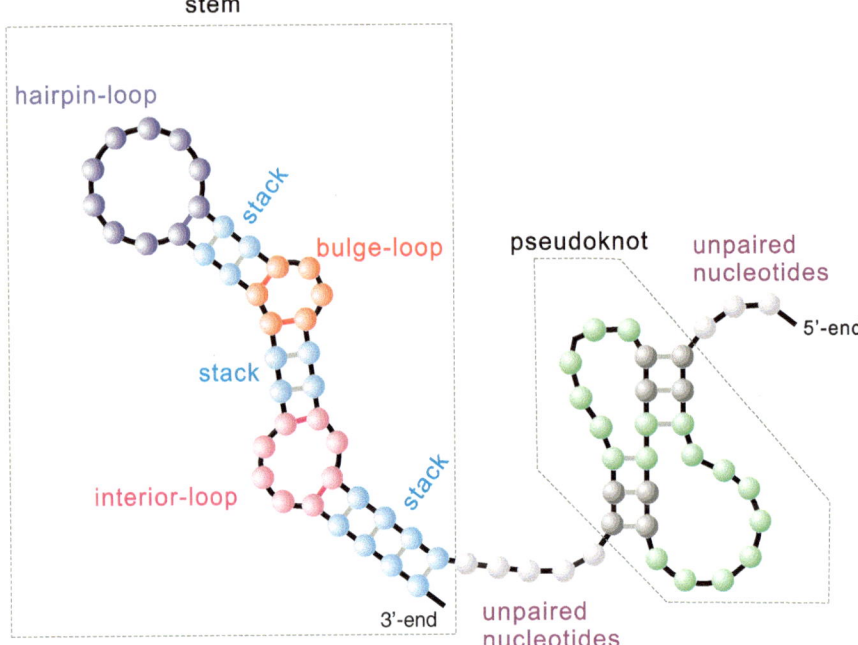

Fig. 1.1. Pseudoknot structures: structural elements and cross-serial interactions (*green*). For details on loops in RNA pseudoknot structures, see Chapter 6.

from comparative sequence analyses imply that there exist additional, cross-serial types of interactions among RNA nucleotides [145]; see Fig. 1.1. These are called pseudoknots and are functionally important in tRNAs, RNAseP [86], telomerase RNA [128], and ribosomal RNAs [84]. Pseudoknots are abundant in nature: in plant virus RNAs they mimic tRNA structures, and in vitro selection experiments have produced pseudoknotted RNA families that bind to the HIV-1 reverse transcriptase [136]. Important general mechanisms such as ribosomal frameshifting are dependent upon pseudoknots [23]. They are conserved in the catalytic core of group I introns. As a result, RNA pseudoknot structures have drawn a lot of attention [119], over the last years.

Despite their biological importance pseudoknots are typically excluded from large-scale computational studies as it is still unknown how to derive them reliably from their primary sequences. Although the problem has attracted considerable attention over the last decade, and several software tools [91, 109, 111, 130] have become available, the required resources have remained prohibitive for applications beyond individual molecules. The problem is that the prediction of general RNA pseudoknot structures is NP-complete [87]. To make matters worse, the algorithmic difficulties are confounded by the fact that the thermodynamics of pseudoknots is poorly understood.

In the literature, oftentimes some variant of the dynamic programming (DP) paradigm is used [111], where certain subclasses of pseudoknots are considered. In Chapter 6 we discuss that the DP paradigm is ideally suited for an inductive, or context-free, structure class. However, due to the existence of cross-serial bonds, RNA pseudoknot structures cannot be recursively generated. Consequently, the DP paradigm is only of limited applicability. Besides these conceptual shortcomings, DP-based approaches are oftentimes not even particularly time efficient. Therefore, staying within the DP paradigm, it is unlikely that folding algorithms can be substantially improved.

Here we introduce the mathematical framework for a completely different view on pseudoknotted structures, that is not based on recursive decomposition, i.e., parsing with respect to (some extension of) context-free grammars (CFG).

1.1 RNA secondary structures

Let us begin by discussing RNA secondary structures. An RNA secondary structure [82, 97, 143] is a contact structure, identified with a set of Watson–Crick (**A-U**, **G-C**), and (**U-G**), base pairs without considering any notion of spatial embedding. In other words, a secondary structure is a graph over n nucleotides whose arcs are the base pairs; see Fig. 1.2. One important feature of the secondary structure is that the energies involved in its formation are large compared to those of tertiary contacts [43]. Our first objective will be to introduce the most commonly used representations.

The first representation interprets a secondary structure as a diagram: a labeled graph over the vertex set $[n] = \{1, \ldots, n\}$ with vertex degrees ≤ 1, represented by drawing its vertices $1, \ldots, n$ in a horizontal line and its arcs

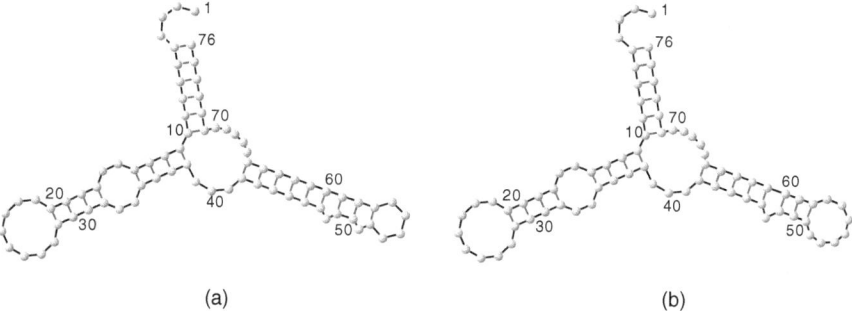

(a)   (b)

Fig. 1.2. The phenylalanine tRNA secondary structure: (**a**) the structure of phenylalanine tRNA, as folded by the loop-based DP-routine ViennaRNA [67, 68]. (**b**) Phenylalanine structure as folded by the loop-based folding algorithm `cross`, see Chapter 6. Due to the fact that `cross` does not consider stacks, i.e., arc sequences of the form $(i, j), (i-1, j+1), \ldots, (i-\ell, j+\ell)$ of length smaller than 3, (**b**) differs from (**a**) with respect to the sequence segment between nucleotides 48 and 60.

(i, j), where $i < j$, in the upper half plane. Obviously, vertices and arcs correspond to nucleotides and Watson–Crick (**A-U**, **G-C**) and (**U-G**) base pairs, respectively. With foresight we categorize diagrams via the maximum number of mutually crossing arcs, $k-1$, the minimum arc length, λ, and the minimum stack length, σ. Here, the length of an arc (i, j) is $j - i$ and a stack of size σ is a sequence of "parallel" arcs of the form

$$((i, j), (i + 1, j - 1), \ldots, (i + (\sigma - 1), j - (\sigma - 1)));$$

see Fig. 1.3. We call a diagram with at most $k - 1$ mutually crossing arcs a k-noncrossing diagram and an arc of length λ is called a λ-arc.

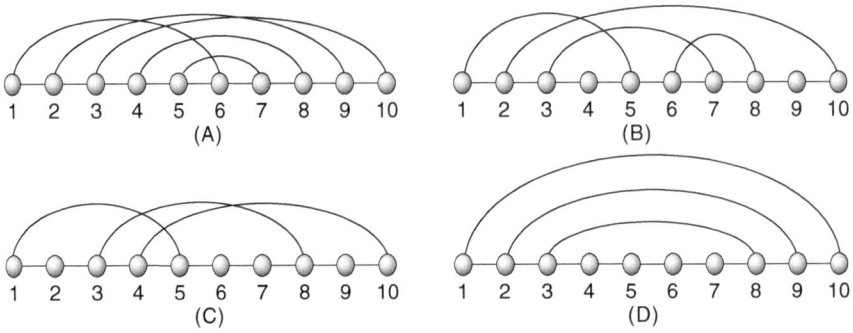

Fig. 1.3. Diagrams: the horizontal line corresponds to the primary sequence of backbone of the RNA molecule, the arcs in the upper half plane represent the nucleotide interactions.

A k-noncrossing, σ-canonical structure is a diagram in which

- there exist no k-mutually crossing arcs
- any stack has at least size σ, see Fig. 1.3 (D), and
- any arc (i, j) has a minimum arc length $j - i \geq 2$.

In the language of k-noncrossing structures, RNA secondary structures are simply *noncrossing* structures.[1] We remark that diagrams have a "raison d'étre" as purely combinatorial objects [27] besides offering a very intuitive representation of k-noncrossing structures.

A second interpretation of secondary structures is that of certain Motzkin paths. A Motzkin path is a path composed by *up*, *down*, and *horizontal* steps. The path starts at the origin and stays in the upper half plane and ends on the x-axis. We shall see that Motzkin paths are not "abstract nonsense", they are well suited to understand the genuine inductiveness of RNA secondary structures. It is easy to see that any RNA secondary structure corresponds uniquely to a peak-free Motzkin path, i.e., a path in which an up-step is

---

[1] That is without arcs of the form $(i, i + 1)$ (also referred to as 1-arcs).

not immediately followed by a down-step. This correspondence is derived as follows: each vertex of the diagram is either an origin or terminus of an arc (i, j) or isolated (unpaired). Mapping each origin into an up-$((1, 1))$, each terminus into a down-$((1, -1))$ and each isolated vertex into an horizontal-$((1, 0))$ step encodes the diagram uniquely into a Motzkin-path. Clearly, the minimum arc length ≥ 2 translates into the peak-freeness. Given a peakfree Motzkin-path it is clear how to recover its associated diagram, see Fig. 1.4. One equivalent presentation is the *point-bracket* notation where we write each up-step as "(", each down-step as ")" and each horizontal step as "$\bullet$".

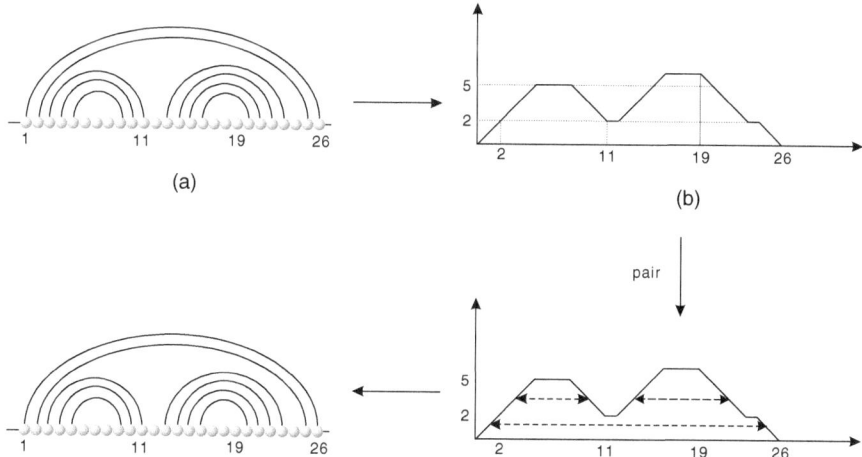

Fig. 1.4. From noncrossing diagrams to Motzkin paths and back. Origins correspond to up-, termini to down- and isolated vertices to horizontal steps, respectively. Labeling the up- and down-steps and subsequent pairing allows to uniquely recover the base pairings as well as the unpaired nucleotides.

Third we may draw a secondary structure as a planar graph. This graph can be viewed as a result of the "folding" of the primary sequence of nucleotides such that pairing nucleotides come close and chemically interact. This interpretation is particularly suggestive when decomposing a structure into loops, an important concept which arises in the context of free energy of RNA structures. This representation, however, is not canonical at all.

In Fig. 1.5 we summarize all three representations of RNA secondary structures.

One first question about RNA secondary structures is how to enumerate them. This means, given $[n] = \{1, \ldots, n\}$ in how many different ways can one draw noncrossing arcs with arc length ≥ 2 over $[n]$? Let $\mathsf{T}_2^{[\lambda]}(n)$ denote the number of RNA secondary structures with arc length $\geq \lambda$ over $[n]$. According to Waterman [142] we have the following recursion:

$$\mathsf{T}_2^{[\lambda]}(n) = \mathsf{T}_2^{[\lambda]}(n-1) + \sum_{j=0}^{n-(\lambda+1)} \mathsf{T}_2^{[\lambda]}(n-2-j)\mathsf{T}_2^{[\lambda]}(j), \qquad (1.1)$$

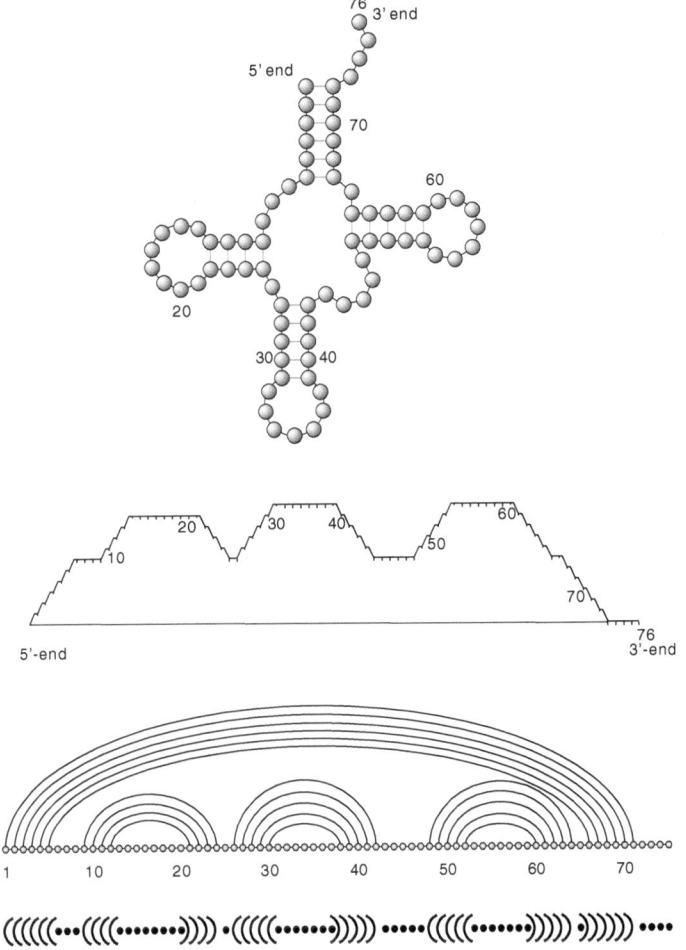

Fig. 1.5. RNA secondary structures: as (outer)-planar graphs, Motzkin path, diagram, and abstract word over the alphabet "(," ")," and "•".

where $\mathsf{T}_2^{[\lambda]}(n) = 1$ for $0 \leq n \leq \lambda$. Equation (1.1) becomes evident when employing the Motzkin-path interpretation of secondary structures. Since each Motzkin path starts and ends on the x-axis, the concatenation of any two Motzkin paths is again a Motzkin path. Indeed, Motzkin paths form an associate monoid with respect to path concatenation. In light of this eq. (1.1) has the following interpretation: a Motzkin path with n-steps starts either with a horizontal step or with an up-step, otherwise. In the latter case there must be a down-step after which one has again a Motzkin path with j-steps. If one shifts down the "elevated" path (i.e., right *after* the up-step and *before* the down-step), one observes that this is again a Motzkin path with $(n-2-j)$ steps; see Fig. 1.6. Since there is always the path consisting only of horizontal

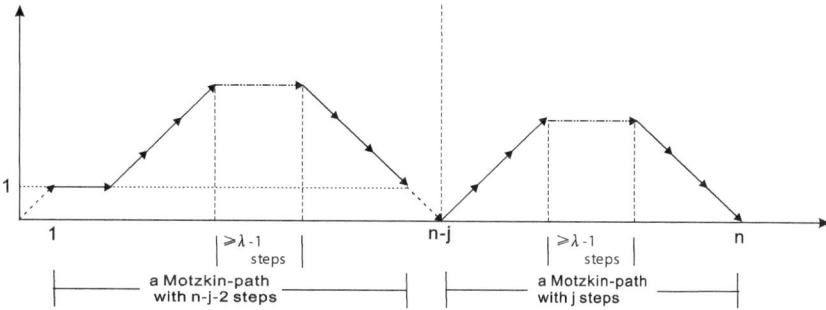

Fig. 1.6. Equation (1.1) interpreted via Motzkin paths.

steps, this path can only be nontrivial for $n - 2 - j \geq \lambda - 1$ steps. It would otherwise produce an arc of length $< \lambda$, which is impossible.

Combinatorialists now evoke an – at first view – abstract object, called the generating function. In our case this generating function reads

$$\mathbf{T}_2^{[\lambda]}(z) = \sum_{n \geq 0} \mathsf{T}_2^{[\lambda]}(n)\, z^n,$$

i.e., a formal power series, whose coefficients are exactly the number of RNA secondary structures for all n. While skepticism is in order whether this leads to deeper understanding, multiplying eq. (1.1) by z^n for all $n > \lambda$ and subsequent calculation imply for the generating function $\mathbf{T}_2^{[\lambda]}(z)$ the simple functional equation

$$z^2\, \mathbf{T}_2^{[\lambda]}(z)^2 - (1 - z + z^2 + \cdots + z^\lambda)\mathbf{T}_2^{[\lambda]}(z) + 1 = 0.$$

Thus we derive a quadratic equation for $\mathbf{T}_2^{[\lambda]}(z)$! Computer algebra systems like MAPLE immediately give the explicit solution. Therefore the "complicated" object $\mathbf{T}_2^{[\lambda]}(z)$, containing the information about all numbers of RNA secondary structures, is easily seen to be a square root – for some a convincing argument for the usefulness of the concept of generating functions.

In fact we want more: ideally we would like to obtain simple formulas for $\mathsf{T}_2^{[\lambda]}(n)$, for large n, for instance, $n = 100$ or 200, say. Not surprisingly, the answer to such formulas lies again in the generating function $\mathbf{T}_2^{[\lambda]}(z)$. We have learned in complex analysis that power series have a radius of convergence, i.e., there exists some real number $r \geq 0$ (possibly zero!) such that $\mathbf{T}_2^{[\lambda]}(z)$ is holomorphic for $|z| < r$. Therefore singular points can only arise for $|z| \geq r$. A classic theorem of Pfringsheim [134] now asserts that if the coefficients of this power series are positive (as it is the case for enumerative generating functions), then r itself is a singular point. We shall show in Section 2.3 that it is the behavior of the power series $\mathbf{T}_2^{[\lambda]}(z)$ close to this singularity that determines the asymptotics of its coefficients. Again the generating function is the key for deriving the asymptotics.

1.2 RNA pseudoknot structures

RNA pseudoknot structures [119, 145] are structures which exhibit cross-ing arcs in the diagram representation discussed in the previous section. We observe that we are not interested in the total number of crossings, but the maximal number of *mutually* crossing arcs. In Fig. 1.7 we display a 4- and a 3-noncrossing diagram and highlight the particular 3- and 2-crossings, respectively.

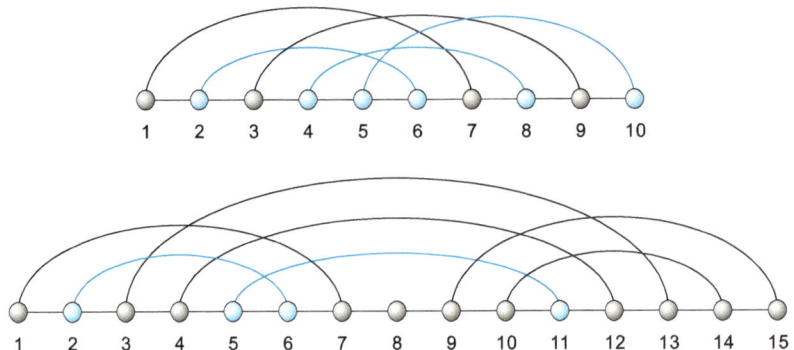

Fig. 1.7. k-noncrossing diagrams: we display a 4-noncrossing, arc length $\lambda \geq 4$ and $\sigma \geq 1$ (*upper*) and 3-noncrossing, $\lambda \geq 4$ and $\sigma \geq 2$ (*lower*) diagram. In both diagrams we highlight one particular 3- and 2-crossings (*blue*).

We stipulate that it is intuitive to consider pseudoknot structures with low crossing numbers as less complex. Point in case are Stadler's bisecondary structures [63], intuitively obtained by drawing a one secondary structure in the upper half-plane and another in the lower half plane such that each vertex has degree at most 1. The bisecondary structure is then derived by "flipping" the arcs contained the lower half plane "up".

It is not difficult to see that bisecondary structures are exactly the planar 3-noncrossing RNA structures. At present time, bisecondary structures are still a combinatorial mystery: no generating function is known. According to Stadler [63] most natural RNA structures exhibit low crossing numbers. How-ever, relatively high numbers of pairwise crossing bonds are also observed in natural RNA structures, for instance, the *gag-pro* ribosomal frameshift signal of the simian retrovirus-1 [131], which is a 10-noncrossing RNA structural motif; see Fig. 1.8.

As for the combinatorics of RNA pseudoknot structures, Stadler and Haslinger [63] suggested a classification of their knot types based on a notion of inconsistency graphs and gave an upper bound for bisecondary structures. What constitutes the main difficulty here is the lack of an inductive recurrence relation, as, for instance, eq. (1.1).

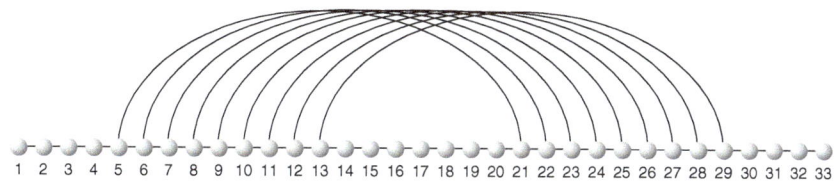

Fig. 1.8. The proposed SRV-1 frameshift [131]: A 10-noncrossing RNA structure motif.

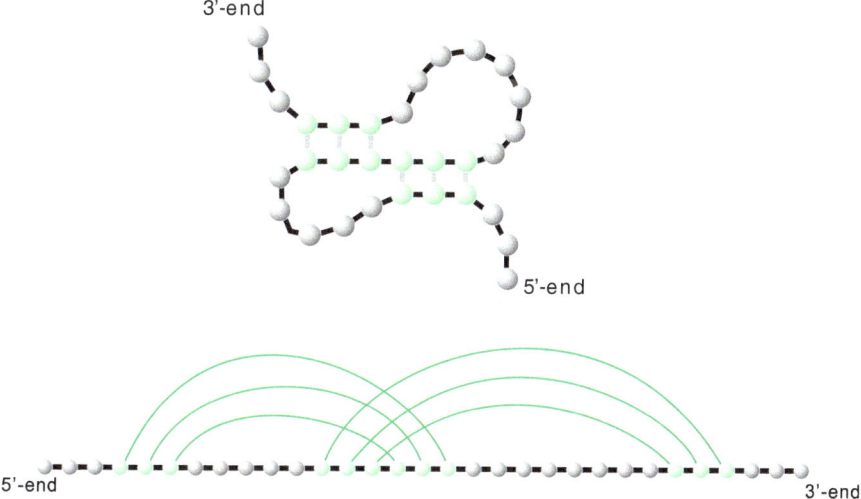

Fig. 1.9. Cross-serial dependencies in k-noncrossing RNA pseudoknot structures. We display a 3-noncrossing structure as planar graph (*top*) and as diagram (*bottom*).

The inherent non-inductiveness of pseudoknot structures, see Fig. 1.9, requires a suite of new ideas developed in Chapter 4. In the course of our study we will discover that k-noncrossing structures share several features of utmost importance with secondary structures:

- As for RNA secondary structures, k-noncrossing structures have a unique loop decomposition, see Proposition 6.2. This result forms the basis for any minimum free energy-based folding algorithm of k-noncrossing structures. For details see Chapter 6. In Fig. 1.10 we give an overview on the different loop types in k-noncrossing structures.
- Their generating functions are D-finite, i.e., their *numbers* satisfy a recursion of finite length with polynomial coefficients, see Theorem 2.13 and Corollary 2.14.

The D-finiteness of the generating function of k-noncrossing structures implies simple asymptotic expressions for the numbers of k-noncrossing and k-noncrossing, canonical structures; see Propositions 4.14 and 4.16. Further-

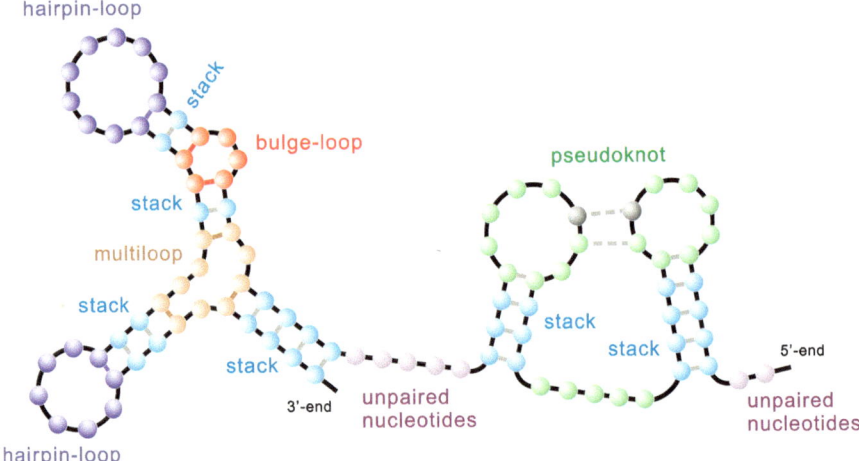

Fig. 1.10. Loop types in k-noncrossing structures; see Chapter 6 for details.

more it facilitates the uniform generation of k-noncrossing structures after $O(n^{k+1})$ preprocessing time in linear time; see Theorem 5.4.

1.3 Sequence to structure maps

The results presented here have been derived in the context of studying the evolution of RNA sequences.

The combinatorics developed in the following chapters has profound implications for the latter. Combined with random graph theory [105, 106] it guarantees the existence of neutral networks and nontrivial sequence to structure maps into RNA pseudoknot structures. To be precise, the induced subgraph of set of sequences, which fold into a particular k-noncrossing pseudoknot structure, exhibits an unique giant component and is exponential in size. Furthermore, for any sequence to structure map into pseudoknot structures, there exist exponentially many distinct k-noncrossing structures.

While the statements about neutral networks of RNA pseudoknot structures are new, neutral networks of RNA secondary structures, see Fig. 1.11, have been studied on different levels:

- Via exhaustive enumeration [50, 55, 56], employing computer folding algorithms, like ViennaRNA [68], which derive for RNA sequences their minimum free energy (mfe) secondary structure.
- Via structural analysis, considering the embedding of neutral networks into sequence space. This line of work has led to the intersection theorem [106], see Chapter 7, which implies that for any two secondary or pseudoknot structures there exists at least one sequence which is compatible to

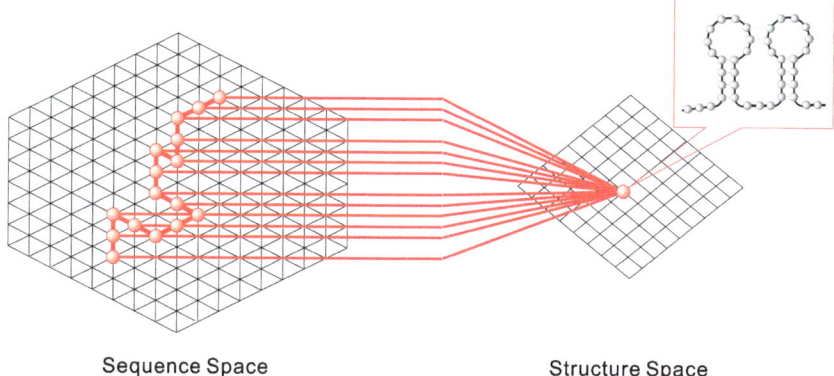

Sequence Space   Structure Space

Fig. 1.11. Neutral networks: sequence space (*left*) and structure space (*right*) represented as lattices. Edges between two sequences are drawn *bold* if they both map into the given structure. Two key properties of neutral nets are connectivity and percolation.

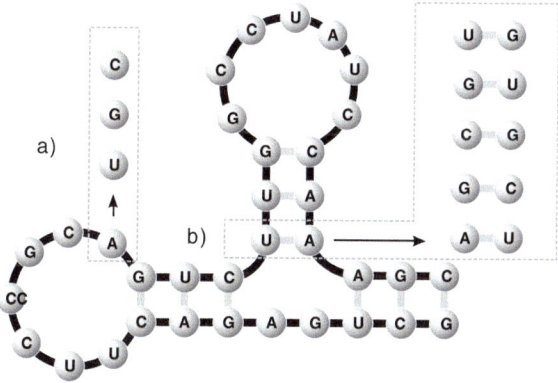

Fig. 1.12. Compatible mutations: here we represent a secondary structure as a planar graph. The *gray edges* correspond to the arcs in the upper half plane of its diagram representation, while the *black edges* represent the backbone of the underlying sequence. We illustrate the different alphabets for compatible mutations in unpaired **a)** and paired **b)** positions, respectively.

both. Here a compatible sequence is a sequence that satisfies all base pair requirements implied by the underlying structure, s, but for which s may not be an mfe structure; see Fig. 1.12. The intersection theorem shows that neutral networks come "close" in sequence space and has motivated exciting experimental work; see, for instance, [117].

■ Via random graphs, where neutral networks have been modeled as random subgraphs of n-cubes [102, 103, 105, 106]. Two important notions originated from this approach: the concepts of connectivity and density of neutral networks. A neutral network is connected if between any two of its sequences there exists a neutral path [118] and r-dense if a Hamming ball of radius r, centered at a compatible sequence (see Chapter 7 for details), contains at least one sequence contained in the neutral network.

A key result in the context of neutral networks for secondary structures has been derived in [69]. For biophysical reasons (folding maps produce typically conformations of low free energy) canonical structures, i.e., structures having *no* isolated base pairs and arc length greater than 4, are of particular relevance. Based on some variant of Waterman's basic recursion, eq. (1.1), and Darboux-type theorems [148], it was proved in [69] that there are asymptotically

$$1.4848 \cdot n^{-3/2} \cdot 1.84877^n \tag{1.2}$$

canonical secondary structures with arc length greater than 4. Clearly, since there are 4^n sequences over the natural alphabet this proves the existence of (exponentially large) neutral networks for sequence to structure maps into RNA secondary structures.

One motivation for our analysis in Chapters 4 and 5 is to generalize and extend the results known for RNA secondary structures to pseudoknot structures. More precisely, we will show that sequence to structure mappings in RNA pseudoknot structures realize an *exponential number* of distinct pseudoknot structures having exponentially large neutral networks.

While the existence of neutral networks for k-noncrossing structures follows from the exponential growth rates, the fact that there are exponentially many of these is a consequence of the statistics of the number of base pairs in k-noncrossing structures in combination with two biophysical facts. First, only 6 out of the 16 possible combinations of 2 nucleotides over the natural alphabet satisfy the Watson–Crick and **G-U** base-pairing rules (**A-U**, **U-A**, **G-C**, **C-G**, and **G-U**, **U-G**) and second the mfe structures generated by folding maps exhibit $O(n)$ base pairs.

Let us have a closer look at the argument for the existence of neutral networks of RNA pseudoknot structures. We present in Table 1.1 the exponential growth rates, computed via singularity analysis in Chapter 4. One important observation here is the drop of the exponential growth rate from arbitrary to canonical structures. For instance, for $k = 3$ we have $\gamma_{3,1} = 4.7913$ for structures with arbitrary stack-length, while canonical structures exhibit $\gamma_{3,2} = 2.5881$. Accordingly, the set of thermodynamically stable conformations is much smaller than the set of all sequences. In the context of inverse folding, this is a relevant feature of a well-suited target-class of folding algorithms. One further consequence is that for k smaller than 7 there exist some canonical structures with exponentially large neutral networks. In the last row of Table 1.1 we present the exponential growth rates, $\gamma_{k,2}^{[4]}$, obtained via the equivalent of eq. (1.2) for k-noncrossing, canonical structures having arc length at least 4; see Theorem 4.25. Table 1.1 shows that these growth rates are only marginally smaller than those of structures with minimum arc length

k	2	3	4	5	6	7	8	9
$\gamma_{k,1}$	2.6180	**4.7913**	6.8541	8.8875	10.9083	12.9226	14.9330	16.941
$\gamma_{k,2}$	1.9680	**2.5881**	3.0382	3.4138	3.7438	4.0420	4.3162	4.5715
$\gamma_{k,2}^{[4]}$	1.8487	2.5410	3.0132	3.3974	3.7319	4.0327	4.3087	4.5654

Table 1.1. Exponential growth rates, $\gamma_{k,\sigma}$, for various classes of k-noncrossing, σ-canonical structures; see Proposition 4.16. $\sigma = 1$ corresponds to structures with isolated arcs and $\sigma = 2$ are canonical structures. Note the drop from $\gamma_{3,1}$ to $\gamma_{3,2}$ (bold entries). $\gamma_{k,2}^{[4]}$ represents the growth rate of k-noncrossing, canonical structures having arc length at least 4; see Theorem 4.25.

2. This leads to the conclusion that minimum arc length requirements do not have a significant impact on the number of RNA pseudoknot structures.

In particular, one generic target for a folding algorithm into RNA pseudoknot structures is the class of 3-noncrossing, canonical structures having arc length at least 4. The equivalent of eq. (1.2) then reads

$$5089.47 \cdot n^{-5} \cdot 2.5410^n, \tag{1.3}$$

see eq. (4.59). Equation (1.3) supports the hypothesis that sequence to structure maps into pseudoknot structures exhibits features reminiscent of those of maps into secondary structures.

Second, we proceed by showing that there exist exponentially many pseudoknot structures with large neutral networks. To this end we analyze the size of a neutral network. Clearly, any neutral network is contained in the set of compatible sequences. As mentioned above, there are 16 pairings of 2 nucleotides; only 6 of which being consistent with the Watson–Crick and **G-U** base-pairing rules. Consequently, if structures contain sufficiently many base-pairs, their compatible sequences and also the neutral networks contained therein are exponentially small compared to sequence space. More precisely, if a sequence to structure map realizes structures in which there are $O(n)$ base pairs, then the neutral networks of any structure are exponentially smaller than sequence space. We can conclude from this that there exist exponentially many such structures.

It is therefore of interest to compute the distribution of the number of base pairs in k-noncrossing structures. This is new even for RNA secondary structures and can – somewhat surprisingly – be answered directly via the generating functions derived in Chapter 4. The idea is to put a combinatorial "label" on any arc in the generating function, thereby passing to a bivariate version. It is then a result of the supercritical paradigm discussed in Chapter 2 that the limit distribution of base pairs in k-noncrossing structures is a Gaussian distribution, governed by Theorem 5.14. To be exact, the mean of any such Gaussian is of the form $\mu_{k,\tau} n$, where $\mu_{k,\tau} > 0$; see Table 1.2 and Fig. 1.13. According to Table 1.2, the numbers of base pairs of

τ	$\mu_{k,\tau}$ ($k=2$)	$\sigma^2_{k,\tau}$	$\mu_{k,\tau}$ ($k=3$)	$\sigma^2_{k,\tau}$	$\mu_{k,\tau}$ ($k=4$)	$\sigma^2_{k,\tau}$	$\mu_{k,\tau}$ ($k=5$)	$\sigma^2_{k,\tau}$
1	0.276393	0.0447214	0.390891	0.0415653	0.425464	0.0314706	0.443020	0.0251601
2	0.317240	0.0643144	0.381701	0.0559928	0.403574	0.0470546	0.416068	0.0413361
3	0.336417	0.0791378	0.383555	0.0670987	0.400288	0.0559818	0.410087	0.0517052
4	0.348222	0.0916871	0.386408	0.0767872	0.400412	0.0667094	0.408701	0.0603242
5	0.356484	0.1028563	0.389134	0.0855937	0.401402	0.0748305	0.408741	0.0680229

Table 1.2. Central limit theorem for the numbers of base pairs in k-noncrossing, τ-canonical structures. We list $\mu_{k,\tau}$ and $\sigma^2_{k,\tau}$ computed via Theorem 5.14.

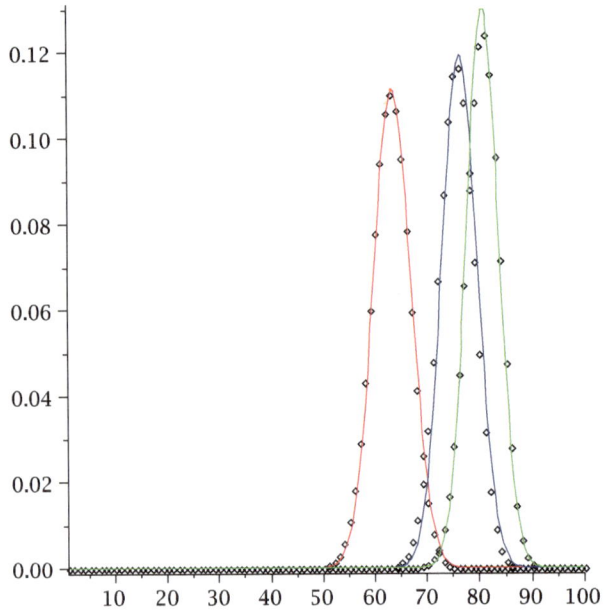

Fig. 1.13. The central limit theorem for arcs in k-noncrossing structures. We display the limit distribution (*solid curves: red/blue/green*) versus exact enumeration data (*dots*) for canonical 2-, 3-, and 4-noncrossing structures of length $n = 200$. The x-axis displays the number of base pairs. See, Theorem 5.14 for details.

2- and 3-noncrossing, canonical RNA structures are concentrated at $0.32\,n$ and $0.38\,n$, respectively. As a result, sequence to structure maps into k-noncrossing, canonical structures exhibits a plethora of structures with large neutral networks. Due to combinatorial as well as biophysical reasons, these maps appear to be ideally suited to facilitate evolutionary optimization based on random point mutations.

1.4 Folding

In light of these RNA functionalities the question of RNA structure prediction becomes important. In the context of folding we employ the notion of minimum free energy structure. As mentioned before, it is possible that there exist nonnative and native conformations exhibiting comparable energies. During the folding one observes the formation of non-native-like secondary structures which in turn imply structural reorganizations in order to realize ultimately the native state. This can result in misfolded molecules – the so-called folding traps, delaying the global fold [99, 135]. As a result the folding does not necessarily lead to a unique structure. In fact, for a fixed RNA primary sequence there can be alternative structures with different biochemical functionalities [10, 12, 13, 100]. The capability of RNA molecules to exhibit a number of meta-stable conformations is used in nature in form of molecular switches in the context of regulating and controlling biochemical processes [8, 39, 57, 65, 85, 138]. However, also artificial molecular switches [117, 123] were designed. In the following we mean by "folding" the generation of a particular mfe conformation, regardless of folding path or kinetic considerations.

The first mfe-folding algorithms for RNA secondary structures are due to [29, 46, 81] and the first dynamic programming (DP) folding routines for secondary structures were derived by Waterman et al. [96, 142, 144, 150]. The DP routines predict the loop-based mfe secondary structure [132] in $O(n^3)$-time and $O(n^2)$-space.

In the following we use the term pseudoknot synonymous with cross-serial dependencies between pairs of nucleotides [21, 120]. We have the following situation: the problem of predicting general RNA pseudoknot structures under the widely used thermodynamic model is NP-complete [87]. There exist, however, polynomial time folding algorithms, capable of the energy-based prediction of certain pseudoknots: Rivas et al. [111], Uemura et al. [137], Akutsu [3], and Lyngsø [87].

For the ab initio folding of pseudoknot RNA, there exist two paradigms: Rivas and Eddy's [111] gap-matrix variant of Waterman's DP-folding routine for secondary structures [70, 96, 142–144] and maximum weighted matching algorithms [22, 35, 47, 130]; see Fig. 1.14. The former method folds into a somewhat "mysterious" class of pseudoknots [112] in polynomial time. Algorithms along these lines have been developed by Dirks and Pierce [31], Reeder and Giegerich [101], and Ren et al. [109]. Additional ideas for pseudoknot folding involve the iterated loop matching approach [113] and the sampling of RNA structures via the Markov chain Monte Carlo method [91]. In Chapter 6 we discuss a different approach via the algorithm, **cross**, which is a priori generating 3-noncrossing structures; see Fig. 1.15. In difference to the DP paradigm, where optimal configurations of a very large, oftentimes unspecified, class can be constructed "locally" in polynomial time, **cross** is build

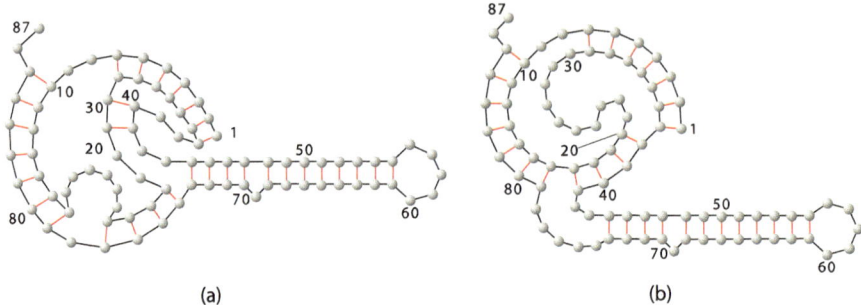

Fig. 1.14. The HDV-pseudoknot structure: (**a**) the structure as folded by Rivas and Eddy's algorithm [111] and (**b**) the structure as folded by `cross`, see Chapter 6, which folds 3-noncrossing, 3-canonical structures with arc length ≥ 4.

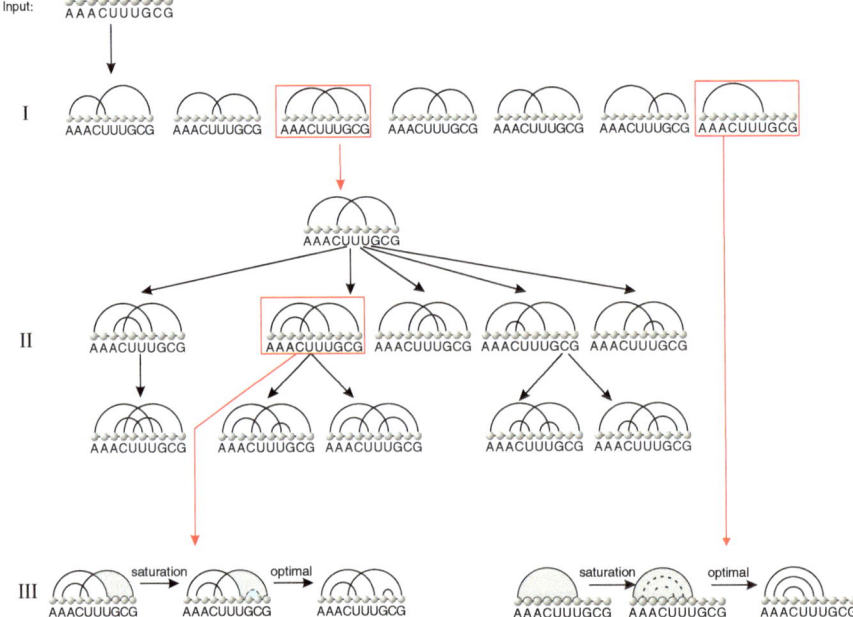

Fig. 1.15. An outline of `cross`: the generation of motifs (I); the construction of skeleta trees, that are rooted in irreducible shadows (II); and the saturation (III). During the latter we derive via DP routines optimal fillings of intervals of skeleta. The *red arrows* represent the processing of two motifs, one of which leads to the generation of a skeleton tree, while the other leads directly to the saturation routine. For details, see Chapter 6.

around an a priori known target class. The key feature of this class is that its cardinality is much smaller than that of sequence space; see eq. (1.3).

So much for the key ideas in RNA folding. We next discuss the DP routine for folding RNA secondary structures [141] in detail. In order to present the basic idea we may begin by considering additive energy contributions of Watson–Crick base pairs [96] given as follows: **G-U** $\equiv$ 1, **A-U** $\equiv$ 2, and **G-C** $\equiv$ 3.

Let $S(i, j)$ denote the optimal score for an mfe structure over $[i, j]$. The key observation is that $S(i, j)$ can be inductively derived. According to Waterman [141], see also eq. (1.1) and Fig. 1.16, the computation of $S(i, j)$ is obtained inductively distinguishing the following cases:

- (i, j) form a base pair, in which case there exists a nested substructure over $(i + 1, j - 1)$.
- i is unpaired, then there exists a substructure over $(i + 1, j)$.
- j is unpaired, then there exists a substructure over $(i, j - 1)$.
- i, j are paired but not to each other, then there are two substructures, over (i, k) and $(k + 1, j)$, respectively.

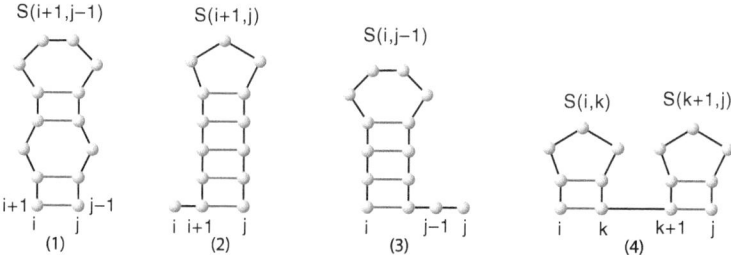

Fig. 1.16. Recursive computation of $S(i, j)$ by distinguishing the cases: (1) (i, j) are a base pair; (2) i is unpaired; (3) j is unpaired; and (4) i, j are paired but do not form the base pair (i, j).

Accordingly, we can inductively generate $S(i, j)$ via

$$
S(i, j) = \text{optimal} \begin{cases} P_{i,j} + S(i + 1, j - 1) \\ S(i + 1, j) \\ S(i, j - 1) \\ \text{optimal}_{i < k < j} S(i, k) + S(k + 1, j) \end{cases}, \tag{1.4}
$$

where $P_{i,j}$ is the energy contribution of base pair (i, j).

Now we are in position to recursively compute the optimal score, $S(1, n)$. For this purpose, we recursively construct the upper triangular matrix $S = (S(i, j))_{i,j}$, where $S(i, j) = 0$ for $j < i$. Evidently, for subsequences of length 0 or 1, there exists no base pair, whence $S(i, i) = S(i, i - 1) = 0$. According to the recursion given in eq. (1.4), we can inductively build the matrix S and eventually compute $S(1, n)$; see Fig. 1.17.

Let us make explicit how to find an optimal structure using dynamic programming. For the sequence **AGGACCUCUU**, we initialize the matrix and

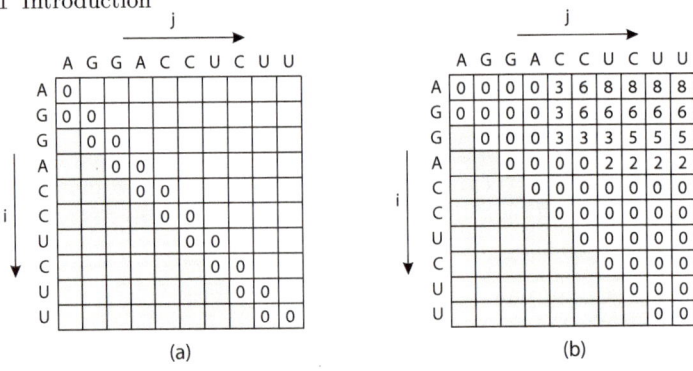

Fig. 1.17. The score matrix: (a) initialization of the matrix and (b) recursive computation of all matrix entries.

subsequently fill it recursively. Let us show how $S(1, 10)$ is obtained via equation (1.4). Since positions 1 and 10 can pair ((**A-U**) pair), we have

$$S(1,10) = \max \begin{cases} 2 + S(2,9) = 8 \\ S(1,9) = 8 \\ S(2,10) = 6 \\ \max_{1<k<10} S(1,k) + S(k+1,10) = 8 \quad (k = 7,8) \end{cases}$$

$$= 8.$$

In general, there exist more than one structure leading to the same score $S(1, n)$. In our example, $S(1, 10)$ can be obtained from $2 + S(2, 8)$, $S(1, 9)$, or $S(1, k) + S(k + 1, 10)$, where $k = 7, 8$. Given S, we can construct an optimal structure by tracing back. We showcase two different structures derived from different traces in Fig. 1.18.

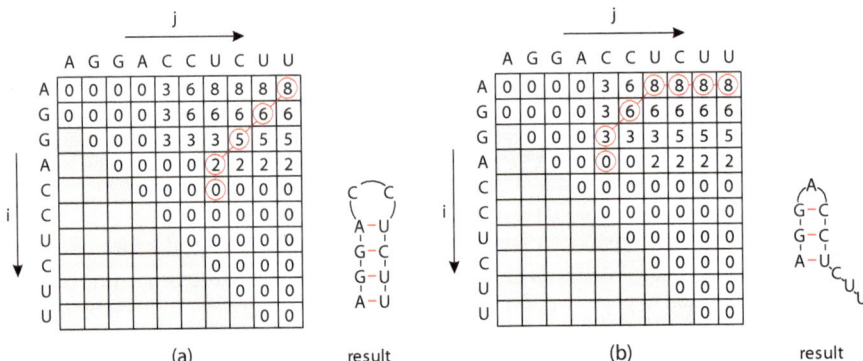

Fig. 1.18. Two traces: different traces induce different structures.

1.5 RNA tertiary interactions: a combinatorial perspective

Like proteins, RNA adopts complex three-dimensional folds for the precise presentation of chemical moieties that are essential for its function as a biological catalyst, translator of genetic information, and structural scaffold. With the techniques for structural analysis such as X-ray crystallography and nuclear magnetic resonance (NMR) spectroscopy, it is tertiary interactions that play a dominant role in establishing the global fold of the molecule. Atomic resolution structures of several large RNA molecules, determined by X-ray crystallography, have elucidated some of the means by which a global fold is achieved. Within these RNAs are tertiary structural motifs that enable the highly anionic double-stranded helices to tightly pack together to create a globular architecture. Base stacking, participation of the ribose 2'-hydroxyl groups in hydrogen-bonding interactions, binding of divalent metal cations, non-canonical base pairing, and backbone topology all serve to stabilize the global structure of RNA and play critical roles in guiding the folding process.

Tertiary interactions have been classified into three general categories: interactions between two double-stranded helical regions, between a helical region and a non-double-stranded region, and between two non-helical regions.

A pseudoknot is a tertiary interaction between unpaired regions defined as a motif in which nucleotides of a hairpin loop base pair with a complementary single-stranded sequence. Furthermore, base triples are tertiary interactions between helical and unpaired motifs. Single-stranded nucleotides can interact with base-paired nucleotides via either the major groove or the minor groove of duplex regions. Nucleotide triples have been shown or proposed to form at junctions of coaxially stacked RNA helices that have adjacent single-stranded regions [9, 122]. Several major groove triples are present in tRNA where they function to stabilize its L-shaped three-dimensional structure.

A first step toward RNA tertiary structures beyond pseudoknot interactions consists in considering base triples. These interactions function to orient regions of secondary structures in large RNA molecules and to stabilize RNA three-dimensional structures. The three dimensional structure, including pseudoknot and RNA triples, is often the key to its function; see Fig. 1.19.

In Chapter 3 we present the combinatorial framework for expressing RNA tertiary interactions. This framework consists of tangled diagrams or "tangles" defined as follows: a tangled diagram is a labeled graph over the vertices $1, \ldots, n$, with vertices of degree at most 2, drawn in increasing order in a horizontal line. The arcs are drawn in the upper half plane. In general, a tangled diagram has isolated points and the types of nonisolated vertices dis-

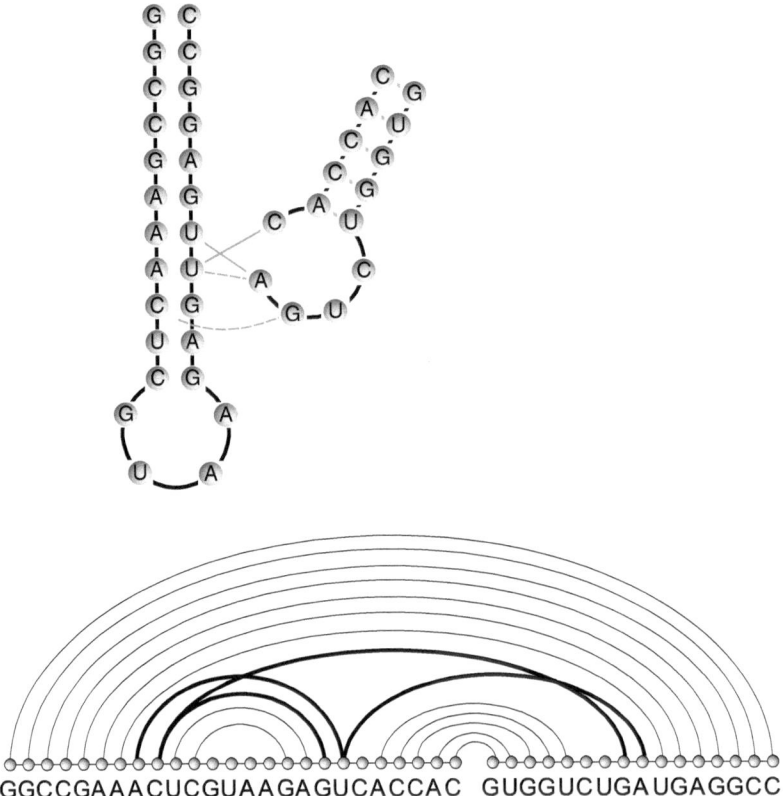

Fig. 1.19. The hammerhead ribozyme [9, 32]. Its two tertiary interactions are shown in the above graph representation as *dashed arcs*. The *bold arcs* in the diagram representation correspond to the arcs of the tangle displayed in Fig. 1.21, Section 3.1. The gap after **C**25 indicates that some nucleotides are omitted, which are involved in an unrelated structural motif. In the lower diagram, the *bold line* is used to denote the arcs which are related with base triples.

played in Fig. 1.20. Tangles are a generalization of diagrams by allowing for vertices of degree 2. They enable us to express all types of tertiary interactions; see Fig. 1.21. As for diagrams we introduce a notion of crossings for tangles: a tangled diagram is k-noncrossing if it does not contain k mutually (geometrically) crossing arcs and k-nonnesting if it does not contain k mutually nesting arcs. In contrast to k-noncrossing diagrams, crossings in tangles are intrinsically geometric. Fig. 1.20 shows that for each vertex of degree 2 we have perfect symmetry: there is exactly one nesting and one crossing configuration. In Fig. 1.21, we display the arc configurations of the hammerhead ribozyme [32] and the catalytic core region of group I self-splicing intron [24] via tangled diagrams, respectively. In Chapter 3 we study tangled diagrams.

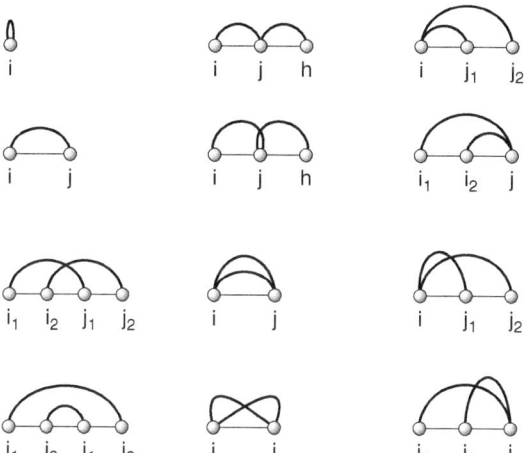

Fig. 1.20. All types of vertices with degree ≥ 1 in tangled diagrams. The figure illustrates the idea behind the notion of geometric crossings.

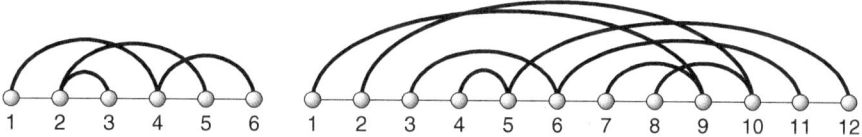

Fig. 1.21. The hammerhead and a catalytic core region as tangles: we represent the bonds of the hammerhead ribozyme (*left*), as displayed in Fig. 1.19 in form of a tangled diagram. In addition we represent the bonds of the catalytic core region of the group I self-splicing intron (*right*) [24].

Basic concepts

Not so many words—just the reason   *a simple mathematician*
                                                       Los Alamos 1996

In this chapter we provide the mathematical foundation for the following results. One main objective here is the self-contained derivation of the generating function of k-noncrossing matchings, which will play a central role for RNA pseudoknot structures.

We begin with the combinatorial framework needed for the reflection principle, facilitating the enumeration of nonrecursive combinatorial objects. The reflection principle [49] requires some understanding of group actions and familiarity with formal power series. This combinatorial section concludes with the discussion of D-finite generating functions [125–127].

Next we discuss the basic ideas behind the singularity analysis which are due to [42]. The proofs of the main theorems there can be found in [42]. We then discuss the implications of singularity analysis for k-noncrossing matchings in the context of Theorems 2.7 and 2.8 [80]. Finally we provide a case study for secondary structures in order to familiarize the reader with the new concepts.

We then conclude this chapter by introducing random-induced subgraphs of n-cubes [102, 105, 106]. Aside from providing the basic terminology we present the key tool needed in Chapter 7: vertex boundaries, branching processes, and Janson's inequality.

2.1 k-Noncrossing partial matchings

A diagram is a labeled graph over the vertex set $[n] = \{1, 2, \ldots, n\}$ with degree smaller or equal than 1. A diagram is represented by drawing the vertices $1, 2, \ldots, n$ in a horizontal line and the arcs (i, j), where $i < j$, in the upper half plane. The length of an arc (i, j) is $s = j - i$ and an arc of length s is called

C. Reidys, *Combinatorial Computational Biology of RNA*,   23
DOI 10.1007/978-0-387-76731-4_2,
© Springer Science+Business Media, LLC 2011

1 2 3 4 5 6 7 8 9
2-noncrossing
1 2 3 4 5 6 7 8 9
3-noncrossing
1 2 3 4 5 6 7 8 9
4-noncrossing

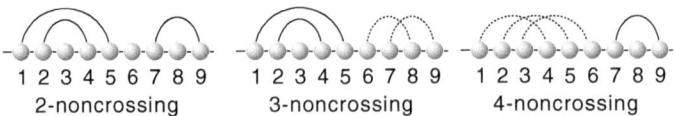

Fig. 2.1. k-Noncrossing diagrams: a 2-noncrossing (*left*), 3-noncrossing (*middle*), and 4-noncrossing diagram (*right*). The *dashed arcs* represent the maximal mutually crossing arcs.

an s-arc. A k-crossing is a set of k distinct arcs $(i_1, j_1), (i_2, j_2), \ldots (i_k, j_k)$ such that

$$i_1 < i_2 < \cdots < i_k < j_1 < j_2 < \cdots < j_k.$$

A diagram without any k-crossings is called k-noncrossing diagram or k-noncrossing partial matching (Fig. 2.1). A k-noncrossing diagram without any isolated points is called a k-noncrossing matching. A k-nesting is a set of k distinct arcs such that

$$i_1 < i_2 < \cdots < i_k < j_k < \cdots j_2 < j_1.$$

A diagram without any k-nestings is called a k-nonnesting diagram. Note that partial matchings can have arcs of *any* length, while the diagram representation of RNA structures assumes a minimum arc length of 2 or 4, respectively.

2.1.1 Young tableaux, RSK algorithm, and Weyl chambers

A Young diagram (shape) is a collection of squares arranged in left-justified rows with weakly decreasing number of boxes in each row. A Young tableau, or tableau, is a filling of the squares by numbers which is weakly increasing in each row and strictly decreasing in each column. A tableau is called standard if each entry occurs exactly once; see Fig. 2.2.

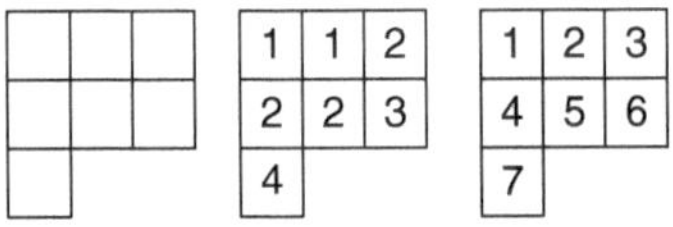

1 1 2
2 2 3
4
1 2 3
4 5 6
7

Fig. 2.2. Shape (*left*), Young tableau (*middle*), and standard Young tableau (*right*).

An oscillating tableau is a sequence

$$\varnothing = \mu^0, \mu^1, \ldots, \mu^n = \varnothing$$

of standard Young diagrams, such that for $1 \leq i \leq n$, μ^i is obtained from μ^{i-1} by either adding one square or removing one square. For instance, the sequence is an oscillating tableaux; see Fig. 2.3. In the following we consider a specific generalization by allowing for hesitation steps, i.e., we consider $*$-tableaux

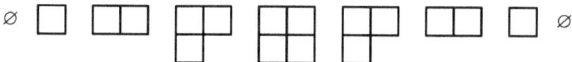

Fig. 2.3. Oscillating tableaux: two subsequent shapes, μ^{i-1} and μ^i, differ by exactly one square.

being sequences $\varnothing = \mu^0, \mu^1, \ldots, \mu^n = \varnothing$ such that for $1 \le i \le n$, μ^i is obtained from μ^{i-1} by either adding/removing one square or doing nothing; see Fig. 2.4. Let μ^{i-1} and μ^i be two shapes. If μ^i contains the shape μ^{i-1} we write $\mu^{i-1} \subseteq \mu^i$ and if, in particular, the shape μ^i is obtained by adding a square to the shape μ^{i-1} we write $\mu^i \setminus \mu^{i-1} = \square$.

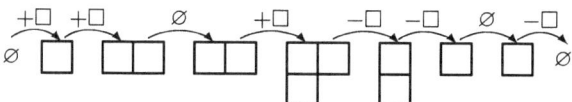

Fig. 2.4. $*$-tableaux: μ^{i-1} and μ^i either differ by one square or are equal.

We next come to a procedure via which elements can be row-inserted into Young tableaux, called RSK algorithm. Suppose we want to insert k into a standard Young tableau λ. Let $\lambda_{i,j}$ denote the element in the ith row and jth column of the Young tableau. Let j be the largest integer such that $\lambda_{1,j-1} \le k$. (If $\lambda_{1,1} > k$, then $j = 1$.) If $\lambda_{1,j}$ does not exist, then simply add k at the end of the first row. Otherwise, if $\lambda_{1,j}$ exists, then replace $\lambda_{1,j}$ by k. Next insert $\lambda_{1,j}$ into the second row following the above procedure and continue until an element is inserted at the end of a row. As a result we obtain a new standard Young tableau with k included. For instance, inserting the sequence of integers $(5, 2, 4, 1, 6, 3)$, see Fig. 2.5, starting with an empty shape yields the following sequence of standard Young tableaux:

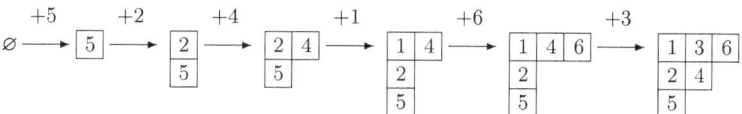

Fig. 2.5. RSK insertion: the sequence of integers $(5, 2, 4, 1, 6, 3)$ is RSK inserted, starting with an empty shape. The labeling of the *arrows* by "$+x$" indicates the RSK insertion of the integer x.

One key observation with respect to the RSK algorithm is that it can be, in some sense, "inverted" [27]. To be precise, we have

Lemma 2.1. *Suppose we are given two shapes μ^{i-1}, μ^i such that $\mu^{i-1} \setminus \mu^i = \square$ and a standard Young tableaux T_{i-1} of shape μ^{i-1}. Then there exists a unique j contained in T_{i-1} and a standard Young tableaux T_i of shape μ^i such that T_{i-1} is obtained from T_i by inserting j via the RSK algorithm.*

Proof. Indeed, suppose μ^{i-1} differs from μ^i in the first row. Then j is the element at the end of the first row in T_{i-1}. Otherwise suppose ℓ is the row of the square being removed from T_{i-1}. Remove the square and insert its element x into the $(\ell-1)$th row at precisely the position, where the removed element y would push it down via the RSK algorithm. That is, y is maximal subject to $y < x$. Since each column is strictly increasing y always exists. Iterating this process results in exactly one element j being removed from T_i and a new filling of the shape μ^{i-1}, i.e., a unique tableau T_{i-1}. By construction, inserting j with the RSK algorithm produces T_{i-1}.

In Fig. 2.6 we give an illustration of Lemma 2.1. We shall furthermore see that $*$-tableaux can be interpreted as lattice walks. This interpretation allows for the application of powerful principles tailored for their enumeration. For this purpose we provide next some basic background on lattice walks.

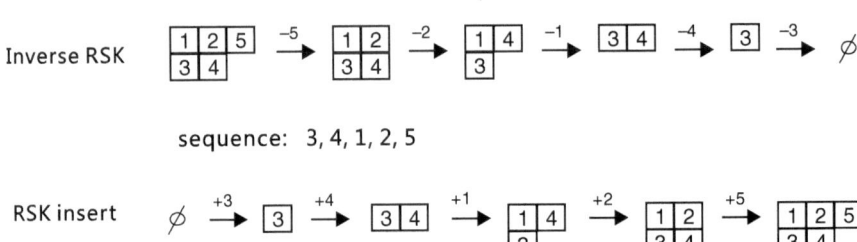

Fig. 2.6. The RSK algorithm and its inverse. First we extract via the inverse RSK and then reinsert using RSK, recovering the original Young tableau. The *arrows* are labeled by "$+x$" and "$-x$" in case of RSK insertion and extraction, respectively.

Let $\mathbb{Z}^{k-1}$ denote the $(k-1)$-dimensional lattice. We consider walks in $\mathbb{Z}^{k-1}$ having the steps s contained in $\{\pm e_i, 0 \mid 1 \leq i \leq k-1\}$, where e_i denotes the ith unit vector and 0 corresponds to a hesitation step. That is for $a, b \in \mathbb{Z}^{k-1}$ a walk from a to b, $\gamma_{a,b}$, of length n is an n tuple $(s_1, \dots, s_n)$ where $s_h \in \{\pm e_i, 0 \mid 1 \leq i \leq k-1\}$ such that $b = a + \sum_{h=1}^{n} s_h$; see Fig. 2.7. We set $\gamma_{a,b}(s_r) = a + \sum_{h=1}^{r} s_h \in \mathbb{Z}^{k-1}$, i.e., the element at which the walk resides at step r.

2.1.2 The Weyl group

We next introduce the Weyl group B_{k-1}. For this purpose we consider the abelian group

$$E_{k-1} \cong \langle -1 \rangle^{k-1},$$

whose elements are $(k-1)$-tuples with coordinates being ± 1. E_{k-1} is generated by the elements ϵ_i, having coordinates "1" everywhere except of the ith coordinate which is "-1." We note that the symmetric group S_{k-1} and E_{k-1} act on $\mathbb{Z}^{k-1}$ via

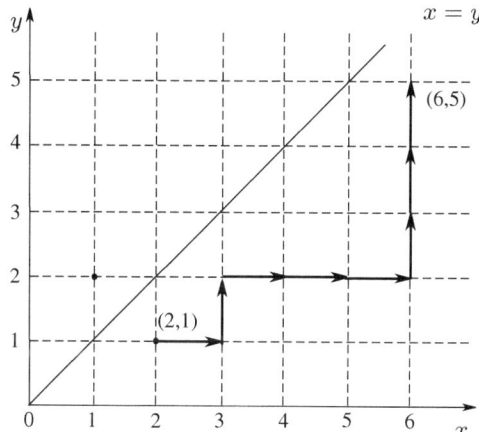

Fig. 2.7. Lattice walks: a walk from $(2, 1)$ to $(6, 5)$ of length 8 inside the fundamental Weyl chamber $C_0 = \{(x_1, x_2) \in \mathbb{Z}^2 \mid 0 \le x_2 \le x_1\}$. See the text for the definition of C_0.

$$\sigma(x_i)_{1 \le i \le k-1} = (x_{\sigma^{-1}(i)})_{1 \le i \le k-1},$$
$$\epsilon_i(x_1, \ldots, x_i, \ldots, x_{k-1}) = (x_1, \ldots, -x_i, \ldots, x_{k-1}).$$

It is straightforward to verify that $\{(\epsilon, \sigma) \mid \sigma \in S_{k-1}, \epsilon \in E_{k-1}\}$ carries a natural group structure via

$$(\epsilon_i, \sigma) \cdot (\epsilon_j, \sigma') = (\epsilon_i \cdot (\sigma \epsilon_j \sigma^{-1}), \sigma \sigma') = (\epsilon_i \epsilon_{\sigma(j)}, \sigma \sigma').$$

This is the Weyl group B_{k-1}, i.e., the semidirect product $E_{k-1} \rtimes S_{k-1}$, and generated by the set $M_{k-1} = \{\epsilon_{k-1}\} \cup \{\rho_j \mid 2 \le j \le k-1\}$, where $\rho_j = (j-1\ j)$ denotes the canonical transposition, i.e., ρ_j transposes the coordinates x_{j-1} and x_j. Since B_{k-1} acts on a basis vector $e_1, \ldots, e_n$ as a permutation, followed by some sign changes, the root system of B_{k-1} [54] is given by

$$\Delta_{k-1} = \{\pm e_i \mid 1 \le i \le k-1\} \cup \{e_i \pm e_j \mid 1 \le i, j \le k-1\}.$$

We observe that there exists a bijection between

$$\Delta'_{k-1} = \{e_{k-1}, e_{j-1} - e_j \mid 2 \le j \le k-1\}$$

and the set of generators M_{k-1} which maps each $\alpha \in \Delta'_{k-1}$ into a reflection as follows (in particular, B_{k-1} is generated by reflections):

$$\{e_{k-1}\} \cup \{e_{j-1} - e_j \mid 2 \le j \le k-1\} \longrightarrow \{\epsilon_{k-1}\} \cup \{\rho_j \mid 2 \le j \le k-1\},$$

$$\alpha \mapsto \left(\beta_\alpha : x \mapsto x - 2\frac{\langle \alpha, x \rangle}{\langle \alpha, \alpha \rangle}\alpha\right)$$

$$(2.1)$$

where $\langle x, x' \rangle$ denotes the standard scalar product in $\mathbb{R}^{k-1}$. It is clear that Δ'_{k-1} is a basis of $\mathbb{R}^{k-1}$. We refer to the subspaces $\langle e_i \rangle$ for $1 \le i \le k-1$ and

$\langle e_{j-1} - e_j \rangle$ for $2 \leq j \leq k-1$ as walls. A Weyl chamber is defined as the set of $x \in \mathbb{Z}^{k-1}$ with the property that $\langle \alpha, x \rangle \geq 0$ for all $\alpha \in \Delta'_{k-1}$. We denote Weyl chambers by "C" and refer to the particular Weyl chamber

$$\{x \in \mathbb{Z}^{k-1} \mid 0 \leq x_{k-1} \leq x_{k-2} \leq \cdots \leq x_1\} \tag{2.2}$$

as the fundamental Weyl chamber, C_0. Any element β of B_{k-1} can be expressed in several ways as a product of reflections and the minimal number of M_{k-1}-reflections needed to represent $\beta \in \mathsf{B}_{k-1}$ is called the length of β, denoted by $\ell(\beta)$. According to a theorem of Iwahori [74], multiplication of β by a M_{k-1}-reflection, β_α, changes $\ell(\beta)$ by either $+1$ or -1, respectively. Therefore we have

$$(-1)^{\ell(\beta_\alpha \beta)} = (-1)^{1+\ell(\beta)}. \tag{2.3}$$

We next show how to compute the length of an element of $\beta \in \mathsf{B}_{k-1}$. Such a β can be written as $\beta = \eta\sigma$, where $\sigma \in S_{k-1}$, $\eta = (\eta_1, \ldots, \eta_{k-1})$, $\eta_i \in \{+1, -1\}$. Let furthermore

$$B = \{i \mid \eta_i = -1, \beta = (\eta_i)_{1 \leq i \leq k-1}\sigma\}.$$

Let us make the action of β on an element of $\mathbb{Z}^{k-1}$ explicit

$$\beta(x_i)_{1 \leq i \leq k-1} = \eta\sigma(x_i)_{1 \leq i \leq k-1} = \left(\prod_{i \in B} \epsilon_i\right)(x_{\sigma^{-1}(i)})_{1 \leq i \leq k-1}.$$

Here $\epsilon_i(x_1, \ldots, x_i, \ldots, x_{k-1}) = (x_1, \ldots, -x_i, \ldots, x_{k-1})$ and the product is taken in B_{k-1}. Accordingly we have

$$\ell(\beta) = \ell(\eta\sigma) = \ell\left(\prod_{i \in B} \epsilon_i \circ \sigma\right), \tag{2.4}$$

where $\epsilon_j = \tau_j \epsilon_{k-1} \tau_j$ and $\tau_j = (k-1, k-2) \cdots (j+1, \ j)$. Since the ϵ_{k-1} and the τ_j are M_{k-1}-reflections we can conclude from eq. (2.3)

$$(-1)^{\ell(\beta)} = (-1)^{\ell(\prod_{i \in B}(\tau_i \epsilon_{k-1} \tau_i)\sigma)} = (-1)^{|B|+\ell(\sigma)}$$

and consequently

$$(-1)^{\ell(\beta)} = \operatorname{sgn}(\sigma) \prod_{i \in B} \eta_i = \operatorname{sgn}(\sigma) \prod_{i=1}^{k-1} \eta_i. \tag{2.5}$$

2.1.3 From tableaux to paths and back

In this section we connect the concepts of $*$-tableaux and specific lattice walks contained in Weyl chambers. Our main result is instrumental for the enumeration of RNA structures. It will allow us to interpret k-noncrossing partial matchings as walks in $\mathbb{Z}^{k-1}$ which remain in the interior of the Weyl chamber C_0. The result is due to Chen et al. [25]. The original bijection between oscillating tableaux and matchings is due to Stanley and was generalized by Sundaram [129]. In Chapter 3 we generalize these ideas in the context of tangled diagrams and prove a generalization of Theorem 2.2. This enables us to develop a framework for RNA tertiary structures.

Theorem 2.2. *(Chen et al. [25]) There exists a bijection between k-noncrossing partial matchings and walks of length n in $\mathbb{Z}^{k-1}$ which start and end at $a = (k-1, k-2, \ldots, 1)$, denoted by $\gamma_{a,a}$, having steps $0, \pm e_i$, $1 \leq i \leq k-1$ such that $0 < x_{k-1} < \cdots < x_1$ at any step, i.e., we have a bijection*

$$M_k(n) \longrightarrow \{\gamma_{a,a} \mid \gamma_{a,a} \text{ remains inside the Weyl chamber } C_0\},$$

where $M_k(n)$ denotes the set of k-noncrossing partial matchings over $[n]$.

Proof. **Claim 1.** There exists a bijection between the set of $*$-tableaux of length n and partial matchings over $[n]$.

Given a tableau $(\mu^i)_{i=0}^n$, where μ^i differs from μ^{i-1} by at most one square, we define a sequence $(G_n^0, T_0), (G_n^1, T_1), \ldots, (G_n^n, T_n)$, recursively, where G_n^i is a diagram and T_i is a standard Young tableau. We define G_n^0 to be the diagram with empty arc set and T_0 to be the empty standard Young tableau. The tableau T_i is obtained from T_{i-1} and the diagram G_n^i is obtained from G_n^{i-1} by the following procedure:

1. (Insert origins) For $\mu^i \supsetneq \mu^{i-1}$, then T_i is obtained from T_{i-1} by adding the entry i in the square $\mu^i \backslash \mu^{i-1}$.
2. (Isolated vertices) For $\mu^i = \mu^{i-1}$ then set $T_i = T_{i-1}$
3. (Remove origins) For $\mu^i \subsetneq \mu^{i-1}$, then let T_i be the unique standard Young tableau of shape μ^i and j be the unique number such that T_{i-1} is obtained from T_i by row-inserting j with the RSK algorithm. Then set $E_{G_n^i} = E_{G_n^{i-1}} \cup \{(j, i)\}$, where $E_{G_n^i}$ is the arc set of the diagram G_n^i.

For instance, given the sequence of tableau $(\mu^i)_{i=0}^7$

$$ \qquad\qquad\qquad\qquad\qquad\qquad\qquad\qquad (*)$$

The previous procedure gives rise to the fillings of μ_i and the diagram G_n^i:

$$
\begin{aligned}
E_{G_n^0} &= \varnothing, & E_{G_n^4} &= \{(2,3)\}, \\
E_{G_n^1} &= \varnothing, & E_{G_n^5} &= \{(2,3), (1,5)\}, \\
E_{G_n^2} &= \varnothing, & E_{G_n^6} &= \{(2,3), (1,5)\}, \\
E_{G_n^3} &= \{(2,3)\}, & E_{G_n^7} &= \{(2,3), (1,5), (4,7)\}.
\end{aligned}
$$

The resulting partial matching G_n^7 is given by

Let $G_n = G_n^n$. Obviously, G_n is a diagram, and the set of i where $\mu^i = \mu^{i-1}$ equals the set of isolated vertices of G_n. By construction each entry j is removed exactly once whence no edges of the form (j, i) and (j, i') can be obtained. Therefore G_n has degree ≤ 1 and we have a well-defined mapping

$$\psi \colon \{(\mu^i)_{i=0}^n \mid (\mu^i)_{i=0}^n \text{ is a } *\text{-tableaux}\} \longrightarrow$$
$$\{G_n | G_n \text{ is a partial matching over } [n]\}.$$

It is clear from the above procedure that G_n is a partial matching and then ψ is injective. To prove subjectivity we observe that each diagram G_n induces an $*$-tableaux as follows. We set $\mu_{G_n}^n = \varnothing$ and $T_n = \varnothing$. Starting from vertex $i = n, n-1, \ldots, 1, 0$ we derive a sequence of Young tableaux $(T_n, T_{n-1}, \ldots, T_0)$ as follows:

I. If i is a terminus of a G_n-arc (j, i) add j via the RSK algorithm to T_i set $\mu_{G_n}^{i-1} \supsetneq \mu_{G_n}^i$ to be the shape of T_{i-1} (corresponds to (3)).

II. If i is an isolated G_n-vertex set $\mu_{G_n}^{i-1} = \mu_{G_n}^i$ (corresponds to (2)).

III. If i is the origin of a G_n-arc (i, k) let $\mu_{G_n}^{i-1} \subsetneq \mu_{G_n}^i$ be the shape of T_{i-1}, the standard Young tableau obtained by removing the square containing i (corresponds to (1)).

Then we have *by construction* $\psi((\mu_{G_n}^i)_{i=0}^n) = G_n$, whence ψ is bijective.

Claim 2. G_n is k-noncrossing if and only if all shapes μ^i in the $*$-tableaux have less than k rows.

From Claim 1 we know $\psi^{-1}(G_n) = (\varnothing = \mu^0, \mu^1, \ldots, \mu^n = \varnothing)$, so it suffices to prove that the maximal number of rows in the shape set $\psi^{-1}(G_n)$ is less than k. First we observe that the arcs $(i_1, j_1), \ldots (i_\ell, j_\ell)$ form a ℓ-crossing of G_n if and only if there exists a tableau T_i such that elements $i_1, i_2, \ldots i_\ell$ are in the ℓ squares of T_i and being deleted in increasing order $i_1 < i_2 < \ldots < i_\ell$ afterwards. Next, we will obtain a permutation π_i from the entries in each tableau T_i recursively as follows:

1. If T_{i-1} is obtained from T_i by row-inserting j with the RSK algorithm, then $\pi_{i-1} = \pi_i j$.
2. If $T_{i-1} = T_i$, then $\pi_i = \pi_{i-1}$.
3. If T_{i-1} is obtained from T_i by deleting the entry i, then π_{i-1} is obtained from π_i by deleting i.

If $\pi = r_1 r_2 \ldots r_t$, then the entries being deleted afterward are in the order $r_t, \ldots, r_2, r_1$.

Using the RSK algorithm w.r.t. the permutation π_i, the resulting row-inserting Young tableau is exactly T_i. We prove this by induction in reverse order of the $*$-tableaux. It is trivial for the case $i = n$. Suppose it holds for j, $1 \leq j \leq n$. Consider the above three cases: inserting an element, doing nothing, and deleting an element. In the first case, the assertion is implied by the RSK algorithm in the construction of the $*$-tableaux. In the second case, it holds by the induction hypothesis on step j.

It remains to consider the third case, that is, removing the entry from T_j to get T_{j-1}. We show that also in this case the insertion Young tableau of π_i equals the labeled tableau T_i. Write $\pi_j = x_1 x_2 \ldots x_p j y_1 y_2 \ldots y_q$ and $\pi_{j-1} = x_1 x_2 \ldots x_p y_1 y_2 \ldots y_q$. In view of step 3 j is larger than elements $x_1, x_2, \ldots, x_p, y_1, \ldots y_q$. We need to prove that the insertion tableau S_{j-1} of π_{j-1} by the RSK algorithm is exactly the same as deleting the entry j in T_j. We proceed by induction on q. In the case $q = 0$, T_j is obtained from T_{j-1} by adding j at the end of the first row. Suppose the assertion holds for $q-1$, that is $S_{j-1}(x_1 x_2 \ldots x_p y_1 y_2 \ldots y_{q-1}) = S_j(x_1 x_2 \ldots x_p j y_1 y_2 \ldots y_{q-1}) \setminus \boxed{j}$. Consider inserting y_q into S_{j-1}, via the RSK algorithm. If the insertion track path never touches the position of j, then $S_{j-1}(x_1 x_2 \ldots x_p y_1 y_2 \ldots y_{q-1} y_q) = S_j(x_1 x_2 \ldots x_p j y_1 y_2 \ldots y_{q-1} y_q) \setminus \boxed{j}$. Otherwise, if the insertion path touched j and pushed j into the next row, then since j is greater than any other entry, j must be moved to the end of next row and the push process stops. Accordingly, the insertion path in $S_{j-1}(x_1 x_2 \ldots x_p y_1 y_2 \ldots y_{q-1})$ is the same path as in $S_j(x_1 x_2 \ldots x_p j y_1 y_2 \ldots y_{q-1})$ except the last step moving j to a new position j, so deleting j will get $S_{j-1}(x_1 x_2 \ldots x_p y_1 y_2 \ldots y_{q-1} y_q) = S_j(x_1 x_2 \ldots x_p j y_1 y_2 \ldots y_{q-1} y_q) \setminus \boxed{j}$. According to Schensted's theorem [115], for any permutation π, assume A is the corresponding insertion Young tableau by using the RSK algorithm on π. Then the length of the longest decreasing subsequences of π is the number of rows in A, whence the assertion.

Now we can prove Claim 2. A diagram contains an ℓ-crossing if and only if there exists a π_i which has decreasing subsequence of length ℓ. And the insertion Young tableau of π_i equals the labeled tableau T_i. According to Schensted's theorem, π has a decreasing sequence of length ℓ if and only if rows of T_i is ℓ.

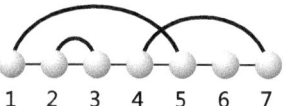

Fig. 2.8. The diagram corresponding to the sequence of tableaux in eq. $(*)$.

For instance, consider the partial matching of Fig. 2.8. We then obtain the sequence $\pi = (\varnothing \leftarrow 1 \leftarrow 12 \leftarrow 1 \leftarrow 1 \leftarrow 41 \leftarrow 4 \leftarrow 4 \leftarrow \varnothing)$. For the segment $\pi_1 = 1 \leftarrow \pi_2 = 12$ we have $j = 2$ and $q = 0$. Since the insertion track path never touches the position of $\boxed{2}$

$$S(1) = \boxed{1} = \boxed{1\ 2} \setminus \boxed{2} = \boxed{1} = S(12) \setminus \boxed{2}.$$

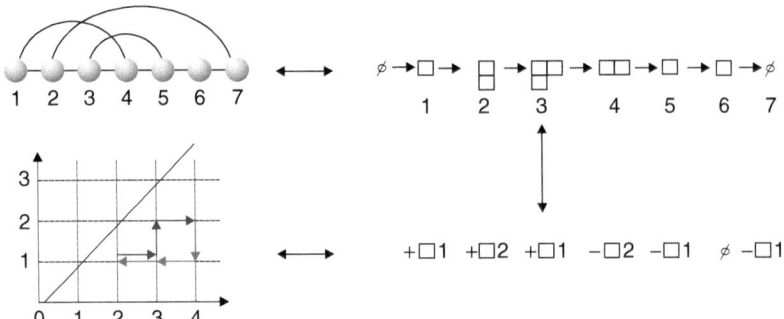

Fig. 2.9. The basic correspondences between partial matchings, $*$-tableaux, and walks inside the Weyl chamber C_0. Here "$\pm\square_i$" denotes the addition and removal of a square in the ith row, respectively.

For the segment $\pi_4 = 1 \leftarrow \pi_5 = 41$ we have $j = 4$ and $q = 1$:

$$S(1) = \boxed{1} = \begin{array}{c}\boxed{1}\\\boxed{4}\end{array} \setminus \boxed{4} = \boxed{1} = S(41) \setminus \boxed{4}.$$

Here the insertion path touches $\boxed{4}$ and 4 moves to the end of the next row, where the push process stops.

Claim 3. There is a bijection between $*$-tableaux with at most $k-1$ rows of length n and walks with steps $\pm e_i, 0$ which stay in the interior of C_0 starting and ending at $(k-1, k-2, \ldots, 1)$ see Fig. 2.9.

This bijection is obtained by setting for $1 \leq \ell \leq k-1$, x_ℓ to be the length of the ℓ-th row. By definition of standard Young tableaux, we have $\lambda_1 \geq \lambda_2 \geq \ldots \geq \lambda_n$, i.e., the length of each row is weakly decreasing. This property also characterizes walks that stay within the Weyl chamber C_0, i.e., where we have $x_1 > x_2 > \cdots > x_{k-1} > 0$ since a walk from $(k-1, \ldots, 2, 1)$ to itself in the interior of C_0 corresponds to a walk from the origin to itself in the region $x_1 \geq x_2 > \cdots \geq x_{k-1} \geq 0$. In an $*$-tableau μ^i differs from μ^{i-1} by at most one square and adding or deleting a square in the ℓth row or doing nothing corresponds to steps $\pm e_\ell$ and 0, respectively. Since the $*$-tableau is of empty shape, we have walks from the origin to itself, whence Claim 3 follows and the proof of the theorem is complete.

To summarize, given an $*$-tableaux of empty shape, $(\varnothing, \lambda^1, \ldots, \lambda^{n-1}, \varnothing)$, reading $\lambda^i \setminus \lambda^{i-1}$ from left to right, at step i, we do the following:

- For a $+\square$-step we insert i into the new square
- For a $\varnothing$-step we do nothing
- For a $-\square$-step we extract the unique entry, $j(i)$, of the tableaux T^{i-1}, which via RSK insertion into T^i recovers it (Fig. 2.6)

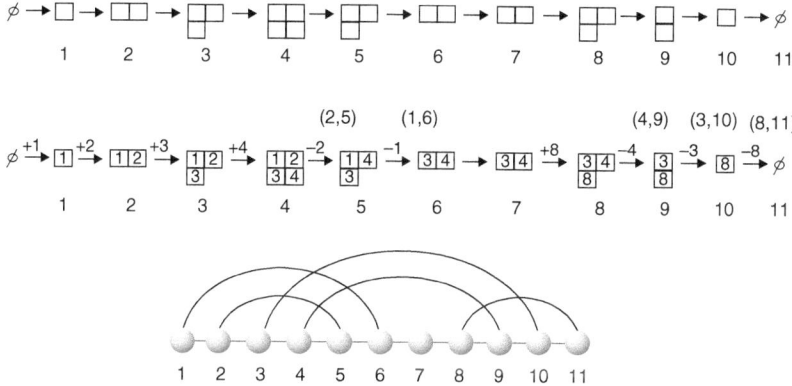

Fig. 2.10. From $*$-tableaux to partial matchings. If $\lambda^i \setminus \lambda^{i-1} = -\square$, then the unique number is extracted, which, if RSK inserted into λ^i, recovers λ^{i-1}. This yields the arc set of a k-noncrossing, partial matching.

The latter extractions generate the arc set $\{(i, j(i)) \mid i \text{ is a } -\square\text{-step}\}$ of a k-noncrossing diagram; see Fig. 2.10. Given a k-noncrossing diagram, starting with the empty shape, consider the sequence $(n, n-1, \ldots, 1)$ and do the following:

- If j is the endpoint of an arc (i, j), then RSK insert i.
- If j is the startpoint of an arc (j, s), then remove the square containing j.
- If j is an isolated point, then do nothing; see Fig. 2.11.

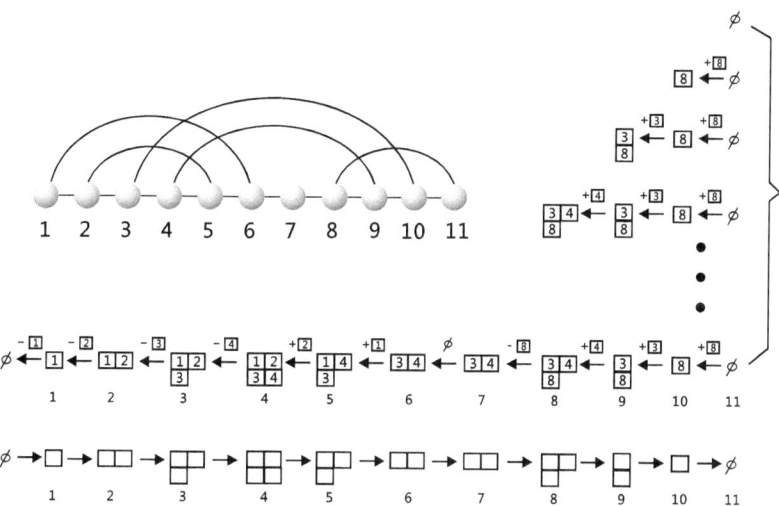

Fig. 2.11. From k-noncrosssing diagrams to $*$-tableaux using RSK insertion of the origins of arcs and removal of squares at the termini.

2.1.4 The generating function via the reflection principle

In this section we compute the enumerative generating function of k-noncrossing partial matchings. Our computation is based on the reflection principle. The key idea behind the reflection principle goes back to André [5, 49] and is to count walks that remain in the interior of a Weyl chamber by observing that all "bad" walks, i.e., those which touch a wall, cancel themselves. The particular method for deriving this pairing is via reflecting the walk choosing a point where it touches a wall. The following observation is essential for the reflection principle, formulated in Theorem 2.4.

Lemma 2.3. *Let* $\Delta'_{k-1} = \{e_{k-1}, e_{j-1} - e_j \mid 2 \le j \le k-1\}$. *Then every walk starting at some lattice point in the interior of a Weyl chamber,* C, *having steps* $\pm e_i, 0$ *that crosses from inside* C *into outside* C *touches a subspace* $\langle e_{j-1} - e_j \mid 2 \le j \le k-1 \rangle$ *or* $\langle e_j \mid 1 \le j \le k-1 \rangle$.

Proof. To prove the lemma we can, without loss of generality, assume

$$C = C_0 = \{(x_1, \ldots, x_{k-1}) \mid x_1 \ge x_2 \ge \cdots \ge x_{k-1} \ge 0\}.$$

Then the assertion is that every walk having steps $\pm e_i, 0$ that crosses from the inside C_0 into outside C_0 intersects either $\langle e_{k-1} \rangle$ or $\langle e_{j-1} - e_j \rangle$ for $2 \le j \le k-1$. This is correct since to leave C_0 is tantamount to the existence of some i such that $x_i < x_{i+1}$. Let s_j be minimal w.r.t. $a + \sum_h^{j+1} s_h \notin C_0$. Since we have steps $\pm e_i, 0$ we conclude $x_{k-1} = 0$ or $x_j = x_{j-1}$ for some $2 \le j \le k-1$, whence the lemma.

Let $\Gamma_n(a, b)$ be the number of walks $\gamma_{a,b}$. For $a, b \in C_0$ (eq. (2.2)) let $\Gamma_n^+(a, b)$ denote the number of walks $\gamma_{a,b}$ that never touches a wall, i.e., remain in the interior of C_0. Finally for $a, b \in \mathbb{Z}^{k-1}$, let $\Gamma_n^-(a, b)$ denote the number of walks $\gamma_{a,b} = (s_1, \ldots, s_n)$ that hit a wall at some step s_r. $\ell(\beta)$ denotes the length of $\beta \in \mathsf{B}_{k-1}$. For $a = b = (k-1, \ldots, 1)$ we have according to Theorem 2.2

$$\Gamma_n^+(a, a) = \mathsf{M}_k(n),$$

where $\mathsf{M}_k(n) = |M_k(n)|$, i.e., the number of all k-noncrossing partial matchings over $[n]$.

Theorem 2.4. (Reflection Principle) *(Gessel and Viennot [49])* *Suppose* $a, b \in C_0$, *then we have*

$$\Gamma_n^+(a, b) = \sum_{\beta \in \mathsf{B}_{k-1}} (-1)^{\ell(\beta)} \, \Gamma_n(\beta(a), b).$$

Theorem 2.4 allows us to compute the exponential generating function for $\Gamma_n^+(a, b)$, which is the number of walks from a to b that remain in the interior of C_0 [53]. Fig. 2.12 gives a simple application of reflection principles in lattice walk.

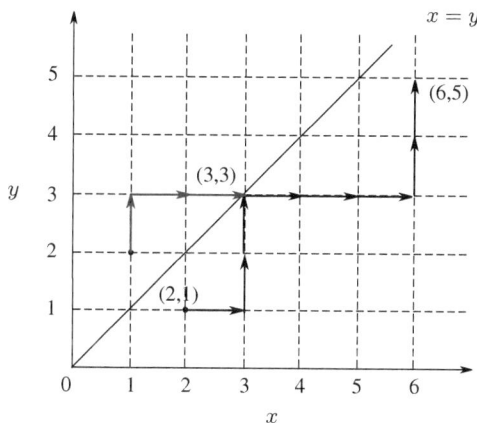

Fig. 2.12. Illustration of the reflection principle: "bad" walks cancel each other. Each lattice walk (here we consider only walks with steps $(1,0)$ or $(0,1)$) from $(2,1)$ to $(6,5)$ that hits the wall $y = x$ can uniquely be reflected into the walk from $(1,2)$ to $(6,5)$. Setting $a = (2,1)$, $b = (n+2, n+1)$, and $\tilde{a} = (1,2)$, the largest root corresponds to the subspace $\langle e_2 - e_1 \rangle$. We display a walk that hits this wall after three steps. Its initial segment (*red*) is then reflected leading to a walk from $(2,1)$ to $(6,5)$. Reflection implies $\Gamma_n^+(a,b) = \Gamma_n(a,b) - \Gamma_n(\tilde{a}, b) = C_n$, where C_n is Catalan number.

Proof. Totally order the roots of Δ. Let $\Gamma_n^-(a,b)$ be the number of walks γ from a to b, $a, b \in \mathbb{Z}^{k-1}$ of length n using the steps s, $s \in \{\pm e_i, 0\}$ such that $\langle \gamma(s_r), \alpha \rangle = 0$ for some $\alpha \in \Delta$ (i.e., the walk intersects with the subspace $\langle \alpha \rangle$). According to Lemma 2.3 every walk that crosses from inside C into outside C touches a wall from which we can draw two conclusions:

$$\Gamma_n(a,b) = \Gamma_n^+(a,b) + \Gamma_n^-(a,b),$$

$$\beta \neq \text{id} \implies \Gamma_n(\beta(a), b) = \Gamma_n^-(\beta(a), b).$$

Claim. $\sum_{\beta \in B_{k-1}} (-1)^{\ell(\beta)} \Gamma_n^-(\beta(a), b) = 0$.

Let $(s_1, \ldots, s_n)$ be a walk from $\beta(a)$ to b. By assumption there exists some step s_r at which we have $\langle \gamma_{\beta(a),b}(s_r), \alpha \rangle = 0$, for $\alpha \in \Delta$, where $\langle\ ,\ \rangle$ denotes the standard scalar product in $\mathbb{R}^{k-1}$. Let α^* be the largest root for which we have $\langle \gamma_{\beta(a),b}(s_r), \alpha^* \rangle = 0$ and $\beta_{\alpha^*}(x) = x - \frac{2\langle \alpha^*, x \rangle}{\langle \alpha^*, \alpha^* \rangle} \alpha^*$ its associated reflection (eq. (2.1)). We consider the walk

$$(\beta_{\alpha^*}(s_1), \ldots, \beta_{\alpha^*}(s_r), s_{r+1}, \ldots, s_n).$$

Now by definition $(\beta_{\alpha^*}(s_1), \ldots, \beta_{\alpha^*}(s_r), s_{r+1}, \ldots, s_n)$ starts at $(\beta_{\alpha^*} \circ \beta)(a)$ and we have according to eq. (2.3)

$$(-1)^{\ell(\beta_{\alpha^*} \circ \beta)} = (-1)^{\ell(\beta)+1}.$$

Therefore to each element $\gamma_{\beta(a),b}$ of $\Gamma_n^-(\beta(a),b)$ having sign $(-1)^{\ell(\beta)}$ there exists a $\gamma_{\beta_{\alpha^*}\beta(a),b} \in \Gamma_n^-(\beta_{\alpha^*}\beta(a),b)$ with sign $(-1)^{\ell(\beta)+1}$ and the claim follows. We immediately derive

$$\sum_{\beta \in B_{k-1}} (-1)^{\ell(\beta)} \, \Gamma_n(\beta(a),b)$$

$$= \Gamma_n(a,b) + \sum_{\beta \in B_{k-1}, \beta \neq \mathsf{id}} (-1)^{\ell(\beta)} \, \underbrace{\Gamma_n(\beta(a),b)}_{=\Gamma_n^-(\beta(a),b)}$$

$$= \Gamma_n^+(a,b) + \Gamma_n^-(a,b) + \underbrace{\sum_{\beta \in B_{k-1}, \beta \neq \mathsf{id}} (-1)^{\ell(\beta)} \, \Gamma_n^-(\beta(a),b),}_{\sum_{\beta \in B_{k-1}} (-1)^{\ell(\beta)} \, \Gamma_n^-(\beta(a),b)=0}$$

whence the theorem. $\square$

We can now achieve our main objective and specify the generating functions of the walks $\Gamma_n^+(a,b)$ having steps $0, \pm e_i$ and $\Gamma_n'^+(a,b)$ having steps $\pm e_i$ as a determinant of Bessel functions [53].

Theorem 2.5. (Grabiner and Magyar [53]) *Let* $I_r(2x) = \sum_{j\geq 0} \frac{x^{2j+r}}{j!(r+j)!}$ *be the hyperbolic Bessel function of the first kind of order* r. *Then the exponential generating functions for* $\Gamma_n^+(a,b)$ *and* $\Gamma_n'^+(a,b)$ *are given by*

$$\sum_{n\geq 0} \Gamma_n^+(a,b) \frac{x^n}{n!} = e^x \det[I_{b_j-a_i}(2x) - I_{a_i+b_j}(2x)]|_{i,j=1}^{k-1},$$

$$\sum_{n\geq 0} \Gamma_n'^+(a,b) \frac{x^n}{n!} = \det[I_{b_j-a_i}(2x) - I_{a_i+b_j}(2x)]|_{i,j=1}^{k-1}.$$

Proof. Let u_i, $1 \leq i \leq k-1$, be indeterminants and $u = (u_i)_1^{k-1}$. We define $u^{b-a} = \prod_{i=1}^{k-1} u_i^{b_i-a_i}$. Let $F(x,u)$ be a generating function, then $F(x,u)|_{u^{b-a}}$ equals the family of coefficients $a_i(u)$ at u^{b-a} of $\sum_{i\geq 0} a_i(u)x^i = F(x,u)$. We first consider unrestricted walks from a to b whose cardinality is given by

$$\Gamma_n(a,b) = \left[1 + \sum_{i=1}^{k-1}(u_i + u_i^{-1}) \right]^n \Bigg|_{u^{b-a}}.$$

The exponential generating function of $\Gamma_n(a,b)$ is

$$\sum_{n\geq 0} \Gamma_n(a,b) \frac{x^n}{n!} = \sum_{n\geq 0} \left[1 + \sum_{i=1}^{k-1}(u_i + u_i^{-1}) \right]^n \Bigg|_{u^{b-a}} \frac{x^n}{n!}$$

$$= \sum_{n \geq 0} \frac{[1 + \sum_{i=1}^{k-1}(u_i + u_i^{-1})]^n}{n!} x^n \bigg|_{u^{b-a}}$$

$$= e^x \cdot \exp[x \sum_{i=1}^{k-1}(u_i + u_i^{-1})]\bigg|_{u^{b-a}}$$

$$= e^x \cdot \prod_{i=1}^{k-1}\left(\exp(x(u_i + u_i^{-1}))\bigg|_{u_i^{b_i-a_i}}\right).$$

According to Theorem 2.4 we have

$$\sum_{n \geq 0} \Gamma_n^+(a,b)\frac{x^n}{n!} = \sum_{n \geq 0}\sum_{\beta \in B_{k-1}} (-1)^{\ell(\beta)}\, \Gamma_n(\beta(a),b)\frac{x^n}{n!}$$

$$= \sum_{\beta \in B_{k-1}} (-1)^{\ell(\beta)} \sum_{n \geq 0} \Gamma_n(\beta(a),b)\frac{x^n}{n!}$$

$$= e^x \sum_{\beta \in B_{k-1}} (-1)^{\ell(\beta)} \prod_{i=1}^{k-1} \exp(x(u_i + u_i^{-1}))\bigg|_{u^{b-\beta(a)}},$$

whereas in case of $\Gamma_n'^+(a,b)$

$$\sum_{n \geq 0} \Gamma_n'^+(a,b)\frac{x^n}{n!} = \sum_{\beta \in B_{k-1}} (-1)^{\ell(\beta)} \prod_{i=1}^{k-1} \exp(x(u_i + u_i^{-1}))\bigg|_{u^{b-\beta(a)}}$$

holds. We continue by analyzing $\sum_{n \geq 0}\Gamma_n^+(a,b)\frac{x^n}{n!}$. Equation (2.5) provides an interpretation of the term $(-1)^{\ell(\beta)}$:

$$(-1)^{\ell(\beta)} = \text{sgn}(\sigma) \prod_{i \in B} \eta_i = \text{sgn}(\sigma) \prod_{i=1}^{k-1} \eta_i,$$

where $\eta_i = \pm 1$. Based on this interpretation we compute

$$\sum_{n \geq 0} \Gamma_n^+(a,b)\frac{x^n}{n!} =$$

$$e^x \sum_{\sigma \in S_{k-1}} \sum_{\eta_i = -1,+1} \text{sgn}(\sigma) \prod_{i=1}^{k-1} \eta_i \left(\exp(x(u_i + u_i^{-1}))\bigg|_{u_i^{b_i - \eta_i a_{\sigma_i}}}\right) =$$

$$e^x \sum_{\sigma \in S_{k-1}} \text{sgn}(\sigma) \prod_{i=1}^{k-1}\left(\exp(x(u_i + u_i^{-1}))\bigg|_{u_i^{b_i - a_{\sigma_i}}} - \exp(x(u_i + u_i^{-1}))\bigg|_{u_i^{b_i + a_{\sigma_i}}}\right).$$

We proceed by analyzing the terms $\exp(x(u_i + u_i^{-1}))$:

$$\exp(x(u_i + u_i^{-1})) = \sum_{n \geq 0} \frac{x^n}{n!}(u_i + u_i^{-1})^n$$

$$= \sum_{n \geq 0} \frac{x^n}{n!} \sum_{j=0}^{n} \binom{n}{j} u_i^{n-2j}$$

$$= \sum_{n \geq 0} x^n \sum_{j=0}^{n} \frac{u_i^{n-2j}}{j!(n-j)!}$$

$$= \sum_{r=-\infty}^{\infty} u_i^r \sum_{j=0}^{\infty} \frac{x^{2j+r}}{j!(j+r)!}$$

$$= \sum_{r=-\infty}^{\infty} u_i^r I_r(2x).$$

Therefore, for any $r \in \mathbb{Z}$, we have

$$\exp(x(u_i + u_i^{-1}))\Big|_{u_i^r} = \sum_{j \geq 0} \frac{x^{2j+r}}{j!(j+r)!} = I_r(2x).$$

As a result we arrive at

$$\sum_{n \geq 0} \Gamma_n^+(a,b) \frac{x^n}{n!} = e^x \sum_{\sigma \in S_{k-1}} \text{sgn}(\sigma) \prod_{i=1}^{k-1} \left(I_{b_i - a_{\sigma_i}}(2x) - I_{b_i + a_{\sigma_i}}(2x) \right), \quad (2.6)$$

that is

$$\sum_{n \geq 0} \Gamma_n^+(a,b) \frac{x^n}{n!} = e^x \det[I_{b_j - a_i}(2x) - I_{a_i + b_j}(2x)]|_{i,j=1}^{k-1},$$

completing the proof of the theorem.

Let $f_k(n, 0)$ denote the number of k-noncrossing matchings without isolated vertices over $[n]$. By abuse of notation we will in later chapters simply write $f_k(n)$ instead of $f_k(n, 0)$. When n is odd, by the definition, $f_k(n, 0) = 0$. Since $\Gamma_n^+(a, a) = M_k(n)$ and $\Gamma_n'^+(a, a) = f_k(n, 0)$ we obtain according to Theorem 2.5 for the generating functions of matchings and partial matchings as follows:

Corollary 2.6. *Let $I_r(2x) = \sum_{j \geq 0} \frac{x^{2j+r}}{j!(r+j)!}$ be the hyperbolic Bessel function of the first kind of order r. Then the generating functions for matchings and partial matchings are given by*

$$\sum_{n \geq 0} f_k(2n, 0) \cdot \frac{x^{2n}}{(2n)!} = \det[I_{i-j}(2x) - I_{i+j}(2x)]|_{i,j=1}^{k-1}, \quad (2.7)$$

$$\sum_{n \geq 0} M_k(n) \cdot \frac{x^n}{n!} = e^x \det[I_{i-j}(2x) - I_{i+j}(2x)]|_{i,j=1}^{k-1}. \quad (2.8)$$

Let

$$\mathbf{H}_k(z) = \sum_{n \geq 0} f_k(2n, 0) \cdot \frac{z^{2n}}{(2n)!}.$$

The main importance of Corollary 2.6 lies in the fact that it implies that $\mathbf{H}_k(z)$ is D-finite; see Corollary 2.14. It does not allow to derive "simple" expressions for $\mathbf{H}_k(z)$ for $k \geq 3$.

By taking the approximation of the Bessel function[1], for $-\frac{\pi}{2} < \arg(z) < \frac{\pi}{2}$, and

$$I_r(z) = \frac{e^x}{\sqrt{2\pi z}} \left(\sum_{h=0}^{H} \frac{(-1)^h}{h! 8^h} \prod_{t=1}^{h} (4r^2 - (2t-1)^2) z^{-h} + O(|z|^{-H-1}) \right)$$

into the determinant given in eq. (2.7), we derive the following asymptotic formula.

Theorem 2.7. (Jin et al. [80]) *For arbitrary $k \in \mathbb{N}$, $k \geq 2$, $\arg(z) \neq \pm\frac{\pi}{2}$ holds*

$$\mathbf{H}_k(z) = \left[\prod_{i=1}^{k-1} \Gamma\left(i+1-\frac{1}{2}\right) \prod_{r=1}^{k-2} r! \right] \left(\frac{e^{2z}}{\pi} \right)^{k-1} z^{-(k-1)^2 - \frac{k-1}{2}} \ (1 + O(|z|^{-1})),$$

where $\Gamma(z)$ denotes the gamma function.

Employing the subtraction of singularities principle [98], in combination with Theorem 2.7, we obtain the following result, which is of central importance for all asymptotic formulas involving k-noncrossing matchings:

Theorem 2.8. (Jin et al. [80]) *For arbitrary $k \in \mathbb{N}$, $k \geq 2$ we have*

$$f_k(2n, 0) \sim c_k \, n^{-((k-1)^2 + (k-1)/2)} \, (2(k-1))^{2n}, \quad \text{where } c_k > 0. \tag{2.9}$$

The proofs of Theorems 2.7 and 2.8 are elementary but involved and beyond the scope of this book. We refer the interested reader to [80]. Note that Theorem 2.8 implies that $\rho_k^2 = (2(k-1))^{-2}$ is a singularity of $\mathbf{F}_k(z)$; see Section 2.3.

Instead, we shall proceed by analyzing the relation between k-noncrossing matchings and k-noncrossing partial matchings. For this purpose we recruit the powerful concept of integral representations [36] in which combinatorial quantities like, for instance, binomial coefficients are replaced by contour integrals.

Lemma 2.9. *Let z be an indeterminate over $\mathbb{C}$. Then we have the identity of power series*

$$\forall |z| < \mu_k; \quad \sum_{n \geq 0} \mathsf{M}_k(n) \, z^n = \left(\frac{1}{1-z} \right) \sum_{n \geq 0} f_k(2n, 0) \left(\frac{z}{1-z} \right)^{2n}. \tag{2.10}$$

Proof. We have

$$\mathsf{M}_k(n) = \sum_{m=0}^{\lfloor \frac{n}{2} \rfloor} \binom{n}{2m} f_k(2m, 0),$$

where $\lfloor a \rfloor$ is the largest integer not larger than a. Expressing the combinatorial terms by contour integrals [36] we obtain

$$\binom{n}{2m} = \frac{1}{2\pi i} \oint_{|u|=\alpha} (1+u)^n u^{-2m-1} du,$$

$$f_k(2m, 0) = \frac{1}{2\pi i} \oint_{|v|=\beta} \mathbf{F}_k(v^2) v^{-2m-1} dv,$$

where α, β are arbitrary small positive numbers and $\mathbf{F}_k(z) = \sum_{n \geq 0} f_k(2n, 0) z^n$. We derive

$$\mathsf{M}_k(n) = \frac{1}{(2\pi i)^2} \sum_m \oint_{|u|=\alpha, |v|=\beta} (1+u)^n u^{-2m-1} \mathbf{F}_k(v^2) v^{-2m-1} du dv$$

$$= \frac{1}{(2\pi i)^2} \oint_{|u|=\alpha, |v|=\beta} (1+u)^n \frac{uv}{(uv)^2 - 1} \mathbf{F}_k(v^2) du dv$$

$$= \frac{1}{(2\pi i)^2} \oint_{|v|=\beta} \mathbf{F}_k(v^2) v^{-1} \left[\oint_{|u|=\alpha} \frac{(1+u)^n u}{(u + \frac{1}{v})(u - \frac{1}{v})} du \right] dv.$$

Since $u = \frac{1}{v}$ and $u = -\frac{1}{v}$ are the only singularities (poles) enclosed by the particular contour, eq. (2.10) implies

$$\oint_{|u|=\alpha} \frac{(1+u)^n u}{(u + \frac{1}{v})(u - \frac{1}{v})} du = 2\pi i \left[\frac{(1+u)^n u}{u - \frac{1}{v}} \Big|_{u=-\frac{1}{v}} + \frac{(1+u)^n u}{u + \frac{1}{v}} \Big|_{u=\frac{1}{v}} \right]$$

$$= \pi i \left(\left[1 - \frac{1}{v} \right]^n + \left[1 + \frac{1}{v} \right]^n \right).$$

Therefore, for $|z| < \mu_k$

$$\sum_{n \geq 0} \mathsf{M}_k(n) z^n$$

$$= \frac{1}{4\pi i} \sum_{n \geq 0} \oint_{|v|=\beta} \mathbf{F}_k(v^2) v^{-1} \left(\left[1 - \frac{1}{v} \right]^n + \left[1 + \frac{1}{v} \right]^n \right) z^n dv$$

$$= \frac{1}{4\pi i} \oint_{|v|=\beta} \mathbf{F}_k(v^2) \frac{1}{v - (v-1)z} dv + \frac{1}{4\pi i} \oint_{|v|=\beta} \mathbf{F}_k(v^2) \frac{1}{v - (v+1)z} dv.$$

The first integrand has its unique pole at $v = -\frac{z}{1-z}$ and the second at $v = \frac{z}{1-z}$, respectively:

$$\frac{1}{v - (v-1)z} = \frac{1}{v + \frac{z}{1-z}} \frac{1}{1-z} \quad \text{and} \quad \frac{1}{v - (v+1)z} = \frac{1}{v - \frac{z}{1-z}} \frac{1}{1-z}.$$

We derive

$$\sum_{n\geq 0} \mathsf{M}_k(n)z^n = \frac{1}{1-z}\left[\frac{1}{2}\mathbf{F}_k\left(\left(\frac{z}{1-z}\right)^2\right) + \frac{1}{2}\mathbf{F}_k\left(\left(\frac{z}{1-z}\right)^2\right)\right]$$

$$= \frac{1}{1-z}\mathbf{F}_k\left(\left(\frac{z}{1-z}\right)^2\right),$$

whence Lemma 2.1.

2.1.5 D-finiteness

The power series, $\mathbf{F}_k(x) = \sum_{n\geq 0} f_k(2n,0)x^n$, [125] is of central importance in Section 2.3 in the context of singularity analysis [42]. It is a D-finite power series and allows for analytic continuation in any simply connected domain containing zero.

Definition 2.10. *(a) A sequence $f(n)$ of complex number is said to be P-recursive, if there are polynomials $p_0(n), \ldots, p_m(n) \in \mathbb{C}[n]$ with $p_m(n) \neq 0$, such that for all $n \in \mathbb{N}$*

$$p_m(n)f(n+m) + p_{m-1}(n)f(n+m-1) + \cdots + p_0(n)f(n) = 0. \quad (2.11)$$

(b) A formal power series $F(x) = \sum_{n\geq 0} f(n)x^n$ is rational, if there are polynomials $A(x)$ and $B(x)$ in $\mathbb{C}[x]$ with $B(x) \neq 0$, such that

$$F(x) = \frac{A(x)}{B(x)}.$$

(c) $F(x)$ is algebraic, if there exist polynomials $q_0(x), \ldots, q_m(x) \in \mathbb{C}[x]$ with $q_m(x) \neq 0$, such that

$$q_m(x)F^m(x) + q_{m-1}(x)F^{m-1}(x) + \cdots + q_1(x)F(x) + q_0(x) = 0.$$

(d) $F(x)$ is D-finite, if there are polynomials $q_0(x), \ldots, q_m(x) \in \mathbb{C}[x]$ with $q_m(x) \neq 0$, such that

$$q_m(x)F^{(m)}(x) + q_{m-1}(x)F^{(m-1)}(x) + \cdots + q_1(x)F'(x) + q_0(x)F(x) = 0, \quad (2.12)$$

where $F^{(i)}(x) = d^i F(x)/dx^i$, and $\mathbb{C}[x]$ is the ring of polynomials in x with complex coefficients.

Let $\mathbb{C}(x)$ denote the rational function field, i.e., the field generated by taking equivalence classes of fractions of polynomials. Let $\mathbb{C}_{\text{alg}}[[x]]$ and $\mathcal{D}$ denote the sets of algebraic power series over $\mathbb{C}$ and D-finite power series, respectively. Clearly, a rational formal power series is in particular algebraic. Furthermore, if $u \in \mathbb{C}_{\text{alg}}[[x]]$, then u is also D-finite[127].

It is well known that a sequence is P-recursive if and only if its generating function is D-finite[125].

Lemma 2.11. *Suppose $F(z) = \sum_{n \geq 0} f(n) z^n$. Then $F(z)$ is D-finite if only if $f(n)$ is P-recursive.*

Proof. Since

$$z^j F^{(i)}(z) = \sum_{n \geq 0} (n + i - j)_i f(n + i - j) z^n, \qquad (2.13)$$

where $(n - j + i)_i = (n - j + i)(n - j + i - 1) \cdots (n - j + 1)$ denotes the falling factorials, combining eqs. (2.13) and (2.12) implies the recurrence of eq. (2.11) for $f(n)$ by equating the coefficients of z^n. Accordingly, we conclude that the coefficients $f(n)$ of the power series $F(z)$ are P-recursive and we can derive the unique recurrence from the differential equation (2.12) of $F(z)$. If a sequence $f(n)$ is P-recursive, then eq. (2.11) holds. Since each $p_i(n) \in \mathbb{C}[n]$ can be represented as $\mathbb{C}$-linear combination of $(n + i)_j$, $j \geq 0$, the term $\sum_{n \geq 0} p_i(n) f(n + i) z^n$ can also be represented as a $\mathbb{C}$-linear combination of series of the form $\sum_{n \geq 0} (n + i)_j f(n + i) z^n$. In view of

$$\sum_{n \geq 0} (n + i)_j f(n + i) z^n = R_i(z) + z^{j-i} F^{(j)}(z),$$

where $R_i(z) \in z^{-1} \mathbb{C}[z^{-1}]$, we can recover eq. (2.12) by multiplying eq. (2.11) with z^n and summing over $n \geq 0$. Thus for a given recurrence of $f(n)$, we can derive a unique differential equation of $F(z)$ in the form (2.12). ∎

Lemma 2.12. *Each P-recursion of $f_k(2n, 0)$, $\mathcal{R}$, having polynomial coefficients with greatest common divisor (gcd) one corresponds to a P-recursion of $e_k(n) = f_k(2n, 0)/(2n)!$, $\epsilon(\mathcal{R})$. Each P-recursion of $e_k(2n, 0)$, $\mathcal{Q}$, corresponds uniquely to a P-recursion of $f_k(2n, 0)$, $\omega(\mathcal{Q})$, having polynomial coefficients with gcd one. Furthermore, we have $\omega(\epsilon(\mathcal{R})) = \mathcal{R}$.*

Proof. Suppose we have a P-recurrence $\sum_{i=0}^{r_k} a_i(n) f_k(2(n + i), 0) = 0$, where $a_i(n)$ are polynomials in n with integer coefficients, having gcd one and $a_0(n) \neq 0$. Then

$$\sum_{i=0}^{r_k} a_i(n) (2(n + i))_{2i} \, e_k(n + i) = 0,$$

i.e., a P-recurrence for $e_k(n)$. Suppose now we have a P-recurrence for $e_k(n)$, $\sum_{i=0}^{r_k} b_i(n) e_k(n + i) = 0$, where the $b_i(n)$ are all polynomials of n with integer coefficients, and $b_0(n) \neq 0$. We then immediately derive

$$\sum_{i=0}^{r_k} c_i(n) f_k(2(n + i), 0) = 0,$$

where $c_i(n) = b_i(n) \frac{(2n)!}{(2(n+i))!}$. $c_i(n)$ are *rational* functions in n. Suppose $d(n)$ is the lcm of the denominators of the $c_i(n)$. Then

$$\sum_{i=0}^{r_k} c'_i(n) f_k(2(n+i), 0) = 0,$$

where the $c'_i(n) = d(n) b_i(n) \frac{(2n)!}{(2(n+i))!}$ are by construction polynomials, having gcd one and $c'_0(n) \neq 0$, whence the lemma.

We proceed by studying closure properties of D-finite power series which are of key importance in the following chapters.

Theorem 2.13. (Stanley [127]) *P-recursive sequences, D-finite, and algebraic power series have the following properties:*

(a) If f, g are P-recursive, then $f \cdot g$ is P-recursive.
(b) If $F, G \in \mathcal{D}$, and $\alpha, \beta \in \mathbb{C}$, then $\alpha F + \beta G \in \mathcal{D}$ and $FG \in \mathcal{D}$.
(c) If $F \in \mathcal{D}$ and $G \in \mathbb{C}_{alg}[[x]]$ with $G(0) = 0$, then $F(G(x)) \in \mathcal{D}$.

Here we omit the proof of (a) and (b) which can be found in [127]. We present, however, a direct proof of (c).

Proof. (c) We assume that $G(0) = 0$ so that the composition $F(G(x))$ is well defined. Let $K = F(G(x))$. Then $K^{(i)}$ is a linear combination of $F(G(x))$, $F'(G(x))$, ..., over $\mathbb{C}[G, G', \ldots]$, i.e., the ring of polynomials in $G, G', \ldots$ with complex coefficients.

Claim. $G^{(i)} \in \mathbb{C}(x, G)$, $i \geq 0$, and therefore $\mathbb{C}[G, G', \ldots] \subset \mathbb{C}(x, G)$, where $\mathbb{C}(x, G)$ denotes the field generated by x and G.
Since G is algebraic, it satisfies

$$q_d(x) G^d(x) + q_{d-1}(x) G^{d-1}(x) + \cdots + q_1(x) G(x) + q_0(x) = 0, \qquad (2.14)$$

where $q_0(x), \ldots, q_d(x) \in \mathbb{C}[x]$, $q_d(x) \neq 0$ and d is minimal, i.e., $(G^i(x))_{i=0}^{d-1}$ is linear independent over $\mathbb{C}[x]$. In other words, for all $(\tilde{q}_i(x))_{i=1}^{d-1} \neq 0$ we have

$$\tilde{q}_{d-1}(x) G^{d-1}(x) + \cdots + \tilde{q}_1(x) G(x) + \tilde{q}_0(x) \neq 0.$$

We consider

$$P(x, G) = q_d(x) G^d(x) + q_{d-1}(x) G^{d-1}(x) + \cdots + q_1(x) G(x) + q_0(x).$$

Differentiating eq. (2.14) once, we derive

$$0 = \frac{d}{dx} P(x, G) = \frac{\partial P(x, y)}{\partial x} \bigg|_{y=G} + G' \frac{\partial P(x, y)}{\partial y} \bigg|_{y=G}.$$

The degree of $\frac{\partial P(x, y)}{\partial y} \big|_{y=G}$ in G is smaller than $d - 1$ and $q_d(x) \neq 0$, whence

$$\frac{\partial P(x, y)}{\partial y} \bigg|_{y=G} \neq 0. \text{ We therefore arrive at}$$

$$G' = -\frac{\frac{\partial P(x,y)}{\partial x}\big|_{y=G}}{\frac{\partial P(x,y)}{\partial y}\big|_{y=G}} \in \mathbb{C}(x,G).$$

Iterating the above argument, we obtain $G^{(i)} \in \mathbb{C}(x,G)$, $i \geq 0$, and therefore $\mathbb{C}[G, G', \ldots] \subset \mathbb{C}(x,G)$, whence the claim.

Let $\widetilde{V}$ be the $\mathbb{C}(x,G)$ vector space spanned by $F(G(x))$, $F'(G(x))$, $\ldots$. Since $F \in \mathcal{D}$, we have $\dim_{\mathbb{C}(x)}\langle F, F', \cdots \rangle < \infty$, immediately implying the finiteness of $\dim_{\mathbb{C}(G)}\langle F(G), F'(G), \cdots \rangle$. Thus, since $\mathbb{C}(G)$ is a subfield of $\mathbb{C}(x,G)$, we derive

$$\dim_{\mathbb{C}(x,G)}\langle F, F', \cdots \rangle < \infty$$

and consequently $\dim_{\mathbb{C}(x,G)} \widetilde{V} < \infty$ and $\dim_{\mathbb{C}(x)} \mathbb{C}(x,G) < \infty$. As a result

$$\dim_{\mathbb{C}(x)} \widetilde{V} = \dim_{\mathbb{C}(x,G)} \widetilde{V} \cdot \dim_{\mathbb{C}(x)} \mathbb{C}(x,G) < \infty$$

follows and since each $K^{(i)} \in \widetilde{V}$, we conclude that $F(G(x))$ is D-finite.

Corollary 2.14. *The generating function of k-noncrossing matchings over $2n$ vertices, $\mathbf{F}_k(z) = \sum_{n\geq 0} f_k(2n,0)\, z^n$, is D-finite.*

Proof. Corollary 2.6 gives the exponential generating function of $f_k(2n,0)$

$$\sum_{n\geq 1} f_k(2n,0)\frac{x^{2n}}{(2n)!} = \det[I_{i-j}(2x) - I_{i+j}(2x)]_{i,j=1}^{k-1}, \qquad (2.15)$$

where $I_m(x)$ is Bessel function of the first order. Recall that the Bessel function of the first kind satisfies $I_n(x) = i^{-n}J_n(ix)$ and $J_n(x)$ is the solution of the Bessel differential equation

$$x^2\frac{d^2y}{dx^2} + x\frac{dy}{dx} + (x^2 - n^2)y = 0.$$

For every fixed $n \in \mathbb{N}$, $J_n(x)$ is D-finite. Let $G(x) = ix$. Clearly, $G(x) \in \mathbb{C}_{\mathrm{alg}}[[x]]$ and $G(0) = 0$, $J_n(ix)$ and $I_n(x)$ are accordingly D-finite in view of the assertion (c) of Theorem 2.13. Analogously we show that $I_n(2x)$ is D-finite for every fixed $n \in \mathbb{N}$. Using eq. (2.15) and assertion (b) of Theorem 2.13, we conclude that

$$\mathbf{H}_k(x) = \sum_{n\geq 0} \frac{f_k(2n,0)}{(2n)!} x^{2n}$$

is D-finite. In other words the sequence $f(n) = \frac{f_k(2n,0)}{(2n)!}$ is P-recursive and furthermore $g(n) = (2n)!$ is, in view of $(2n+1)(2n+2)g(n) - g(n+1) = 0$, P-recursive. Therefore, $f_k(2n,0) = f(n)g(n)$ is P-recursive. This proves that $\mathbf{F}_k(z) = \sum_{n\geq 0} f_k(2n,0)z^n$ is D-finite.

2.2 Symbolic enumeration

In the following we will compute various generating functions via the symbolic enumeration method [42].

Definition 2.15. *A combinatorial class is a set $\mathcal{C}$ together with a size function, $w_{\mathcal{C}} \colon \mathcal{C} \longrightarrow \mathbb{Z}^{+}$, $(\mathcal{C}, w_{\mathcal{C}})$ such that $w_{\mathcal{C}}^{-1}(n)$ is finite for any $n \in \mathbb{Z}^{+}$.*

Suppose $(\mathcal{C}, w_{\mathcal{C}})$ is a combinatorial class and $c \in \mathcal{C}$. We call $w_{\mathcal{C}}(c)$ the size of c and write simply $w(c)$. There are two special combinatorial classes: $\mathcal{E}$ and $\mathcal{Z}$ which contain only one element of sizes 0 and 1, respectively. The subset of $\mathcal{C}$ which contains all the elements of size n, $w_{\mathcal{C}}^{-1}(n)$, is denoted by $\mathcal{C}_n$, and let $C_n = |\mathcal{C}_n|$. The generating function of a combinatorial class $(\mathcal{C}, w_{\mathcal{C}})$ is given by

$$\mathbf{C}(z) = \sum_{c \in \mathcal{C}} z^{w_{\mathcal{C}}(c)} = \sum_{n \geq 0} C_n z^n,$$

where $\mathcal{C}_n \subset \mathcal{C}$. In particular, the generating functions of the classes $\mathcal{E}$ and $\mathcal{Z}$ are

$$\mathbf{E}(z) = 1 \quad \text{and} \quad \mathbf{Z}(z) = z. \tag{2.16}$$

Definition 2.16. *Suppose $\mathcal{C}, \mathcal{D}$ are combinatorial classes. Then $\mathcal{C}$ is isomorphic to $\mathcal{D}$, $\mathcal{C} \cong \mathcal{D}$, if and only if*

$$\forall\, n \geq 0, \quad |\mathcal{C}_n| = |\mathcal{D}_n|.$$

In the following we shall identify isomorphic combinatorial classes and write $\mathcal{C} = \mathcal{D}$ if $\mathcal{C} \cong \mathcal{D}$. We set

- $\mathcal{C} + \mathcal{D} := \mathcal{C} \cup \mathcal{D}$, if $\mathcal{C} \cap \mathcal{D} = \varnothing$ and for $a \in \mathcal{C} + \mathcal{D}$,

$$w_{\mathcal{C}+\mathcal{D}}(a) = \begin{cases} w_{\mathcal{C}}(a) & \text{if } a \in \mathcal{C} \\ w_{\mathcal{D}}(a) & \text{if } a \in \mathcal{D}. \end{cases}$$

- $\mathcal{C} \times \mathcal{D} := \{a = (c, d) \mid c \in \mathcal{C}, d \in \mathcal{D}\}$ and for $a \in \mathcal{C} \times \mathcal{D}$,

$$w_{\mathcal{C} \times \mathcal{D}}(a) = w_{\mathcal{C}}(c) + w_{\mathcal{D}}(d).$$

We furthermore set

- $\mathcal{C}^m := \prod_{h=1}^{m} \mathcal{C}$ and
- $\mathrm{SEQ}(\mathcal{C}) := \mathcal{E} + \mathcal{C} + \mathcal{C}^2 + \cdots.$

In view of eq. (2.16), $\mathrm{SEQ}(\mathcal{C})$ is a combinatorial class if and only if there is no element in $\mathcal{C}$ of size 0.

Theorem 2.17. *Suppose* $\mathcal{A}$, $\mathcal{C}$, *and* $\mathcal{D}$ *are combinatorial classes with generating functions* $\mathbf{A}(z)$, $\mathbf{C}(z)$, *and* $\mathbf{D}(z)$. *Then*

(a) $\mathcal{A} = \mathcal{C} + \mathcal{D} \Longrightarrow \mathbf{A}(z) = \mathbf{C}(z) + \mathbf{D}(z)$,
(b) $\mathcal{A} = \mathcal{C} \times \mathcal{D} \Longrightarrow \mathbf{A}(z) = \mathbf{C}(z) \cdot \mathbf{D}(z)$,
(c) $\mathcal{A} = \text{SEQ}(\mathcal{C}) \Longrightarrow \mathbf{A}(z) = \frac{1}{1 - \mathbf{C}(z)}$.

Proof. Suppose $\mathcal{A} = \mathcal{C} + \mathcal{D}$, then

$$\mathbf{A}(z) = \sum_{a \in \mathcal{A}} z^{w_{\mathcal{A}}(a)} = \sum_{a \in \mathcal{C}} z^{w_{\mathcal{C}}(a)} + \sum_{a \in \mathcal{D}} z^{w_{\mathcal{D}}(a)} = \mathbf{C}(z) + \mathbf{D}(z).$$

In case of $\mathcal{A} = \mathcal{C} \times \mathcal{D}$, we compute

$$\begin{aligned}
\mathbf{A}(z) &= \sum_{a \in \mathcal{A}} z^{w_{\mathcal{A}}(a)} \\
&= \sum_{(c,d) \in \mathcal{C} \times \mathcal{D}} z^{w_{\mathcal{C}}(c) + w_{\mathcal{D}}(d)} \\
&= \left(\sum_{c \in \mathcal{C}} z^{w_{\mathcal{C}}(c)} \right) \cdot \left(\sum_{d \in \mathcal{D}} z^{w_{\mathcal{D}}(d)} \right) \\
&= \mathbf{C}(z) \cdot \mathbf{D}(z).
\end{aligned}$$

Consequently, in case of $\mathcal{A} = \text{SEQ}(\mathcal{C})$,

$$\mathbf{A}(z) = 1 + \mathbf{C}(z) + \mathbf{C}(z)^2 + \cdots = \frac{1}{1 - \mathbf{C}(z)}.$$

In order to keep track of some specific combinatorial class in order to express multivariate generating functions, we introduce the concept of combinatorial markers. A combinatorial marker is a combinatorial class with only one element of size 0 or one element of size 1.

For instance, suppose $\mathcal{F}_{k,h}$ is the combinatorial class of all k-noncrossing matchings with h arcs and its size function is the length of a matching in $\mathcal{F}_{k,h}$, i.e., the number of vertices. Let $\mathcal{P}_{k,h}$ denote the combinatorial class of all the k-noncrossing partial matchings with h arcs and its size function counting the total number of vertices. Let $\mathcal{Z}$ represent the combinatorial class consisting of a single vertex. Then, plainly

$$\mathcal{P}_{k,h} = \mathcal{F}_{k,h} \times (\text{SEQ}(\mathcal{Z}))^{2h+1}.$$

Suppose now we want to keep track of the number of isolated vertices in a k-noncrossing partial matching having h arcs. Then we introduce the combinatorial marker μ in order to keep track of the isolated vertices as follows:

$$\mathcal{P}_{k,h} = \mathcal{F}_{k,h} \times (\text{SEQ}(\mu \times \mathcal{Z}))^{2h+1},$$

whence

$$\mathbf{P}_{k,h}(z, u) = \mathbf{F}_{k,h}(z) \cdot \left(\frac{1}{1 - uz} \right)^{2h+1},$$

where $\mathbf{P}_{k,h}(z, u)$ and $\mathbf{F}_{k,h}(z)$ are the generating functions of the combinatorial classes $\mathcal{P}_{k,h}$ and $\mathcal{F}_{k,h}$ and u is an indeterminant.

2.3 Singularity analysis

Let $f(z) = \sum_n a_n z^n$ be a generating function with radius of convergence, R. In light of the fact that explicit formulas for the coefficients a_n can be very complicated or even impossible to obtain, we shall investigate the generating function $f(z)$ by deriving information about a_n for large n.

In the following we are primarily concerned with the estimation of a_n in terms of the exponential factor γ and the subexponential factor $P(n)$, that is, we have the following situation

$$a_n \sim P(n) \cdot \gamma^n, \tag{2.17}$$

where γ is a fixed number and $P(n)$ is a polynomial in n. While this is, of course, a vast simplification of the original problem (explicit computation of the coefficients a_n), eq. (2.17) extracts key information about the coefficients.

2.3.1 Transfer theorems

The derivation of exponential growth rate and subexponential factors of eq. (2.17) mainly rely on singular expansions and transfer theorems. Transfer theorems realize the translation of error terms from functions to coefficients. The underlying basic tool here is, of course, Cauchy's integral formula

$$a_n = \frac{1}{2\pi i} \oint_C \frac{f(z)}{z^{n+1}} dz,$$

where C is any simple closed curve in the region $0 < |x| < R$, containing 0.

In the following we shall employ a particular integration path; see Fig. 2.13. The contour is a path, slightly "outside" the disc of radius R. This contour is comprised of an inner arc segment 3 and an outer arc segment 1 and two connecting linear part segments 2 and 4. The major contribution to the contour integral stems from segments 2, 3, and 4.

The behavior of $f(z)$ close to the dominant singularity is the determining factor for the asymptotic behavior of its coefficients. Let us get started by specifying a suitable domain for our contours.

Definition 2.18. *Given two numbers ϕ, r, where $r > |\rho|$ and $0 < \phi < \frac{\pi}{2}$, the open domain $\Delta_\rho(\phi, r)$ is defined as*

$$\Delta_\rho(\phi, r) = \{z \mid |z| < r, z \neq \rho, |\mathrm{Arg}(z - \rho)| > \phi\}.$$

A domain is a Δ_ρ-domain at ρ if it is of the form $\Delta_\rho(\phi, r)$ for some r and ϕ.
A function is Δ_ρ-analytic if it is analytic in some Δ_ρ-domain.

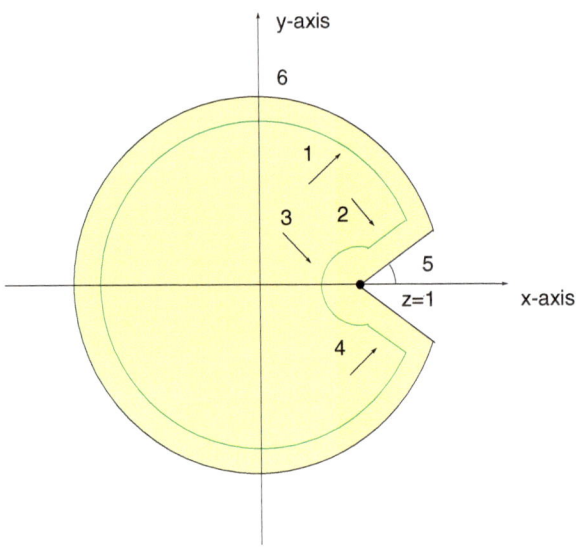

Fig. 2.13. Δ_1-domain enclosing a contour. We assume $z = 1$ to be the unique dominant singularity. The coefficients are obtained via Cauchy's integral formula and the integral path is decomposed into four segments. Segment 1 becomes asymptotically irrelevant since by construction the function involved is bounded on this segment. Relevant are the rectilinear segments 2 and 4 and the inner circle 3. The only contributions to the contour integral are being made here.

Let $[z^n]\, f(z)$ denote the coefficient of z^n of the power series expansion of $f(z)$ at 0. Since the Taylor coefficients have the property

$$\forall \gamma \in \mathbb{C} \setminus 0; \quad [z^n] f(z) = \gamma^n [z^n] f\left(\frac{z}{\gamma}\right),$$

we can, without loss of generality, reduce our analysis to the case where $z = 1$ is the unique dominant singularity. We use $U(a, r) = \{z \in \mathbb{C} \mid |z - a| < r\}$ in order to denote the open neighborhood of a in $\mathbb{C}$. Furthermore, we use the notations

$$(f(z) = O\,(g(z)) \text{ as } z \to \rho) \iff (f(z)/g(z) \text{ is bounded as } z \to \rho),$$
$$(f(z) = o\,(g(z)) \text{ as } z \to \rho) \iff (f(z)/g(z) \to 0 \text{ as } z \to \rho),$$
$$(f(z) = \Theta\,(g(z)) \text{ as } z \to \rho) \iff (f(z)/g(z) \to c \text{ as } z \to \rho),$$
$$(f(z) \sim g(z) \text{ as } z \to \rho) \iff (f(z)/g(z) \to 1 \text{ as } z \to \rho),$$

where c is some constant. If we write $f(z) = O(g(z))$, $f(z) = o(g(z))$, $f(z) = \Theta(g(z))$, or $f(z) \sim g(z)$, it is implicitly assumed that z tends to a (unique) singularity.

Theorem 2.19. (Waterman [41]) *(a) Suppose* $f(z) = (1-z)^{-\alpha}$, $\alpha \in \mathbb{C}\backslash\mathbb{Z}_{\leq 0}$, *then*

$$[z^n] f(z) \sim \frac{n^{\alpha-1}}{\Gamma(\alpha)} \left[1 + \frac{\alpha(\alpha-1)}{2n} + \frac{\alpha(\alpha-1)(\alpha-2)(3\alpha-1)}{24n^2} + \frac{\alpha^2(\alpha-1)^2(\alpha-2)(\alpha-3)}{48n^3} + O\left(\frac{1}{n^4}\right) \right].$$

(b) Suppose $f(z) = (1-z)^r \log(\frac{1}{1-z})$, $r \in \mathbb{Z}_{\geq 0}$, *then we have*

$$[z^n]f(z) \sim (-1)^r \frac{r!}{n(n-1)\dots(n-r)}.$$

Theorems 2.19 and 2.20 are the key tools for the singularity analysis of the generating function of RNA pseudoknot structures.

Theorem 2.20. (Flajolet and Sedgewick [42]) *Let* $f(z)$ *be a* Δ_1-*analytic function at its unique singularity* $z = 1$. *Let* $g(z)$ *be a linear combination of functions in the set* B, *where*

$$B = \{(1-z)^\alpha \log^\beta \left(\frac{1}{1-z}\right) | \alpha, \beta \in \mathbb{R}\},$$

that is, we have in the intersection of a neighborhood of 1 *with the* Δ_1-*domain*

$$f(z) = o(g(z)) \quad \text{for } z \to 1.$$

Then we have

$$[z^n]f(z) = o([z^n]g(z)),$$

where $\circ \in \{O, o, \Theta, \sim\}$.

Let $S(\rho, n)$ denote the subexponential factor of $[z^n] f(z)$ at the dominant singularity ρ. In general [42], if $f(z)$ has multiple dominant singularities, $[z^n] f(z)$ is asymptotically determined by the sum over *all* dominant singularities, i.e.,

$$[z^n] f(z) \sim \sum_i S(\rho_i, n)\rho_i^n.$$

2.3.2 The supercritical paradigm

In this section we discuss an implication of Theorem 2.20. The supercritical paradigm refers to a composition of two functions where the "inner" function is regular at the singularity of the outer function. In this case the singularity type is that of the "outer" function. What happens is that the inner function only "shifts" the singularity of the outer function.

The scenario considered here is tailored for Chapters 4 and 5.

Theorem 2.21. *Let $\psi(z, s)$ be an algebraic, analytic function in a domain $\mathcal{D} = \{(z, s) | |z| \leq r, |s| < \epsilon\}$ such that $\psi(0, s) = 0$. In addition suppose $\gamma(s)$ is the unique dominant singularity of $\mathbf{F}_k(\psi(z, s))$ and unique analytic solution of $\psi(\gamma(s), s) = \rho_k^2$, $|\gamma(s)| \leq r$, $\partial_z \psi(\gamma(s), s) \neq 0$ for $|s| < \epsilon$. Then $\mathbf{F}_k(\psi(z, s))$ has a singular expansion and*

$$[z^n]\mathbf{F}_k(\psi(z, s)) \sim A(s)\, n^{-((k-1)^2 + (k-1)/2)} \left(\frac{1}{\gamma(s)}\right)^n, \qquad (2.18)$$

uniformly in s contained in a small neighborhood of 0 and $A(s)$ is continuous.

We postpone the proof of Theorem 2.21 to Section 2.4.2. The key property of the singular expansion of Theorem 2.21 is the uniformity of eq. (2.18) in the parameter s.

In the following chapters, we will be working with compositions $\mathbf{F}_k(\vartheta(z))$, where $\vartheta(z)$ is algebraic and satisfies $\vartheta(0) = 0$, that is, we apply Theorem 2.21 for fixed parameter s. According to Theorem 2.13, $\mathbf{F}_k(\vartheta(z))$ is D-finite and Theorem 2.21 implies that if ϑ satisfies certain conditions the subexponential factors of $\mathbf{F}_k(\vartheta(z))$ coincide with those of $\mathbf{F}_k(z)$.

2.4 The generating function $\mathbf{F}_k(z)$

While Theorems 2.7 and 2.8 shed light of the generating function $\mathbf{F}_k(z)$, Theorem 2.21 motivates a closer look in particular at its singular expansion. The key to this is to find the ODE that $\mathbf{F}_k(z)$ satisfies. This is not "just" a matter of computation, in Proposition 2.22 we have to prove that the latter are correct.

2.4.1 Some ODEs

In Section 2.1.5, we have shown that $\mathbf{F}_k(z)$ is D-finite, that is, there exists some $e \in \mathbb{N}$ for which $\mathbf{F}_k(z)$ satisfies an ODE of the form

$$q_{0,k}(z)\frac{d^e}{dz^e}\mathbf{F}_k(z) + q_{1,k}(z)\frac{d^{e-1}}{dz^{e-1}}\mathbf{F}_k(z) + \cdots + q_{e,k}(z)\mathbf{F}_k(z) = 0, \qquad (2.19)$$

where $q_{j,k}(z)$ are polynomials. The fact that $\mathbf{F}_k(z)$ is the solution of an ODE implies the existence of an analytic continuation into any simply connected domain [125], i.e., $\Delta_{\rho_k^2}$-analyticity.

Explicit knowledge of the above ODE is of key importance for two reasons:

- Any dominant singularity of a solution is contained in the set of roots of $q_{0,k}(z)$ [125]. In other words the ODE "controls" the dominant singularities that are crucial for asymptotic enumeration.
- Under certain regularity conditions (discussed below) the singular expansion of $\mathbf{F}_k(z)$ follows from the ODE; see Proposition 2.24.

Accordingly, let us first compute for $2 \leq k \leq 9$ the ODEs for $\mathbf{F}_k(z)$.

Proposition 2.22. *For $2 \leq k \leq 9$, $\mathbf{F}_k(z)$ satisfies the ODEs listed in Table 2.1 and we have in particular*

$$q_{0,2}(z) = (4z - 1)\, z, \tag{2.20}$$

$$q_{0,3}(z) = (16z - 1)\, z^2, \tag{2.21}$$

$$q_{0,4}(z) = (144\, z^2 - 40\, z + 1)\, z^3, \tag{2.22}$$

$$q_{0,5}(z) = (1024\, z^2 - 80\, z + 1)\, z^4, \tag{2.23}$$

$$q_{0,6}(z) = (14{,}400\, z^3 - 4144\, z^2 + 140\, z - 1)\, z^5, \tag{2.24}$$

$$q_{0,7}(z) = (147{,}456\, z^3 - 12{,}544\, z^2 + 224\, z - 1)\, z^6, \tag{2.25}$$

$$q_{0,8}(z) = (2{,}822{,}400 z^4 - 826{,}624 z^3 + 31{,}584 z^2 - 336 z + 1) z^7, \tag{2.26}$$

$$q_{0,9}(z) = (37{,}748{,}736 z^4 - 3{,}358{,}720 z^3 + 69{,}888 z^2 - 480 z + 1) z^8, \tag{2.27}$$

Proposition 2.22 immediately implies the following sets of roots:

$$\nabla_2 = \left\{\frac{1}{4}\right\}; \ \nabla_4 = \nabla_2 \cup \left\{\frac{1}{36}\right\}; \ \nabla_6 = \nabla_4 \cup \left\{\frac{1}{100}\right\}; \ \nabla_8 = \nabla_6 \cup \left\{\frac{1}{196}\right\};$$

$$\nabla_3 = \left\{\frac{1}{16}\right\}; \ \nabla_5 = \nabla_3 \cup \left\{\frac{1}{64}\right\}; \ \nabla_7 = \nabla_5 \cup \left\{\frac{1}{144}\right\}; \ \nabla_9 = \nabla_7 \cup \left\{\frac{1}{256}\right\}.$$

Equations (2.20), (2.21), (2.22), (2.23), (2.24), (2.25), (2.26), and (2.27) and Theorem 2.8 show that for $2 \leq k \leq 9$ the unique dominant singularity of $\mathbf{F}_k(z)$ is given by ρ_k^2, where $\rho_k = 1/2(k-1)$.

Proof. The ODEs for $\mathbf{F}_k(z)$, $2 \leq k \leq 9$, listed in Table 2.1, induce according to Lemma 2.11 uniquely respective P-recurrences $\mathcal{R}_k$. For $2 \leq k \leq 9$ the polynomial coefficients of any $\mathcal{R}_k$ have a greatest common divisor (gcd) of 1 and, in addition, the coefficient of the $f_k(2n, 0)$-term in $\mathcal{R}_k$ is nonzero. According to Lemma 2.12, each $\mathcal{R}_k$ corresponds to a unique P-recurrence $\epsilon(\mathcal{R}_k)$ for $f_k(2n, 0)/(2n)!$, which in turn corresponds uniquely to an ODE for the exponential generating function $\mathbf{H}_k(z) = \sum_{n \geq 0} f_k(2n, 0) \cdot \frac{z^{2n}}{(2n)!}$; see Corollary 2.14. We furthermore have according to eq. (2.15)

$$\sum_{n \geq 1} f_k(2n, 0) \frac{x^{2n}}{(2n)!} = \det[I_{i-j}(2x) - I_{i+j}(2x)]_{i,j=1}^{k-1}.$$

According to Lemma 2.11 the P-recurrences $\epsilon(\mathcal{R}_k)$ induce respective ODEs for $\mathbf{H}_k(z)$. The key point is now that for $\mathbf{H}_k(z)$, eq. (2.15) provides an interpretation of $\mathbf{H}_k(z)$ as a determinant of Bessel functions. We proceed by verifying for $2 \leq k \leq 9$ that $\det[I_{i-j}(2x) - I_{i+j}(2x)]_{i,j=1}^{k-1}$ satisfies the $\mathbf{H}_k(z)$-ODEs derived from Table 2.1 via Lemmas 2.11 and 2.12. Consequently we have now established the correctness of the derived $\mathbf{H}_k(z)$-ODEs. These allow us via Lemmas 2.12 and 2.11 to recover the ODEs listed in Table 2.1 and the proposition follows.

2.4.2 The singular expansion of $\mathbf{F}_k(z)$

Let us begin by introducing some concepts: a meromorphic ODE is an ODE of the form

$$f^{(r)}(z) + d_1(z)f^{(r-1)}(z) + \cdots + d_r(z)f(z) = 0, \qquad (2.28)$$

where $f^{(m)}(z) = \frac{d^m}{dz^m}f(z)$, $0 \le m \le r$ and the $d_j(z)$, are meromorphic in some domain Ω. Assuming that ζ is a pole of a meromorphic function $d(z)$, $\omega_\zeta(d)$ denotes the order of the pole ζ. In case $d(z)$ is analytic at ζ we write $\omega_\zeta(d) = 0$.

Meromorphic differential equations have a singularity at ζ if at least one of the $\omega_\zeta(d_j)$ is positive. Such a ζ is said to be a regular singularity if

$$\forall 1 \le j \le r; \quad \omega_\zeta(d_j) \le j$$

and an irregular singularity otherwise. The indicial equation $I(\alpha) = 0$ of a differential equation of the form (2.28) at its regular singularity ζ is given by

$$I(\alpha) = (\alpha)_r + \delta_1(\alpha)_{r-1} + \cdots + \delta_r, \qquad (\alpha)_\ell := \alpha(\alpha-1)\cdots(\alpha-\ell+1),$$

where $\delta_j := \lim_{z \to \alpha}(z-\alpha)^j d_j(z)$.

Theorem 2.23. (Henrici; Wasow [66, 140]) *Suppose we are given a meromorphic differential equation (2.28) with regular singularity ζ. Then, in a slit neighborhood of ζ, any solution of eq. (2.28) is a linear combination of functions of the form*

$$(z-\zeta)^{\alpha_i}(\log(z-\zeta))^{\ell_{ij}} H_{ij}(z-\zeta), \quad \text{for } 1 \le i \le r, \ 1 \le j \le i,$$

where $\alpha_1, \ldots, \alpha_r$ are the roots of the indicial equation at ζ, ℓ_{ij} are non-negative integer, and each H_{ij} is analytic at 0.

According to Proposition 2.22, the ODEs for $\mathbf{F}_k(z)$ for $2 \le k \le 9$ are known. We next proceed by deriving from these ODEs the singular expansion of $\mathbf{F}_k(z)$.

Proposition 2.24. *For $2 \le k \le 9$, the singular expansion of $\mathbf{F}_k(z)$ for $z \to \rho_k^2$ is given by*

$$\mathbf{F}_k(z) = \begin{cases} P_k(z-\rho_k^2) + c_k'(z-\rho_k^2)^{((k-1)^2+(k-1)/2)-1} \log(z-\rho_k^2)\,(1+o(1)) \\ P_k(z-\rho_k^2) + c_k'(z-\rho_k^2)^{((k-1)^2+(k-1)/2)-1}\,(1+o(1)) \end{cases}$$

depending on k being odd or even. Furthermore, the terms $P_k(z)$ are polynomials of degree not larger than $(k-1)^2 + (k-1)/2 - 1$, c_k' is some constant, and $\rho_k = 1/2(k-1)$.

Note the appearance of the logarithmic term for odd k in the singular expansion of $\mathbf{F}_k(z)$.

Proof. Claim 1. The dominant singularity ρ_k^2 of the ordinary differential equation of $\mathbf{F}_k(z)$ is regular.

We express eq. (2.19) as

$$\mathbf{F}_k^{(r_k)}(z) + \frac{q_{1,k}(z)}{q_{0,k}(z)}\mathbf{F}_k^{(r_k-1)}(z) + \frac{q_{2,k}(z)}{q_{0,k}(z)}\mathbf{F}_k^{(r_k-2)}(z) + \cdots + \frac{q_{r_k,k}(z)}{q_{0,k}(z)}\mathbf{F}_k(z) = 0,$$

writing $\mathbf{F}_k^{(m)}(z) = \frac{d^m}{dz^m}\mathbf{F}_k(z)$ for $0 \le m \le r_k$. For $2 \le k \le 9$, see Table 2.1, $q_{0,k}(z)$ has simple nonzero roots. Since all singularities of $\mathbf{F}_k(z)$

- are contained in the roots of $q_{0,k}(z)$ and
- according to Theorem 2.8 we have

$$f_k(2n,0) \sim c_k\, n^{-((k-1)^2+(k-1)/2)}\,(2(k-1))^{2n}, \quad \text{where } c_k > 0$$

and accordingly derive

$$q_{0,k}(z) = (z - \rho_k^2)q_{0,k}'(z),$$

where $q_{0,k}'(z)$ has also simple nonzero roots. Let

$$d_{j,k}(z) = q_{j,k}(z)/q_{0,k}(z), \quad 1 \le j \le k.$$

Then

$$(z - \rho_k^2)^j d_{j,k}(z) = (z - \rho_k^2)^j \frac{q_{j,k}(z)}{q_{0,k}(z)} = (z - \rho_k^2)^{j-1}\frac{q_{j,k}(z)}{q_{0,k}'(z)}. \tag{2.29}$$

We set $\delta_{j,k} = \lim_{z \to \rho_k^2}(z-\rho_k^2)^j d_{j,k}(z)$. Equation (2.29) shows that $\delta_{1,k}$ exists and $\delta_{j,k} = 0$ for $j \ge 2$. Furthermore, the order of the pole of $d_{j,k}(z)$, for $j \ge 1$, at ρ_k^2 is at most 1. Therefore, for $2 \le k \le 9$, the dominant singularity, ρ_k^2, is unique and regular.

According to Claim 1 the singularity ρ_k^2 is regular and Theorem 2.23 implies

$$\mathbf{F}_k(z) = \sum_{i=1}^{k}\sum_{j=1}^{i}\lambda_{ij}(z-\rho_k^2)^{\alpha_i}\log^{\ell_{ij}}(z-\rho_k^2)H_{ij}(z-\rho_k^2), \tag{2.30}$$

where ℓ_{ij} is a non-negative integer, H_{ij} is analytic at 0, and $\alpha_1,\alpha_2,\ldots,\alpha_k$ are the roots of the indicial equation, $\lambda_{ij} \in \mathbb{C}$. For $2 \le k \le 9$ we derive from the indicial equations

$$\alpha_i = \begin{cases} i-1 & \text{for } i \le k-1, \\ (k-1)^2 + \frac{k-1}{2} - 1 & \text{for } i = k. \end{cases}$$

Since H_{ij} is analytic at 0, its Taylor expansion at 0 exists

$$(z-\rho_k^2)^{\alpha_i}\log^{\ell_{ij}}(z-\rho_k^2)H_{i,j}(z-\rho_k^2) = \sum_{t=0}^{\infty}a_{ijt}(z-\rho_k^2)^{\alpha_i+t}\log^{\ell_{ij}}(z-\rho_k^2).$$

Substituting the Taylor expansion into (2.30), we obtain

$$\mathbf{F}_k(z) = \sum_{i=1}^{k} \sum_{j=1}^{i} \sum_{t=0}^{\infty} a_{ijt}(z - \rho_k^2)^{\alpha_i + t} \log^{\ell_{ij}}(z - \rho_k^2). \qquad (2.31)$$

We set

$$M_1 = \{(i, j, t) \mid 1 \le i \le k, 1 \le j \le i, 0 \le t, a_{ijt} \ne 0, \ell_{i,j} > 0\},$$
$$M_2 = \{(i, j, t) \mid 1 \le i \le k, 1 \le j \le i, 0 \le t, a_{ijt} \ne 0, \alpha_i + t \notin \mathbb{N}\},$$

and $M = M_1 \cup M_2$. Clearly, M is not empty since $\mathbf{F}_k(z)$ would be analytic at $z = \rho_k^2$, otherwise. Let

$$m_k = \min\{\alpha_i + t \mid (i, j, t) \in M, a_{ijt} \ne 0\}$$
$$l_k = \max\{\ell_{ij} \mid \alpha_i + t = m_k, (i, j, t) \in M, a_{ijt} \ne 0\}$$

and let c_k' denotes the coefficient of $(z - \rho_k^2)^{m_k} \log^{l_k}(z - \rho_k^2)$ in eq. (2.31). By construction we then arrive at

$$\mathbf{F}_k(z) = P_k(z - \rho_k^2) + c_k'(z - \rho_k^2)^{m_k} \log^{l_k}(z - \rho_k^2)(1 + o(1)), \qquad (2.32)$$

where $P_k(z)$ is a polynomial of degree $\le m_k$ and Theorem 2.20 implies

$$[z^n]\mathbf{F}_k(z) \sim [z^n]c_k'(z - \rho_k^2)^{m_k} \log^{l_k}(z - \rho_k^2). \qquad (2.33)$$

We distinguish the cases of k being odd and even. In case of k being odd, the terms α_i are, for $1 \le i \le k$, all positive integers and the same holds for m_k. This implies $l_k \ne 0$, since $\mathbf{F}_k(z)$ would be analytic at ρ_k^2, otherwise. According to [42], we have

$$[z^n]c_k'(z - \rho_k^2)^{m_k} \log^{l_k}(z - \rho_k^2) \sim c_k'' \left(\rho_k^2\right)^{-n} n^{-m_k - 1} \sum_{j \ge 0} \frac{F_{j,k}(\log n)}{n^j},$$

where the $F_{j,k}(z)$ are polynomials whose degree is $l_k - 1$. In view of eq. (2.9)

$$[z^n]\mathbf{F}_k(z) \sim c_k \left(\rho_k^2\right)^{-n} n^{-(k-1)^2 - \frac{k-1}{2}}, \qquad (2.34)$$

where c_k is some positive constant, whence

$$m_k = (k - 1)^2 + \frac{k - 1}{2} - 1 \quad \text{and} \quad l_k = 1.$$

In case of k being even, $\alpha_k = (k-1)^2 + \frac{k-1}{2} - 1 \notin \mathbb{Z}$ while $\alpha_i \in \mathbb{Z}$ for $1 \le i < k$. Equation (2.34) implies that m_k is not an integer and according to [42] we have

$$[z^n]c_k'(z - \rho_k^2)^{m_k} \log^{l_k}(z - \rho_k^2) \sim c_k'' \left(\rho_k^2\right)^{-n} \frac{n^{-m_k - 1}}{\Gamma(-m_k)} \sum_{j \ge 0} \frac{E_{j,k}(\log n)}{n^j},$$

where $E_{j,k}(z)$ is a polynomial whose degree is l_k. In view of eq. (2.34) we conclude that

$$m_k = (k-1)^2 + \frac{k-1}{2} - 1 \quad \text{and} \quad l_k = 0.$$

Thus we have proved that for $z \to \rho_k^2$,

$$\mathbf{F}_k(z) = \begin{cases} P_k(z - \rho_k^2) + c_k'(z - \rho_k^2)^{(k-1)^2 + \frac{k-1}{2} - 1} \log(z - \rho_k^2)\,(1 + o(1)) \\ P_k(z - \rho_k^2) + c_k'(z - \rho_k^2)^{(k-1)^2 + \frac{k-1}{2} - 1}\,(1 + o(1)), \end{cases}$$

depending on k being odd or even and where $P_k(z)$ is a polynomial of degree $\leq (k-1)^2 + \frac{k-1}{2} - 1$ and c_k' is some constant.

Proposition 2.24 provides for $2 \leq k \leq 9$ the singular expansion of $\mathbf{F}_k(z)$. These particular expansions and a simple scaling property of the Taylor expansion are the key tools for proving Theorem 2.21.

Proof of Theorem 2.21. We consider the composite function $\mathbf{F}_k(\psi(z,s))$. In view of $[z^n]f(z,s) = \gamma^n[z^n]f(\frac{z}{\gamma},s)$ it suffices to analyze the function $\mathbf{F}_k(\psi(\gamma(s)z,s))$ and to subsequently rescale in order to obtain the correct exponential factor. For this purpose we set

$$\widetilde{\psi}(z,s) = \psi(\gamma(s)z,s),$$

where $\psi(z,s)$ is analytic in a domain $\mathcal{D} = \{(z,s) | |z| \leq r, |s| < \epsilon\}$. Consequently $\widetilde{\psi}(z,s)$ is analytic in $|z| < \widetilde{r}$ and $|s| < \widetilde{\epsilon}$, for some $1 < \widetilde{r}, 0 < \widetilde{\epsilon} < \epsilon$, since it is a composition of two analytic functions in $\mathcal{D}$. Taking its Taylor expansion at $z = 1$,

$$\widetilde{\psi}(z,s) = \sum_{n \geq 0} \widetilde{\psi}_n(s)(1-z)^n, \tag{2.35}$$

where $\widetilde{\psi}_n(s)$ is analytic in $|s| < \widetilde{\epsilon}$. According to Proposition 2.24, the singular expansion of $\mathbf{F}_k(z)$, for $z \to \rho_k^2$, is given by

$$\mathbf{F}_k(z) = \begin{cases} P_k(z - \rho_k^2) + c_k'(z - \rho_k^2)^{((k-1)^2 + (k-1)/2) - 1} \log(z - \rho_k^2)\,(1 + o(1)) \\ P_k(z - \rho_k^2) + c_k'(z - \rho_k^2)^{((k-1)^2 + (k-1)/2) - 1}\,(1 + o(1)), \end{cases}$$

depending on whether k is odd or even and where $P_k(z)$ are polynomials of degree $\leq (k-1)^2 + (k-1)/2 - 1$, c_k' is some constant, and $\rho_k = 1/2(k-1)$. By assumption, $\gamma(s)$ is the unique analytic solution of $\psi(\gamma(s),s) = \rho_k^2$ and by construction $\mathbf{F}_k(\psi(\gamma(s)z,s)) = \mathbf{F}_k(\widetilde{\psi}(z,s))$. In view of eq. (2.35), we have for $z \to 1$ the expansion

$$\widetilde{\psi}(z,s) - \rho_k^2 = \sum_{n \geq 1} \widetilde{\psi}_n(s)(1-z)^n = \widetilde{\psi}_1(s)(1-z)(1 + o(1)), \tag{2.36}$$

that is uniform in s since $\widetilde{\psi}_n(s)$ is analytic for $|s| < \widetilde{\epsilon}$ and $\widetilde{\psi}_0(s) = \psi(\gamma(s), s) = \rho_k^2$. As for the singular expansion of $\mathbf{F}_k(\widetilde{\psi}(z, s))$ we derive, substituting the eq. (2.36) into the singular expansion of $\mathbf{F}_k(z)$, for $z \to 1$,

$$\begin{cases} \widetilde{P}_k(z, s) + c_k(s)(1 - z)^{((k-1)^2 + (k-1)/2) - 1} \log(1 - z)\, (1 + o(1)) & \text{for } k \text{ odd,} \\ \widetilde{P}_k(z, s) + c_k(s)(1 - z)^{((k-1)^2 + (k-1)/2) - 1}\, (1 + o(1)) & \text{for } k \text{ even} \end{cases}$$

where $\widetilde{P}_k(z, s) = P_k(\widetilde{\psi}(z, s) - \rho_k^2)$ and $c_k(s) = c'_k \widetilde{\psi}_1(s)^{((k-1)^2 + (k-1)/2) - 1}$ and

$$\widetilde{\psi}_1(s) = \partial_z \widetilde{\psi}(z, s)|_{z=1} = \gamma(s)\partial_z \psi(\gamma(s), s) \neq 0 \quad \text{for } |s| < \epsilon.$$

Furthermore $\widetilde{P}_k(z, s)$ is analytic at $|z| \le 1$, whence $[z^n]\widetilde{P}_k(z, s)$ is exponentially small compared to 1. Therefore, we arrive at

$$[z^n]\mathbf{F}_k(\widetilde{\psi}(z, s)) \sim \begin{cases} [z^n]c_k(s)(1 - z)^{((k-1)^2 + (k-1)/2) - 1} \log(1 - z)\, (1 + o(1)) \\ [z^n]c_k(s)(1 - z)^{((k-1)^2 + (k-1)/2) - 1}\, (1 + o(1)), \end{cases}$$

$$(2.37)$$

depending on k being odd or even and uniformly in $|s| < \widetilde{\epsilon}$. We observe that $c_k(s)$ is analytic in $|s| < \widetilde{\epsilon}$. Note that a dependency in the parameter s is only given in the coefficients $c_k(s)$ that are analytic in s. The transfer Theorem 2.20 and eq. (2.37) imply that

$$[z^n]\mathbf{F}_k(\widetilde{\psi}(z, s)) \sim A(s)\, n^{-((k-1)^2 + (k-1)/2)} \quad \text{for some } A(s) \in \mathbb{C},$$

uniformly in s contained in a small neighborhood of 0. Finally, as mentioned in the beginning of the proof, we use the scaling property of Taylor expansions in order to derive

$$[z^n]\mathbf{F}_k(\psi(z, s)) = (\gamma(s))^{-n}\, [z^n]\mathbf{F}_k(\widetilde{\psi}(z, s))$$

and the proof of the theorem is complete.

2.5 n-Cubes

In this section we deal with a formalization of the space of all sequences. For this purpose we regard the nucleotides an element of an arbitrary finite set (alphabet), A. The existence of the so-called point-mutations, that is mutations of individual nucleotides, see Fig. 2.14, suggests to consider two sequences to be adjacent, if they differ in exactly one position. This point of view gives rise to consider sequence space as a graph. In this graph each $\mathbf{A}, \mathbf{U}, \mathbf{G}, \mathbf{C}$ sequence of n nucleotides has $3n$ neighbors.

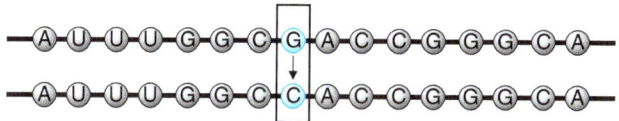

Fig. 2.14. Single point mutations.

k	
2	$(4x - 1) x f''(x) + (10x - 2) f'(x) + 2 f(x) = 0$
3	$(16x^3 - x^2) f^{(3)}(x) + (96x^2 - 8x) f''(x) + (108x - 12) f'(x) + 12 f(x) = 0$
4	$(144x^5 - 40x^4 + x^3) f^{(4)}(x) + (1584x^4 - 556x^3 + 20x^2) f^{(3)}(x)$ $+ (4428x^3 - 1968x^2 + 112x) f''(x) + (3024x^2 - 1728x + 168) f'(x)$ $+ (216x - 168) f(x) = 0$
5	$(1024x^6 - 80x^5 + x^4) f^{(5)}(x) + (20,480x^5 - 2256x^4 + 40x^3) f^{(4)}(x)$ $+ (121,600x^4 - 19,380x^3 + 532x^2) f^{(3)}(x) + (241,920x^3 - 56,692x^2 + 2728x)$ $f''(x) + (130,560x^2 - 46,048x + 4400) f'(x) + (7680x - 4400) f(x) = 0$
6	$(14,400x^8 - 4144x^7 + 140x^6 - x^5) f^{(6)}(x)$ $+ (367,200x^7 - 148,368x^6 + 7126x^5 - 70x^4) f^{(5)}(x)$ $+ (3,078,000x^6 - 1,728,900x^5 + 123,850x^4 - 1792x^3) f^{(4)}(x)$ $+ (10,179,000x^5 - 7,880,640x^4 + 880,152x^3 - 20,704x^2) f^{(3)}(x)$ $+ (12,555,000x^4 - 13,367,880x^3 + 2,399,184x^2 - 106,016x) f''(x)$ $+ (4,374,000x^3 - 6,475,680x^2 + 1,922,736x - 187,200) f'(x)$ $+ (162,000x^2 - 350,640x + 187,200) f(x) = 0$
7	$(147,456x^9 - 12,544x^8 + 224x^7 - x^6) f^{(7)}(x)$ $+ (6,193,152x^8 - 757,760x^7 + 18,816x^6 - 112x^5) f^{(6)}(x)$ $+ (89,800,704x^7 - 16,035,456x^6 + 582,280x^5 - 4872x^4) f^{(5)}(x)$ $+ (561,254,400x^6 - 146,691,840x^5 + 8,254,664x^4 - 104,480x^3) f^{(4)}(x)$ $+ (1,535,708,160x^5 - 585,419,280x^4 + 54,069,792x^3 - 1,151,984x^2) f^{(3)}(x)$ $+ (1,651,829,760x^4 - 916,833,600x^3 + 144,777,216x^2 - 6,094,528x) f''(x)$ $+ (516,741,120x^3 - 421,901,280x^2 + 117,590,208x - 11,797,632) f'(x)$ $+ (17,418,240x^2 - 22,034,880x + 11,797,632) f(x) = 0$
8	$(2,822,400x^{11} - 826,624x^{10} + 31,584x^9 - 336x^8 + x^7) f^{(8)}(x)$ $+ (129,830,400x^{10} - 55,968,384x^9 + 3,026,208x^8 - 43,512x^7 + 168x^6) f^{(7)}(x)$ $+ (2,202,883,200x^9 - 1,363,532,352x^8 + 107,691,912x^7 - 2,188,752x^6$ $+ 11,424x^5)$ $f^{(6)}(x) + (17455132800x^8 - 15,140,260,128x^7 + 1,789,953,376x^6$ $- 54349,728x^5 + 405,200x^4) f^{(5)}(x)$ $+ (67,586,778,000x^7 - 80,551,356,480x^6 + 14,421,855,200x^5$ $- 698,609,104x^4 + 8,035,104x^3) f^{(4)}(x)$ $+ (122,393,376,000x^6 - 197,784,236,160x^5 + 53,661,386,080x^4$ $- 4437573,920x^3 + 88,180,864x^2) f^{(3)}(x)$

Table 2.1. The differential equations for $\mathbf{F}_k(z)(2 \leq k \leq 9)$, obtained by Maple package `gfun`.

$$-4437573,920\,x^3 + 88,180,864\,x^2)f^{(3)}\,(x)$$
$$+(90,239,184,000\,x^5 - 196,676,000,640\,x^4 + 80,758,975,680\,x^3$$
$$-11,973,419,104\,x^2 + 488,846,272\,x)f''\,(x)$$
$$+(19,559,232,000\,x^4 - 57,892,907,520\,x^3 + 35,467,753,520\,x^2$$
$$-9,969,500,032\,x + 1,033,305,728)f'\,(x)$$
$$+\left(444,528,000\,x^3 - 1,852,865,280\,x^2 + 186,993,760\,x - 1,033,305,728\right)f\,(x) = 0$$

9 | $\left(37,748,736\,x^{12} - 3,358,720\,x^{11} + 69,888\,x^{10} - 480\,x^9 + x^8\right)f^{(9)}\,(x)$
$$+\left(2,717,908,992\,x^{11} - 351,387,648\,x^{10} + 10,065,408\,x^9 - 90,912\,x^8\right.$$
$$\left.+240\,x^7\right)f^{(8)}\,(x)$$
$$+(72,873,934,848\,x^{10} - 1,378,440,8064\,x^9$$
$$+563,449,728\,x^8 - 6,950,616\,x^7 + 24,024\,x^6)f^{(7)}\,(x)$$
$$+(940,566,380,544\,x^9 - 258,478,202,880\,x^8 + 15,638,941,312\,x^7$$
$$-2,368,505,160\,x^6 + 1,304,336\,x^5)f^{(6)}\,(x)$$
$$+(6,273,464,795,136\,x^8 - 2,467,959,432,192\,x^7 + 227,994,061,392\,x^6$$
$$-18,674,432,128\,x^5 + 41,782,224\,x^4)f^{(5)}\,(x)$$
$$+(21,523,928,186,880\,x^7 - 119,317,461,350,40\,x^6 + 17,131,29,509,184\,x^5$$
$$-75,115,763,872\,x^4 + 802,970,368\,x^3)f^{(4)}\,(x)$$
$$+(35,583,374,131,200\,x^6 - 27,454,499,6659,20\,x^5 + 614,7724,228,704\,x^4$$
$$-475,182,777,504\,x^3 + 8,956,331,968\,x^2)f^{(3)}\,(x)$$
$$+(24,400,027,975,680\,x^5 - 26,056,335,882,240\,x^4 + 9,086,553,292,608\,x^3$$
$$-1,308,864,283,488\,x^2 + 52,313,960,192\,x)f''\,(x)$$
$$+(4,976,321,495,040\,x^4 - 740,2528,051,200\,x^3 + 4,051,342,551,744\,x^2$$
$$-1,122,348,764,928\,x + 120,086,385,408)f'\,(x)$$
$$+\left(107,017,666,560\,x^3 - 230,051,819,520\,x^2 + 208,033,076,736\,x - 120,\right.$$
$$\left.086,385,408\right)f\,(x) = 0$$

Table 2.1. continued

2.5.1 Some basic facts

The n-cube, Q_α^n, is a combinatorial graph with vertex set A^n, where A is some finite alphabet of size $\alpha \geq 2$. Without loss of generality we will assume $\mathbb{F}_2 \subset A$ (here $\mathbb{F}_2$ denotes the field having the two elements $0, 1$) and call Q_2^n the binary n-cube. In an n-cube two vertices are adjacent if they differ in exactly one coordinate; see Fig. 2.15.

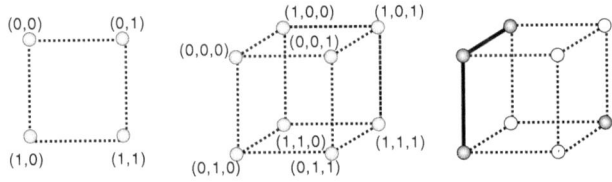

Fig. 2.15. The n-cubes Q_2^n for $n = 2$ (*left*) and $n = 3$ (*middle*). On the RHS we display an induced Q_2^3-subgraph, induced by the gray vertices.

Let $d(v, v')$ be the number of coordinates by which v and v' differ. $d(v, v')$ is oftentimes referred to as Hamming metric. We set $\forall\, C \subset A^n$, $j \le n$

$$\mathsf{B}(C, j) = \{v \in \mathbb{A}^n \mid \exists\, a \in C;\ d(v, a) \le j\}$$
$$\mathsf{S}(C, j) = \mathsf{B}(C, j) \setminus \mathsf{B}(C, j - 1)$$
$$\mathsf{d}(C) = \mathsf{B}(C, 1) \setminus C$$

and call $\mathsf{B}(C, j)$ and $\mathsf{d}(C)$ the ball of radius j around C and the vertex boundary of C in Q_α^n, respectively. If $C = \{v\}$, we simply write $\mathsf{B}(v, j)$. Let $B, C \subset A^n$, we call B ℓ-dense in C if $\mathsf{B}(v, \ell) \cap B \ne \varnothing$ for any $v \in C$.

Q_2^n can also be viewed as the Cayley graph $\mathsf{Cay}(\mathbb{F}_2^n, \{e_i \mid i = 1, \ldots, n\})$, where e_i is the canonical base vector. We will view $\mathbb{F}_2^n$ as a $\mathbb{F}_2$-vectorspace and denote the linear hull over $\{v_1, \ldots, v_h\}$, $v_j \in \mathbb{F}_2^n$ by $\langle v_1, v_2, \ldots, v_h \rangle$.

There exists a natural linear order $\le$ over Q_2^n given by

$$v \le v' \quad \Longleftrightarrow \quad (d(v, 0) < d(v', 0)) \vee (d(v, 0) = d(v', 0) \wedge v \le_{\mathrm{lex}} v'), \quad (2.38)$$

where $\le_{\mathrm{lex}}$ denotes the lexicographical order. Any notion of minimal element or smallest element in $A \subset Q_2^n$ is considered with respect to the linear order $\le$ of eq. (2.38).

Each $B \subset A^n$ induces a unique induced subgraph in Q_α^n, denoted by $Q_\alpha^n[B]$, in which $b_1, b_2 \in B$ are adjacent iff b_1, b_2 are adjacent in Q_α^n.

We next prove a combinatorial lemma, which is a slightly stronger version of a result in [14].

Lemma 2.25. *Let* $d \in \mathbb{N}$, $d \ge 2$ *and let* v, v' *be two* Q_2^n-*vertices where* $d(v, v') = d$. *Then any* Q_2^n-*path from* v *to* v' *has length* $2\ell + d$ *and there are at most*

$$\binom{2\ell + d}{\ell + d} \binom{\ell + d}{\ell} n^\ell\, \ell!\, d!$$

Q_2^n-*paths from* v *to* v' *of length* $2\ell + d$.

Proof. Without loss of generality, we can assume $v = (0, \ldots, 0)$ and $v' = (x_i)_i$, where $x_i = 1$ for $1 \le i \le d$ and $x_i = 0$, otherwise. Each path of length m induces the family of steps $(\epsilon_s)_{1 \le s \le m}$, where $\epsilon_s \in \{e_j \mid 1 \le j \le n\}$. Since each path ends at v', we have for fixed $1 \le i \le n$

$$\sum_{\{\epsilon_s \mid \epsilon_s = e_i\}} \epsilon_s = \begin{cases} 1 & \text{for } 1 \le i \le d, \\ 0 & \text{otherwise.} \end{cases}$$

Hence the families induced by these paths contain necessarily the set $\{e_1, \ldots, e_d\}$. Let $(\epsilon'_s)_{1 \le s \le m'}$ be the family obtained from $(\epsilon_s)_{1 \le s \le m}$ by removing the steps $e_1, \ldots, e_d$, at the smallest index at which they occur. Then $(\epsilon'_s)_{1 \le s \le m'}$ represents a cycle starting and ending at v. Furthermore, we have for all i; $\sum_{\{\epsilon'_s \mid \epsilon'_s = e_i\}} \epsilon'_s = 0$, i.e., all steps must come in up-step/down-step pairs. As a result we derive $m = 2\ell + d$ and there are exactly ℓ steps of the form e_j that can be freely chosen (free up-steps). We proceed by counting the number of the $(2\ell + d)$-tuples $(\epsilon_s)_{1 \leq s \leq 2\ell+d}$. There are exactly $\binom{2\ell+d}{\ell+d}$ ways to select the $(\ell+d)$ indices for the up-steps within the set of all $2\ell+d$ indices. Furthermore, there are at most $\binom{\ell+d}{\ell}$ ways to select the positions for the ℓ up-steps and at most n^ℓ ways to choose the free up-steps themselves (once their positions are fixed). Since a free up-step is paired with a unique down-step reversing it, the ℓ free up-steps determine all ℓ down-steps. Clearly, there are at most $\ell!$ ways to assign the down-steps to their ℓ indices. Finally, there are at most $d!$ ways to assign the fixed up-steps and the lemma follows.

2.5.2 Random subgraphs of the n-cube

Let Q^n_{α,λ_n} be the random graph consisting of Q^n_α-subgraphs, Γ_n, induced by selecting each Q^n_α-vertex with independent probability λ_n; see Fig. 2.16. Q^n_{α,λ_n} is the finite probability space

$$(\{Q^n_\alpha[B] \mid B \subset \mathbb{A}^n\}, \mathbb{P}_n),$$

with the probability measure $\mathbb{P}_n(B) = \lambda_n^{|B|}(1 - \lambda_n)^{\alpha^n - |B|}$.

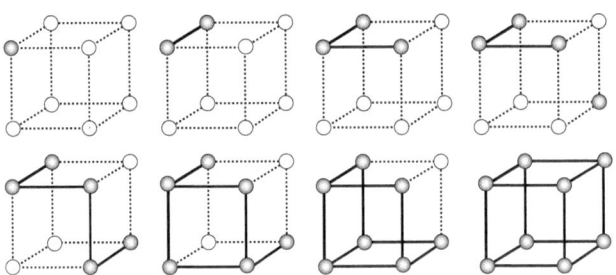

Fig. 2.16. Eight random-induced subgraphs of Q^3_2

A property M_n is a subset of induced subgraphs of Q^n_α closed under graph isomorphisms. The terminology "M_n holds a.s." is equivalent to

$$\lim_{n \to \infty} \mathbb{P}(\mathsf{M}_n) = 1.$$

We use the notation

$$B_m(\ell, \lambda_n) = \binom{m}{\ell} \lambda_n^\ell (1 - \lambda_n)^{m-\ell}$$

and write $g(n) = O(f(n))$ and $g(n) = o(f(n))$ for $g(n)/f(n) \to \kappa$ as $n \to \infty$ and $g(n)/f(n) \to 0$ as $n \to \infty$, respectively.

A component of Γ_n is a maximal connected induced Γ_n-subgraph, C_n. The largest Γ_n-component is denoted by $C_n^{(1)}$. Analogously, the second largest component is denoted by $C_n^{(2)}$. The largest Γ_n-component $C_n^{(1)}$ is called a giant component or giant if and only if

$$|C_n^{(2)}| = o(|C_n^{(1)}|).$$

Furthermore, we write $x_n \sim y_n$ if and only if (a) $\lim_{n\to\infty} x_n/y_n$ exists and (b) $\lim_{n\to\infty} x_n/y_n = 1$.

Let $Z_n = \sum_{i=1}^n \xi_i$ be a sum of mutually independent indicator random variables (r.v.), ξ_i having values in $\{0,1\}$. Then we have, [58], for $\eta > 0$ and $c_\eta = \min\{-\ln(e^\eta[1+\eta]^{-[1+\eta]}), \frac{\eta^2}{2}\}$

$$\mathsf{Prob}(\,|Z_n - \mathbb{E}[Z_n]| > \eta\,\mathbb{E}[Z_n]) \le 2e^{-c_\eta \mathbb{E}[Z_n]}. \tag{2.39}$$

n is always assumed to be sufficiently large and ϵ is a positive constant satisfying $0 < \epsilon < 1$.

2.5.3 Vertex boundaries

In this section we present some generic results on vertex boundaries, which are instrumental for our analysis of connectivity, large components, and distances in n-cubes. The first result is due to [7] used for Sidon sets in groups in the context of Cayley graphs. In the following G denotes a finite group and M a finite set acted upon by G.

Proposition 2.26. *Suppose G act transitively on M and let $A \subset M$, then we have*

$$\frac{1}{|G|}\sum_{g\in G}|A \cap gA| = |A|^2/|M|. \tag{2.40}$$

Proof. We prove eq. (2.40) by induction on $|A|$. For $A = \{x\}$ we derive $\frac{1}{|G|}\sum_{gx=x} 1 = |G_x|/|G|$, since $|M| = |G|/|G_x|$. We next prove the induction step. We write $A = A_0 \cup \{x\}$ and compute

$$\frac{1}{|G|}\sum_g |A \cap gA| = \frac{1}{|G|}\sum_g (|A_0 \cap gA_0| + |\{gx\} \cap A_0| +$$

$$|\{x\} \cap gA_0| + |\{gx\} \cap \{x\}|$$

$$= \frac{1}{|G|}(|A_0|^2|G_x| + 2|A_0||G_x| + |G_x|)$$

$$= \frac{1}{|G|}((|A_0|+1)^2|G_x|) = \frac{|A|^2}{|M|}.$$

Aldous [4, 6] observed how to use Proposition 2.26 for deriving a very general lower bound for vertex boundaries in Cayley graphs:

Theorem 2.27. *Suppose G acts transitively on M and let $A \subset M$, and let S be a generating set of the Cayley graph $\mathsf{Cay}(G,S)$ where $|S| = n$. Then we have*

$$\exists\, s \in S; \quad |sA \setminus A| \geq \frac{1}{n}|A|\left(1 - \frac{|A|}{|M|}\right).$$

Proof. We compute

$$|A| = \frac{1}{|G|}\sum_g (|gA \setminus A| + |A \cap gA|) = \frac{1}{|G|}\sum_g |gA \setminus A| + |A|\frac{|A|}{|M|}$$

and hence $|A|(1 - \frac{|A|}{|M|}) = \frac{1}{|G|}\sum_g |gA \setminus A|$. From this we can immediately conclude

$$\exists\, g \in G; \quad |gA \setminus A| \geq |A|\left(1 - \frac{|A|}{|M|}\right).$$

Let $g = \prod_{j=1}^k s_j$. Since each element of $gA \setminus A$ is contained in at least one set $s_j A \setminus A$ we obtain

$$|gA \setminus A| \leq \sum_{j=1}^k |s_j A \setminus A|.$$

Hence there exists some $1 \leq j \leq k$ such that $|s_j A \setminus A| \geq \frac{1}{k}|gA \setminus A|$ and the lemma follows.

2.5.4 Branching processes and Janson's inequality

Let us next recall some basic facts about branching processes [62, 83]. Suppose ξ is a random variable and $(\xi_i^{(t)})$, $i, t \in \mathbb{N}$ are random variables that count the number of offspring of the ith individual at generation $t - 1$. We consider the family of r.v. $Z = (Z_i)_{i \in \mathbb{N}_0}$, given by

$$Z_0 = 1 \quad \text{and} \quad Z_t = \sum_{i=1}^{Z_{t-1}} \xi_i^{(t)}, \text{ for } t \geq 1$$

and interpret Z_t as the number of individuals "alive" in generation t. We will be interested in the limit probability $\lim_{t\to\infty} \mathsf{Prob}(Z_t > 0)$, i.e., the probability of infinite survival.

In the following, we distinguish three branching processes:

- Suppose the r.v.s ξ and $\xi_i^{(t)}$ are all $B_m(\ell, p)$-distributed. We denote this process by Z^* and its survival probability by

$$\pi_m(p) = \lim_{t\to\infty} \mathsf{Prob}(Z_t^* > 0).$$

- Let Z^0 denote the branching process in which ξ is $B_m(\ell, p)$-distributed and all subsequent r.vs. $\xi_i^{(t)}$ are $B_{m-1}(\ell, p)$-distributed and

$$\pi_0(p) = \lim_{t \to \infty} \mathsf{Prob}(Z_t^0 > 0).$$

- Let Z^P denote the branching process in which the individuals generate offspring according to the Poisson distribution, i.e.,

$$\mathsf{Prob}(\xi_i^{(t)} = j) = \frac{\lambda^j}{j!} e^{-\lambda},$$

where $\lambda > 0$ and let

$$\pi_P(\lambda) = \lim_{t \to \infty} \mathsf{Prob}(Z_t^P > 0).$$

Lemma 2.28. (Bollobás et al. [14])

(1) *For all $0 \le p \le 1$, we have $\pi_{n-1}(p) \le \pi_0(p) \le \pi_n(p)$.*

(2) *If $\lambda > 1$ is fixed, then $\pi_P(\lambda)$ is the unique solution of $x + e^{-\lambda x} = 1$ in the interval $0 < x < 1$.*

(3) *Let $p = \frac{\lambda_n}{n}$ where $\lambda_n = 1 + \epsilon_n$ and $0 < \epsilon_n = o(1)$. Then*

$$\pi_n(p) = \frac{2n\epsilon_n}{n-1} + O(\epsilon_n^2).$$

In particular, if $r = n - s$ then

$$\pi_r(p) = 2\epsilon_n + O(\epsilon_n/n) + O(s/n) + O(\epsilon_n^2);$$

and hence if $s = o(\epsilon_n n)$ then $\pi_r(p) = (1 + o(1))\pi_0(p)$.

Corollary 2.29. *Let $p = \lambda/n$.*

(1) *If $\lambda > 1$ is fixed, then $\pi_0(p) = (1 + o(1))\pi_P(\lambda)$.*

(2) *Let $\lambda_n = 1 + \epsilon_n$, where $0 < \epsilon_n = o(1)$. Then, if $r = n - s$ and $s = o(n\epsilon_n)$,*

$$\pi_0(p) = (1 + o(1))\pi_r(p) = (2 + o(1))\epsilon_n.$$

In Chapter 7 we need the following particular formulation of Corollary 2.29.

Corollary 2.30. *Let $u_n = n^{-\frac{1}{3}}$, $\lambda_n = \frac{1 + \chi_n}{n}$, $m = n - \lfloor \frac{3}{4} u_n n \rfloor$, and*

$$\mathsf{Prob}(\xi = \ell) = B_m(\ell, \lambda_n).$$

Then for $\chi_n = \epsilon$ the r.v. ξ becomes asymptotically Poisson, i.e., $\mathbb{P}(\xi = \ell) \sim \frac{(1+\epsilon)^\ell}{\ell!} e^{-(1+\epsilon)}$ and

$$0 < \lim_{t \to \infty} \mathsf{Prob}(Z_t > 0) = \alpha(\epsilon) < 1,$$

where $0 < \alpha(\epsilon) < 1$ is the unique solution of the equation $x + e^{-(1+\epsilon)x} = 1$. For $o(1) = \chi_n \geq n^{-\frac{1}{3}+\delta}$, $\delta > 0$ we have

$$\lim_{t \to \infty} \mathsf{Prob}(Z_t > 0) = (2 + o(1))\, \chi_n.$$

The next theorem, used in Chapter 7, is Janson's inequality [75]. It facilitates the proof of Theorem 7.15 and Theorem 7.13. Intuitively, Janson's inequality can be viewed as a large deviation result in the presence of correlation.

Theorem 2.31. *Let R be a random subset of some set $[V] = \{1, \ldots, V\}$ obtained by selecting each element $v \in V$ independently with probability λ. Let $S_1, \ldots, S_s$ be subsets of $[V]$ and X be the r.v. counting the number of S_i for which $S_i \subset R$. Let furthermore*

$$\Omega = \sum_{(i,j);\, S_i \cap S_j \neq \varnothing} \mathbb{P}(S_i \cup S_j \subset R),$$

where the sum is taken over all ordered pairs (i,j). Then for any $\gamma > 0$, we have

$$\mathbb{P}(X \leq (1-\gamma)\mathbb{E}[X]) \leq e^{-\frac{\gamma^2 \mathbb{E}[X]}{2 + 2\Omega/\mathbb{E}[X]}}.$$

2.6 Exercises

2.1. Prove Lemma 2.9 via symbolic enumeration. Consider the mapping that assigns to each partial k-noncrossing matching a k-noncrossing matching by removing all isolated vertices. Note that given a k-noncrossing matching, there are exactly $2n+1$ positions in which an arbitrary sequence of isolated vertices can be inserted.

2.2. Compute the generating function of secondary structures with minimum arc length λ and minimum stack-length σ. Hint: Compute the bivariate generating function of noncrossing matchings in which each stack has size exactly one, having exactly m 1-arcs (i.e., arcs of the form $(i, i+1)$). Then use symbolic enumeration and the fact that each secondary structure is mapped into exactly one such matching.

2.3. We analyze the case $k = 2$, i.e., RNA secondary structures. Here the generating function itself coincides with its singular expansion. The particular approach offers a great simplification of the proof in [69] and easily extends to all subclasses of secondary structures, considered there. Prove: The number of RNA secondary, i.e., 2-noncrossing RNA, structures is asymptotically given by

$$T_2^{[2]}(n) \sim \frac{1.9572}{\sqrt{n}} \left(\frac{1}{n+1} - \frac{1}{8n(n+1)} + \frac{1}{128n^2(n+1)} + O(n^{-4}) \right)$$
$$\times \left(\frac{3 + \sqrt{5}}{2} \right)^n.$$

2.4. An *-tableaux is called irreducible if its only two empty shapes are λ^0 and λ^n. Let $\mathbf{Irr}_k^*(z)$ denote the generating function of irreducible *-tableaux. Prove

$$\mathbf{Irr}_k^*(z) = 1 - z - \frac{1}{\frac{1}{1-z} \mathbf{F}_k \left(\frac{z}{1-z} \right)}.$$

Furthermore, prove that

$$[z^n] \mathbf{Irr}_k^*(z) \sim \tilde{c}_k n^{-\mu-1} \left(\frac{\rho_k}{1 - \rho_k} \right)^{-n} (1 + o(1)),$$

where $\tilde{c}_k$ is some computable positive constant, $\mu = (k-1)^2 + \frac{k-1}{2} - 1$, and ρ_k is the real positive dominant singularity of $\mathbf{F}_k(z)$.

2.5. Show: suppose $\lambda > 1$, then $\pi_P(\lambda)$ is the unique solution of $x + e^{-\lambda x} = 1$ in the interval $0 < x < 1$.

2.6. Prove: The number of isolated vertices is asymptotically Poisson distributed in $Q_{2,\lambda}^n$, where $0 < \lambda$.

2.7. Let S_n be the symmetric group and $T_n \subset S_n$ be a minimal generating set of transpositions. We consider the Cayley graph $\Gamma(S_n, T_n)$, having vertex set S_n and edges (v, v') where $v^{-1}v' \in T_n$. Suppose one selects permutations with probability $\frac{1+\epsilon}{n}$. Compute the probability of a cycle of length ℓ, $\mathcal{O}_\ell$, that contains a given permutation.

Tangled diagrams

Most of the material presented in this chapter is derived from [27, 28].

3.1 Tangled diagrams and vacillating tableaux

A tangled diagram, or tangle, is a labeled graph over the vertex set $[n] = \{1, \ldots, n\}$, with vertices of degree at most 2, drawn in increasing order in a horizontal line. Their arcs are drawn in the upper half plane. In general, a tangled diagram has isolated points and other types of degree 2 vertices, as displayed in Fig. 3.1.

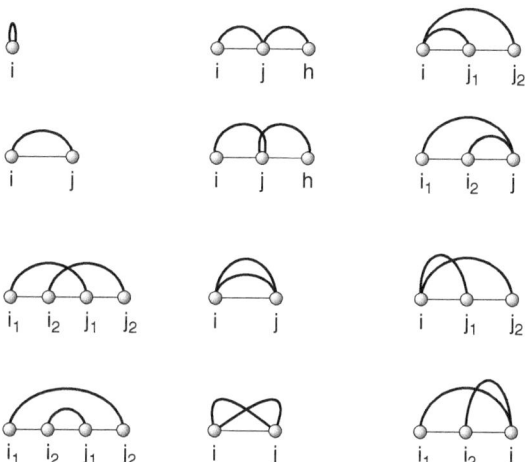

Fig. 3.1. All types of vertices with degree ≥ 1 in tangled diagrams.

Important subclasses of tangles are given as follows: (1) partial matchings, i.e., tangles in which each vertex has degree at most 1; (2) partitions,

C. Reidys, *Combinatorial Computational Biology of RNA*,
DOI 10.1007/978-0-387-76731-4_3,
© Springer Science+Business Media, LLC 2011

i.e., tangles in which any vertex of degree 2, j, is incident to the arcs (i, j) and (j, s), where $i < j < s$. Furthermore, partitions without arcs of the form $(i, i + 1)$ are called 2-regular partitions. (3) Braids, i.e., tangled diagrams in which all vertices of degree 2, j, are either incident to loops (j, j), or crossing arcs (i, j) and (j, h), where $i < j < h$; see Fig. 3.2.

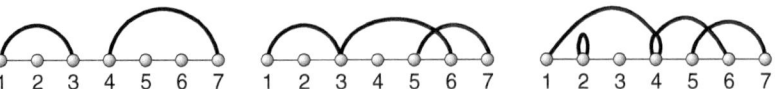

Fig. 3.2. From *left* to *right*: a partial matching, a partition, and a braid, respectively.

In order to describe the geometric crossings in tangled diagrams we map a tangled diagram into a partial matching. This mapping is called inflation. The inflation "splits" each vertex of degree 2, j, into two vertices j and j' having degree 1; see Fig. 3.3.

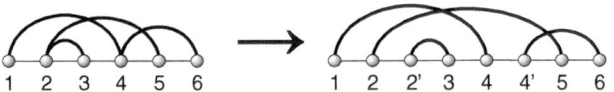

Fig. 3.3. The inflation of the first tangled diagram in Fig. 1.21 into its corresponding partial matching over eight vertices.

Accordingly, a tangle with ℓ vertices of degree 2 over n vertices is expanded into a diagram over $n + \ell$ vertices via inflation. The inflation map has a unique inverse, obtained by simply identifying the vertices j, j'. As RSK insertion refers implicitly a linear order, for this purpose, we consider the following linear ordering on $\{1, 1', \ldots, n, n'\}$:

$$1 < 1' < 2 < 2' < \cdots < n < n'.$$

Let G_n be a tangled diagram with exactly ℓ vertices of degree 2. Then the inflation of G_n, $\eta(G_n)$, is a labeled graph on $\{1, \ldots, n + \ell\}$ vertices with degree less than or equal to 1, obtained as follows:

Suppose first we have $i < j_1 < j_2$. If the arcs (i, j_1), (i, j_2) are crossing, then we map $((i, j_1), (i, j_2))$ into $((i, j_1), (i', j_2))$ and if (i, j_1), (i, j_2) are nesting then $((i, j_1), (i, j_2))$ is mapped into $((i, j_2), (i', j_1))$; see Fig. 3.4.

Second, let $i_1 < i_2 < j$. If (i_1, j), (i_2, j) are crossing, then we map $((i_1, j), (i_2, j))$ into $((i_1, j), (i_2, j'))$. If (i_1, j), (i_2, j) are nesting then we map $((i_1, j), (i_2, j))$ into $((i_1, j'), (i_2, j))$; see Fig. 3.5

Third suppose $i < j$. If $(i, j), (i, j)$ are crossing arcs, then $((i, j), (i, j))$ is mapped into $((i, j), (i', j'))$. If $(i, j), (i, j)$ are nesting arcs, then we map

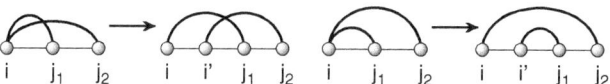

Fig. 3.4. The case $i < j_1 < j_2$: crossing (*left*) and nesting (*right*).

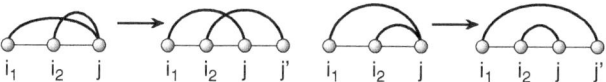

Fig. 3.5. The case $i_1 < i_2 < j$: crossing (*left*) and nesting (*right*).

$((i,j),(i,j))$ into $((i,j'),(i',j))$. Finally, if (i,i) is a loop we map (i,i) into (i,i'); see Fig. 3.6.

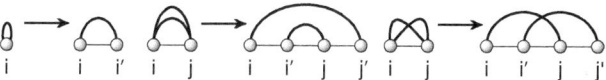

Fig. 3.6. The cases (i,i) and $i < j$: we resolve loops as arcs (*left*) and in case of $i < j$ we distinguish nesting (*middle*) and crossing (*right*).

Lastly, suppose we have $i < j < h$. If (i,j), (j,h) are crossing, then we map $((i,j),(j,h))$ into $((i,j'),(j,h))$ and we map $((i,j),(j,h))$ into $((i,j),(j',h))$, otherwise, see Fig. 3.7.

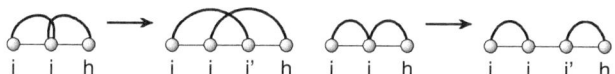

Fig. 3.7. The case $i < j < h$: crossing (*left*) and nesting (*right*).

As mentioned above, identifying all vertex-pairs (i,i') recovers the original tangle, whence we have the bijection

$$\eta: \mathcal{G}_n \longrightarrow \eta(\mathcal{G}_n).$$

The mapping η preserves by definition the maximal number of crossing and nesting arcs, respectively. Equivalently, a tangle G_n is k-noncrossing or k-nonnesting if and only if its inflation $\eta(G_n)$ is k-noncrossing or k-nonnesting, respectively. We have accordingly shown that the notion of crossings and nestings in tangles coincides with the notation of crossings and nestings in partial matchings.

A vacillating tableau V_λ^{2n} of shape λ and length $2n$ is a sequence of shapes $(\lambda^0, \lambda^1, \ldots, \lambda^{2n})$ such that (i) $\lambda^0 = \varnothing$ and $\lambda^{2n} = \lambda$ and (ii) $(\lambda^{2i-1}, \lambda^{2i})$ is derived from λ^{2i-2}, for $1 \le i \le n$, by one of the following operations. $(\varnothing, \varnothing)$:

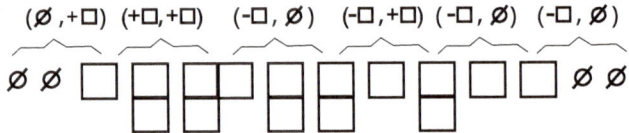

Fig. 3.8. A vacillating tableaux of shape $\varnothing$ and length 12.

do nothing twice; $(-\square, \varnothing)$: first remove a square then do nothing; $(\varnothing, +\square)$: first do nothing then add a square; $(\pm\square, \pm\square)$: add/remove a square at the odd and even steps, respectively. We denote the set of vacillating tableaux by $\mathcal{V}_\lambda^{2n}$; see Fig. 3.8.

3.2 The bijection

Lemma 3.1. *Any vacillating tableaux of shape $\varnothing$ and length $2n$, $V_\varnothing^{2n}$, induces a unique inflation of some tangled diagram on $[n]$, $\phi(V_\varnothing^{2n})$, namely, we have the mapping*

$$\phi \colon V_\varnothing^{2n} \longrightarrow \eta(\mathcal{G}_n).$$

Proof. In order to define ϕ, we recursively define a sequence of triples

$$((P_0, T_0, V_0), (P_1, T_1, V_1), \dots, (P_{2n}, T_{2n}, V_{2n})),$$

where P_i is a set of arcs, T_i is a tableau of shape λ^i, and

$$V_i \subset \{1, 1', 2, 2', \dots, n, n'\}$$

is a set of vertices. $P_0 = \varnothing$, $T_0 = \varnothing$, and $V_0 = \varnothing$. We assume that the left and right endpoints of all P_i-arcs and the entries of the tableau T_i are contained in $\{1, 1', \dots, n, n'\}$. Once given $(P_{2j-2}, T_{2j-2}, V_{2j-2})$, we derive $(P_{2j-1}, T_{2j-1}, V_{2j-1})$ and (P_{2j}, T_{2j}, V_{2j}) as follows:

(I) $(+\square, +\square)$. If $\lambda^{2j-1} \supsetneq \lambda^{2j-2}$ and $\lambda^{2j} \supsetneq \lambda^{2j-1}$, we set $P_{2j-1} = P_{2j-2}$, and T_{2j-1} is obtained from T_{2j-2} by adding the entry j in the square $\lambda^{2j-1} \setminus \lambda^{2j-2}$. Furthermore we set $P_{2j} = P_{2j-1}$ and T_{2j} is obtained from T_{2j-1} by adding the entry j' in the square $\lambda^{2j} \setminus \lambda^{2j-1}$, $V_{2j-1} = V_{2j-2} \cup \{j\}$, and $V_{2j} = V_{2j-1} \cup \{j'\}$; see Fig. 3.9.

(II) $(\varnothing, +\square)$. If $\lambda^{2j-1} = \lambda^{2j-2}$ and $\lambda^{2j} \supsetneq \lambda^{2j-1}$, then $(P_{2j-1}, T_{2j-1}) = (P_{2j-2}, T_{2j-2})$, $P_{2j} = P_{2j-1}$, and T_{2j} is obtained from T_{2j-1} by adding the

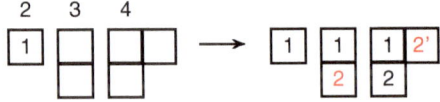

Fig. 3.9. From vacillating tableaux to tangles: in case of $\{+\square, +\square\}$, we have $V_3 = V_2 \cup \{2\}$ and $V_4 = V_3 \cup \{2'\}$.

Fig. 3.10. $(\varnothing, +\square)$: here we have $V_1 = V_0 = \varnothing$ and $V_2 = V_1 \cup \{1\}$.

entry j in the square $\lambda^{2j} \setminus \lambda^{2j-1}$, $V_{2j-1} = V_{2j-2}$, and $V_{2j} = V_{2j-1} \cup \{j\}$; see Fig. 3.10.

(III) $(+\square, -\square)$. If $\lambda^{2j-2} \subsetneq \lambda^{2j-1}$ and $\lambda^{2j} \subsetneq \lambda^{2j-1}$ then T_{2j-1} is obtained from T_{2j-2} by adding the entry j in the square $\lambda^{2j-1} \setminus \lambda^{2j-2}$ and the tableau T_{2j} is the unique tableau of shape λ^{2j} such that T_{2j-1} is obtained from T_{2j} by RSK inserting the unique number i. We then set $P_{2j-1} = P_{2j-2}$, $P_{2j} = P_{2j-1} \cup \{(i, j')\}$, $V_{2j-1} = V_{2j-2} \cup \{j\}$, and $V_{2j} = V_{2j-1} \cup \{j'\}$; see Fig. 3.11.

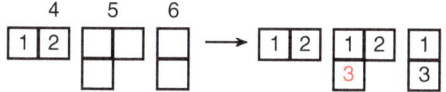

Fig. 3.11. $(+\square, -\square)$: here we have $P_5 = P_4$, $P_6 = P_5 \cup \{(2, 3')\}$, $V_5 = V_4 \cup \{3\}$, and $V_6 = V_5 \cup \{3'\}$.

(IV) $(-\square, \varnothing)$. If $\lambda^{2j-1} \subsetneq \lambda^{2j-2}$ and $\lambda^{2j} = \lambda^{2j-1}$, then T_{2j-1} is the unique tableau of shape λ^{2j-1} such that T_{2j-2} is obtained by RSK inserting the unique number i into T_{2j-1}, $P_{2j-1} = P_{2j-2} \cup \{(i, j)\}$, $(P_{2j}, T_{2j}) = (P_{2j-1}, T_{2j-1})$, $V_{2j-1} = V_{2j-2} \cup \{j\}$, and $V_{2j} = V_{2j-1}$; see Fig. 3.12.

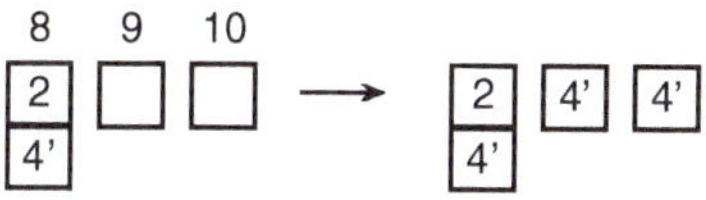

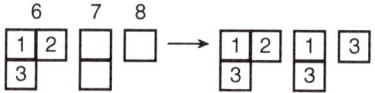

Fig. 3.12. $(-\square, \varnothing)$: here we have $P_5 = P_4 \cup \{(2, 5)\}$, $P_6 = P_5$, $V_5 = V_4 \cup \{5\}$, and $V_6 = V_5$.

(V) $(-\square, -\square)$. If $\lambda^{2j-1} \subsetneq \lambda^{2j-2}$ and $\lambda^{2j} \subsetneq \lambda^{2j-1}$, let T_{2j-1} be the unique tableau of shape λ^{2j-1} such that T_{2j-2} is obtained from T_{2j-1} by RSK inserting i_1 and T_{2j} be the unique tableau of shape λ^{2j} such that T_{2j-1} is obtained from T_{2j} by RSK inserting i_2, $P_{2j-1} = P_{2j-2} \cup \{(i_1, j)\}$, $P_{2j} = P_{2j-1} \cup \{(i_2, j')\}$, $V_{2j-1} = V_{2j-2} \cup \{j\}$, and $V_{2j} = V_{2j-1} \cup \{j'\}$; see Fig. 3.13.

(VI) $(-\square, +\square)$. If $\lambda^{2j-1} \subsetneq \lambda^{2j-2}$ and $\lambda^{2j} \supsetneq \lambda^{2j-1}$, then T_{2j-1} is the unique tableau of shape λ^{2j-1} such that T_{2j-2} is obtained from T_{2j-1} by

Fig. 3.13. $(-\square, -\square)$: here we have $P_7 = P_6 \cup \{(2, 4)\}$, $P_8 = P_7 \cup \{(1, 4')\}$, $V_7 = V_6 \cup \{4\}$, and $V_8 = V_7 \cup \{4'\}$.

RSK inserting the unique number i. Then we set $P_{2j-1} = P_{2j-2} \cup \{(i,j)\}$, $P_{2j} = P_{2j-1}$, and T_{2j} is obtained from T_{2j-1} by adding the entry j' in the square $\lambda^{2j} \setminus \lambda^{2j-1}$, $V_{2j-1} = V_{2j-2} \cup \{j\}$, and $V_{2j} = V_{2j-1} \cup \{j'\}$; see Fig. 3.14.

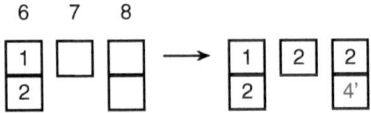

Fig. 3.14. $(-\Box, +\Box)$: we have $P_7 = P_6 \cup \{(1,4)\}$, $P_8 = P_7$, $V_7 = V_6 \cup \{4\}$, and $V_8 = V_7 \cup \{4'\}$.

(VII) $(\varnothing, \varnothing)$. If $\lambda^{2j-1} = \lambda^{2j-2}$ and $\lambda^{2j} = \lambda^{2j-1}$, we have $(P_{2j-1}, T_{2j-1}) = (P_{2j-2}, T_{2j-2})$, $(P_{2j}, T_{2j}) = (P_{2j-1}, T_{2j-1})$, $V_{2j-1} = V_{2j-2} \cup \{j\}$, and $V_{2j} = V_{2j-1}$.

Claim. The image $\phi(V_\varnothing^{2n})$ is the inflation of a tangled diagram.
First, if $(i,j) \in P_{2n}$, then $i < j$. Second, any vertex j can occur only as either a left or right endpoint of an arc, whence $\phi(V_\varnothing^{2n})$ is a 1-diagram. Each step $(+\Box, +\Box)$ induces a pair of arcs of the form (i, j_1), (i', j_2) and each step $(-\Box, -\Box)$ induces a pair of arcs of the form (i_1, j), (i_2, j'). Each step $(-\Box, +\Box)$ corresponds to a pair of arcs (h, j), (j', s) where $h < j < j' < s$, and each step $(+\Box, -\Box)$ induces a pair of arcs of the form (j, s), (h, j'), where $h < j < j' < s$ or a 1-arc of the form (i, i').

Let ℓ be the number of steps not containing $\varnothing$. By construction each of these steps adds the 2-set $\{j, j'\}$, whence (V_{2n}, P_{2n}) corresponds to the inflation of a unique tangled diagram with ℓ vertices of degree 2 and the claim follows.

We remark that, if squares are added, then the corresponding numbers are inserted. If squares are deleted Lemma 2.1 is used to extract a unique number, which then forms the left endpoint of the so-derived arcs; see Fig. 3.15. We proceed by explicitly constructing the inverse of ϕ.

Lemma 3.2. *Any inflation of a tangled diagram on n vertices, $\eta(G_n)$, induces the vacillating tableaux of shape $\varnothing$ and length $2n$, $\psi(\eta(G_n))$, namely, we have the mapping*

$$\psi: \eta(G_n) \longrightarrow V_\varnothing^{2n}. \tag{3.1}$$

Proof. We define ψ as follows. Let $\eta(G_n)$ be the inflation of the tangle G_n. We set

$$\eta_i = \begin{cases} (i, i'), & \text{iff } i \text{ has degree 2 in } G_n, \\ i, & \text{otherwise.} \end{cases}$$

Let $T_{2n} = \varnothing$ be the empty tableau. We will construct a sequence of tableaux T_h of shape $\lambda^h_{\eta(G_n)}$, where $h \in \{0, 1, \dots 2n\}$ by considering η_i for $i = n, n-1, n-2, \dots, 1$. For each η_j we inductively define the pair of tableaux (T_{2j}, T_{2j-1}):

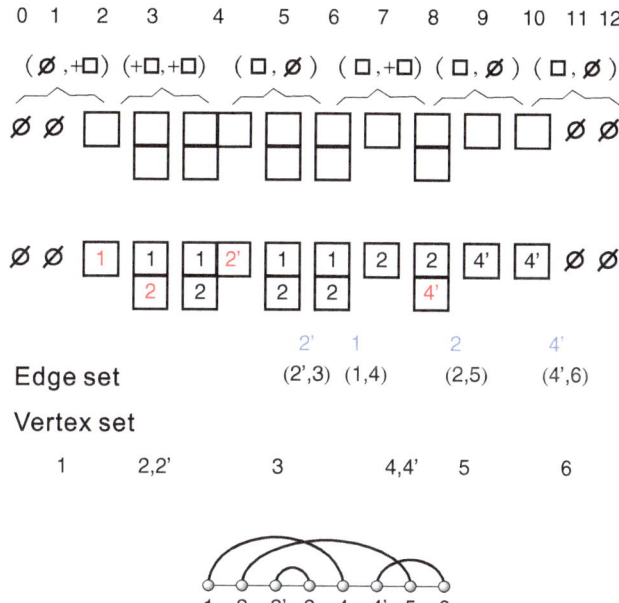

Edge set

Vertex set

Fig. 3.15. Lemma 3.1: from vacillating tableaux to inflated tangles.

(I) j is a left endpoint of degree 2, then we have the two $\eta(G_n)$-arcs (j,r) and (j',h). T_{2j-1} is obtained by removing the square with entry j' from the tableau T_{2j} and T_{2j-2} is obtained by removing the square with entry j from T_{2j-1}. Then we have $\lambda_{\eta(G_n)}^{2j-1} \subsetneq \lambda_{\eta(G_n)}^{2j}$ and $\lambda_{\eta(G_n)}^{2j-2} \subsetneq \lambda_{\eta(G_n)}^{2j-1}$ (left to right: $(+\square, +\square)$); see Fig. 3.16.

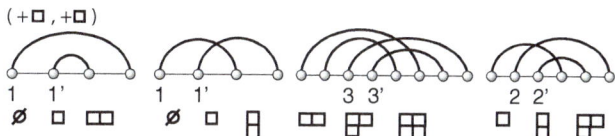

Fig. 3.16. All the possible cases for $(+\square, +\square)$ in case of 3-noncrossing tangles.

(II) j is the left endpoint of exactly one arc (j,k) but not a right endpoint, then first set T_{2j-1} to be the tableau obtained by removing the square with entry j from T_{2j} and let $T_{2j-2} = T_{2j-1}$. Therefore $\lambda_{\eta(G_n)}^{2j-1} \subsetneq \lambda_{\eta(G_n)}^{2j}$ and $\lambda_{\eta(G_n)}^{2j-2} = \lambda_{\eta(G_n)}^{2j-1}$ (left to right: $(\varnothing, +\square)$).

(III) j is a left and right endpoint of crossing arcs or a loop, then we have the two $\eta(G_n)$-arcs (j,s) and (h,j'), $h < j < j' < s$ or an arc of the form (j,j'), respectively. T_{2j-1} is obtained by RSK-inserting h into the tableau T_{2j} and T_{2j-2} is obtained by removing the square with entry j from the T_{2j-1}

$(+\square, -\square)$

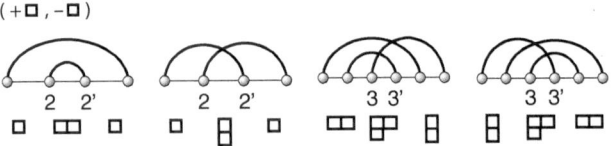

Fig. 3.17. All the possible cases for $(+\square, -\square)$ in case of 3-noncrossing tangles.

or T_{2j-1} is obtained by RSK-inserting j into the tableau T_{2j} and T_{2j-2} is obtained by removing the square with entry j from the T_{2j-1}, respectively (left to right: $(+\square, -\square)$); see Fig. 3.17.

(IV) $\eta_j = j$ is the right endpoint of exactly one arc (i, j) but not a left endpoint, then we set $T_{2j-1} = T_{2j}$ and obtain T_{2j-2} by RSK-inserting i into T_{2j-1}. Consequently we have $\lambda_{\eta(G_n)}^{2j-1} = \lambda_{\eta(G_n)}^{2j}$ and $\lambda_{\eta(G_n)}^{2j-2} \supsetneq \lambda_{\eta(G_n)}^{2j-1}$ (left to right: $(-\square, \varnothing)$).

(V) j is a right endpoint of degree 2, then we have the two $\eta(G_n)$-arcs (i, j) and (h, j'). T_{2j-1} is obtained by RSK-inserting h into T_{2j} and T_{2j-2} is obtained by RSK-inserting i into T_{2j-1}. We derive $\lambda_{\eta(G_n)}^{2j-1} \supsetneq \lambda_{\eta(G_n)}^{2j}$ and $\lambda_{\eta(G_n)}^{2j-2} \supsetneq \lambda_{\eta(G_n)}^{2j-1}$ (left to right: $(-\square, -\square)$); see Fig. 3.18.

$(-\square, -\square)$

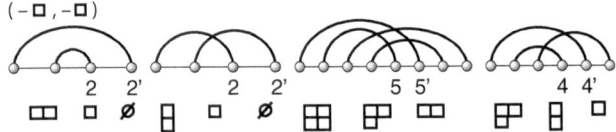

Fig. 3.18. All the possible cases for $(-\square, -\square)$ in case of 3-noncrossing tangles.

(VI) j is a left and right endpoint, then we have the two $\eta(G_n)$-arcs (i, j) and (j', h), where $i < j < j' < h$. First, the tableaux T_{2j-1} is obtained by removing the square with entry j' in T_{2j}. Second, the RSK insertion of i into T_{2j-1} generates the tableau T_{2j-2}. Accordingly, we derive the shapes $\lambda_{\eta(G_n)}^{2j-1} \subsetneq \lambda_{\eta(G_n)}^{2j}$ and $\lambda_{\eta(G_n)}^{2j-2} \supsetneq \lambda_{\eta(G_n)}^{2j-1}$ (left to right: $(-\square, +\square)$); see Fig. 3.19.

(VII) $\eta_j = j$ is an isolated vertex in $\eta(G_n)$, then we set $T_{2j-1} = T_{2j}$ and $T_{2j-2} = T_{2j-1}$. Accordingly, $\lambda_{\eta(G_n)}^{2j-1} = \lambda_{\eta(G_n)}^{2j}$ and $\lambda_{\eta(G_n)}^{2j-2} = \lambda_{\eta(G_n)}^{2j-1}$ (left to right: $(\varnothing, \varnothing)$).

$(-\square, +\square)$

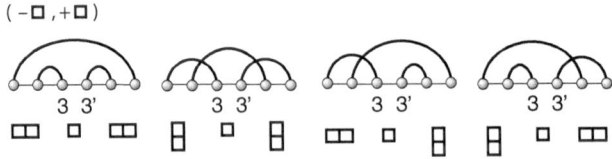

Fig. 3.19. All the possible cases for $(-\square, +\square)$ when restricted to 3-noncrossing tangles.

Therefore, ψ maps the inflation of a tangled diagram into a vacillating tableau and the lemma follows.

As an illustration of Lemma 3.2, see Fig. 3.20: starting from right to left the vacillating tableaux is obtained via the RSK algorithm as follows: if j is a

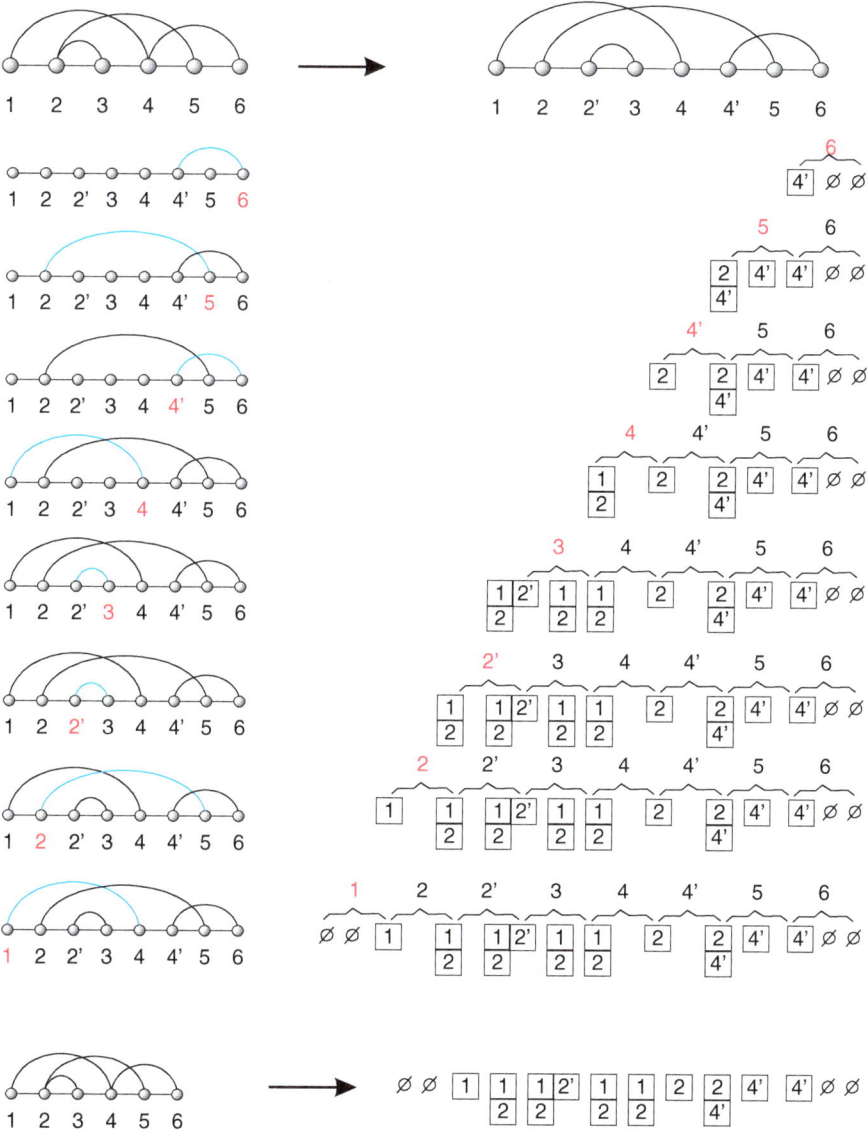

Fig. 3.20. An illustration of Lemma 3.2: how to map a tangle into a vacillating tableaux via ψ.

right endpoint it gives rise to RSK insertion of its (unique) left endpoint and if j is a left endpoint the square filled with j is removed.

Theorem 3.3. *There exists a bijection between the set of vacillating tableaux of shape $\varnothing$ and length $2n$, $\mathcal{V}_\varnothing^{2n}$, and the set of tangles on n vertices, $\mathcal{G}_n$,*

$$\beta\colon \mathcal{V}_\varnothing^{2n} \longrightarrow \mathcal{G}_n.$$

Proof. According to Lemmas 3.1 and 3.2, we have the mappings $\phi\colon \mathcal{V}_\varnothing^{2n} \longrightarrow \eta(\mathcal{G}_n)$ and $\psi\colon \eta(\mathcal{G}_n) \longrightarrow \mathcal{V}_\varnothing^{2n}$. We next show that ϕ and ψ are indeed inverses of each other. By definition, the mapping ϕ generates arcs whose left endpoints, when RSK inserted into T_i, recover the tableaux T_{i-1}. We observe that by definition, the mapping ψ reverses this extraction: it is constructed via the RSK insertion of the left endpoints. Therefore we have the following relations:

$$\phi\circ\psi(\eta(\mathcal{G}_n)) = \phi((\lambda^h_{\eta(\mathcal{G}_n)})_0^{2n}) = \eta(\mathcal{G}_n) \quad \text{and} \quad \psi\circ\phi(\mathcal{V}_\varnothing^{2n}) = \mathcal{V}_\varnothing^{2n},$$

from which we conclude that ϕ and ψ are bijective. Since $\mathcal{G}_n$ is in one-to-one correspondence with $\eta(\mathcal{G}_n)$, the proof of the theorem is complete.

By construction, the bijection $\eta\colon \mathcal{G}_n \longrightarrow \eta(\mathcal{G}_n)$ preserves the maximal number crossing and nesting arcs, respectively. Equivalently, a tangled diagram G_n is k-noncrossing or k-nonnesting if and only if its inflation $\eta(G_n)$ is k-noncrossing or k-nonnesting [25]. Indeed, this follows immediately from the definition of the inflation. Accordingly the next result is directly implied by Theorem 2.2:

Theorem 3.4. *A tangled diagram G_n is k-noncrossing if and only if all shapes λ^i in the corresponding vacillating tableau have less than k rows, i.e., $\phi\colon \mathcal{V}_\varnothing^{2n} \longrightarrow \mathcal{G}_n$ maps vacillating tableaux having less than k rows into k-noncrossing tangles. Furthermore, there is a bijection between the set of k-noncrossing and k-nonnesting tangles.*

Restricting the steps for vacillating tableaux produces the bijection of Chen et al. [25]. Let $\mathcal{M}_k^\dagger(n)$, $\mathcal{P}_k(n)$, and $\mathcal{B}_k(n)$ denote the set of k-noncrossing matchings, partitions, and braids. Theorem 3.3 implies that the tableaux sequences of $\mathcal{M}_k^\dagger(n)$, $\mathcal{P}_k(n)$, and $\mathcal{B}_k(n)$ are composed by the elements in $S_{\mathcal{M}_k^\dagger}, S_{\mathcal{P}_k}$, and $S_{\mathcal{B}_k}$, respectively, where $1 \le h, l \le k-1$ and

$$S_{\mathcal{M}_k^\dagger} = \{(-\Box_h, \varnothing), (\varnothing, +\Box_h)\},$$
$$S_{\mathcal{P}_k} = \{(\varnothing, \varnothing), (-\Box_h, \varnothing), (\varnothing, +\Box_h), (-\Box_h, +\Box_l)\},$$
$$S_{\mathcal{B}_k} = \{(\varnothing, \varnothing), (-\Box_h, \varnothing), (\varnothing, +\Box_h), (+\Box_h, -\Box_l)\},$$

where we use the following notation: if λ_{i+1} is obtained from λ_i by adding, removing a square from the jth row, or doing nothing we write $\lambda_{i+1}\setminus\lambda_i = +\Box_j$, $\lambda_{i+1}\setminus\lambda_i = -\Box_j$ or $\lambda_{i+1}\setminus\lambda_i = \varnothing$, respectively; see Fig. 3.21.

The enumeration of 3-noncrossing partitions and 3-noncrossing enhanced partitions has been studied by Xin and Bousquet-Mélou [17]. The authors obtain their results by solving a functional equation of walks in the first quadrant using the reflection principle [149] and the kernel method [92].

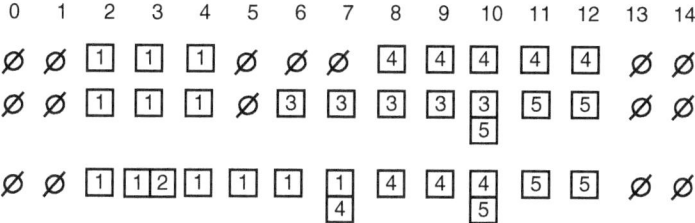

Fig. 3.21. The corresponding tableaux sequences for the partial matching, partition, and braid shown in Fig. 3.2.

A 2-regular, k-noncrossing partition is a k-noncrossing partition without arcs of the form $(i, i+1)$. We denote the set of 2-regular, k-noncrossing partitions by $\mathcal{P}_{k,2}(n)$. There exists a bijection between 2-regular, k-noncrossing partitions and k-noncrossing braids without isolated points, denoted by $\mathcal{B}_k^\dagger(n)$, i.e., k-noncrossing enhanced partitions[25]. This bijection is obtained as follows: for $\delta \in \mathcal{B}_k^\dagger(n)$, we identify loops with isolated points and crossing arcs (i, j) and (j, h), where $i < j < h$, by noncrossing arcs. This identification produces a mapping from $\mathcal{P}_{k,2}(n)$ into a subset of partitions $\mathcal{P}_k^*(n)$, which we refer to as ϑ; see Fig. 3.22.

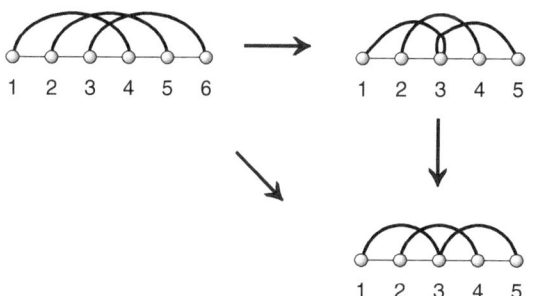

Fig. 3.22. An illustration of Theorem 3.5: the bijection ϑ

Theorem 3.5. *Let $k \in \mathbb{N}$, $k \geq 3$. Then we have a bijection*

$$\vartheta \colon \mathcal{P}_{k,2}(n) \longrightarrow \mathcal{B}_k^\dagger(n-1), \qquad \vartheta((i,j)) = (i, j-1).$$

Proof. By construction, ϑ maps tangled diagrams on $[n]$ to tangled diagrams on $[n-1]$. Since there does not exist any arc of the form $(i, i+1)$, for any $\pi \in \mathcal{P}_{k,2}(n)$, $\vartheta(\pi)$ is loop free. By construction, ϑ preserves the orientation of arcs, whence $\vartheta(\pi)$ is a partition.

Claim 1. $\vartheta \colon \mathcal{P}_{k,2}(n) \longrightarrow \mathcal{B}_k^\dagger(n-1)$ is well defined.
We first prove that $\vartheta(\pi)$ is k-noncrossing. Suppose there exist k mutually crossing arcs, $\{(i_s, j_s)\}_{s=1}^{s=k}$ in $\vartheta(\pi)$. Since $\vartheta(\pi)$ is a partition, we have

$i_1 < \cdots < i_k < j_1 < \cdots < j_k$. So, we obtain for the partition $\pi \in \mathcal{P}_{k,2}(n)$ the k arcs $(i_s, j_s + 1)$, $s = 1, \ldots, k$, where $i_1 < \cdots < i_k < j_1 + 1 < \cdots < j_k + 1$, which is impossible since π is k-noncrossing. We next show that $\vartheta(\pi)$ is a k-noncrossing braid. If $\vartheta(\pi)$ is not a k-noncrossing braid, then it contains k arcs of the form $(i_1, j_1), \ldots, (i_k, j_k)$ such that $i_1 < \cdots < i_k = j_1 < \cdots < j_k$. Then π contains the arcs $(i_1, j_1 + 1)$, $(i_k, j_k + 1)$ where $i_1 < \cdots < i_k < j_1 + 1 < \cdots < j_k + 1$, which is impossible since these arcs are a set of k mutually crossing arcs and Claim 1 follows.

Claim 2. ϑ is bijective.
Clearly ϑ is injective and it remains to prove surjectivity. For any k-noncrossing braid δ there exists 2-regular partition π such that $\vartheta(\pi) = \delta$. We have to show that π is k-noncrossing. Suppose that there exists some partition π with k mutually crossing arcs such that $\vartheta(\pi) = \delta$. Let $M' = \{(i_1, j_1), \ldots, (i_k, j_k)\}$ be a set of k mutually crossing arcs in the standard representation of π, i.e., $i_1 < \cdots < i_k < j_1 < \cdots < j_k$. Then we have in $\vartheta(\pi)$ the arcs $(i_s, j_s - 1)$, $s = 1, \ldots, k$, such that

$$i_1 < \cdots < i_k \leq j_1 - 1 < \cdots < j_k - 1.$$

Since $M = \{(i_1, j_1 - 1), \ldots, (i_k, j_k - 1)\}$ is k-noncrossing, we conclude $i_k = j_1 - 1$. This is impossible in k-noncrossing braids. By transposition, we have proved that any ϑ-preimage is necessarily a k-noncrossing partition, whence Claim 2 and the proof of the theorem is complete.

In Fig. 3.22 we give an illustration of the bijection $\vartheta \colon \mathcal{P}_{k,2}(n) \longrightarrow \mathcal{B}_k^\dagger(n-1)$.

3.3 Enumeration

Let $t_k(n)$ and $\tilde{t}_k(n)$ denote the numbers of k-noncrossing tangles and k-noncrossing tangles without isolated points on $[n]$, respectively. Recall that $f_k(2n, 0)$ is the number of k-noncrossing matchings on $2n$ vertices. In the following we will illustrate that the enumeration of tangles could be reduced to the enumeration of matchings via the inflation map. Without loss of generality we can restrict our analysis to the case of tangles without isolated points since the number of tangled diagrams on $[n]$ is given by

$$t_k(n) = \sum_{i=0}^n \binom{n}{i} \tilde{t}_k(n - i). \tag{3.2}$$

Theorem 3.6. *The number of k-noncrossing tangles without isolated points on $[n]$ is given by*

$$\tilde{t}_k(n) = \sum_{\ell=0}^n \binom{n}{\ell} f_k(2n - \ell, 0).$$

In particular, for $k = 3$ we have

$$\tilde{t}_3(n) = \sum_{\ell=0}^{n} \binom{n}{\ell} \left(C_{\frac{2n-\ell}{2}} \, C_{\frac{2n-\ell}{2}+2} - C_{\frac{2n-\ell}{2}+1}^2 \right),$$

where C_m denotes the mth Catalan number $\frac{1}{m+1}\binom{2m}{m}$.

Proof. Let $\tilde{\mathcal{T}}_k(n, V)$ be the set of tangles without isolated points where $V = \{i_1, \ldots, i_h\}$ is the set of vertices of degree 1 (where $h \equiv 0 \mod 2$ by definition of $\tilde{\mathcal{T}}_k(n, V)$) and let $\mathcal{M}_k^{\dagger}(\{1, 1', \ldots, n, n'\} \setminus V')$, where $V' = \{i'_1, \ldots, i'_h\}$ denotes the set of matchings on $\{1, 1', \ldots, n, n'\} \setminus V'$. By construction, the inflation $\eta \colon \mathcal{G}_n \longrightarrow \eta(\mathcal{G}_n)$ induces a well-defined mapping

$$\hat{\eta} \colon \tilde{\mathcal{T}}_k(n, V) \longrightarrow \mathcal{M}_k^{\dagger}(\{1, 1', \ldots, n, n'\} \setminus V')$$

with inverse κ defined by identifying all pairs (y, y'), where $y, y' \in \{1, 1', \ldots, n, n'\} \setminus V'$. Obviously, we have $|\mathcal{M}_k^{\dagger}(\{1, 1', \ldots, n, n'\} \setminus V')| = f_k(2n - h, 0)$ and

$$\tilde{t}_k(n) = \sum_{V \subset [n]} \tilde{t}_k(n, V) = \sum_{\ell=0}^{n} \binom{n}{\ell} f_k(2n - \ell, 0). \tag{3.3}$$

Suppose $n \equiv 0 \mod 2$. Let C_m denote the mth Catalan number. Then we have [53]

$$f_3(n, 0) = C_{\frac{n}{2}} \, C_{\frac{n}{2}+2} - C_{\frac{n}{2}+1}^2,$$

and the theorem follows.

The first five numbers of 3-noncrossing tangles are given by 2, 7, 39, 292, 2635.

In eq. (3.3) we relate the generating functions of k-noncrossing tangles $\mathbf{T}_k(z) = \sum_n t_k(n) z^n$ and k-noncrossing matchings $\mathbf{F}_k(z) = \sum_n f_k(2n, 0) z^n$. We derive the functional equation which is instrumental to prove eq. (3.6) for $2 \leq k \leq 9$.

For this purpose we employ Cauchy's integral formula: let D be a simply connected domain and let C be a simple closed positively oriented contour that lies in D. If f is analytic inside C and on C, except at the vertices $z_1, z_2, \ldots, z_n$ that are in the interior of C, then we have Cauchy's integral formula

$$\int_C f(z) dz = 2\pi i \sum_{k=1}^{n} Res[f, z_k]. \tag{3.4}$$

In particular, if f has a simple pole at z_0, then $Res[f, z_0] = \lim_{z \to z_0} (z - z_0) f(z)$.

Lemma 3.7. *Let $k \in \mathbb{N}$, $k \geq 2$. Then we have*

$$\mathbf{T}_k \left(\frac{z^2}{1 + z + z^2} \right) = \frac{1 + z + z^2}{z + 2} \, \mathbf{F}_k(z^2). \tag{3.5}$$

Proof. The relation between the number of k-noncrossing tangles, $t_k(n)$, and k-noncrossing matchings, $f_k(2n,0)$, given in eq. (2.7), which implies

$$t_k(n) = \sum_{r,\ell} \binom{n}{r}\binom{n-r}{\ell} f_k(2n-2r-\ell,0).$$

Substituting the combinatorial terms with the contour integrals we derive

$$\binom{n}{r} = \frac{1}{2\pi i} \oint_{|u|=\alpha} (1+u)^n u^{-r-1} du,$$

$$f_k(2n-2r-\ell,0) = \frac{1}{2\pi i} \oint_{|z|=\beta_3} \mathbf{F}_k(z^2) z^{-(2n-2r-\ell)-1} dz,$$

$$t_k(n) = \sum_{r,\ell} \binom{n}{r}\binom{n-r}{\ell} f_k(2n-2r-\ell,0)$$

$$= \frac{1}{(2\pi i)^3} \sum_{r,\ell} \oint_{\substack{|v|=\beta_1 \\ |z|=\beta_2 \\ |u|=\beta_3}} (1+u)^n u^{-r-1} (1+v)^{n-r} v^{-\ell-1} \times$$

$$\mathbf{F}_k(z^2)\, z^{-(2n-2r-\ell)-1} dv\, du\, dz,$$

where $\alpha, \beta_1, \beta_2, \beta_3$ are arbitrary small positive numbers. Since the series are absolute convergent, we obtain

$$t_k(n) = \frac{1}{(2\pi i)^3} \sum_{r} \oint_{\substack{|v|=\beta_1 \\ |z|=\beta_2 \\ |u|=\beta_3}} (1+u)^n u^{-r-1} \mathbf{F}_k(z^2)\, z^{-2n+2r-1} (1+v)^{n-r} v^{-1} \times$$

$$\sum_{\ell} \left(\frac{z}{v}\right)^\ell dv\, du\, dz,$$

which gives rise to

$$t_k(n) = \frac{1}{(2\pi i)^3} \sum_{r} \oint_{\substack{|u|=\beta_3 \\ |z|=\beta_2}} (1+u)^n u^{-r-1} \mathbf{F}_k(z^2)\, z^{-2n+2r-1} \times$$

$$\left(\oint_{|v|=\beta_1} \frac{(1+v)^{n-r}}{v-z} dv\right) du\, dz.$$

Since $v = z$ is the unique (simple) pole in the integral domain, eq. (3.4) implies

$$\oint_{|v|=\beta_1} \frac{(1+v)^{n-r}}{v-z} dv = 2\pi i\,(1+z)^{n-r}.$$

We accordingly have

$$t_k(n) = \frac{1}{(2\pi i)^2} \sum_{r} \oint_{\substack{|u|=\beta_3 \\ |z|=\beta_2}} (1+u)^n u^{-r-1} \mathbf{F}_k(z^2)\, z^{-2n+2r-1} (1+z)^{n-r} du\, dz.$$

Proceeding analogously with respect to the summation over r yields

$$t_k(n) = \frac{1}{(2\pi i)^2} \oint_{\substack{|u|=\beta_3 \\ |z|=\beta_2}} (1+u)^n \mathbf{F}_k(z^2) \, z^{-2n-1}(1+z)^n u^{-1} \sum_r \frac{z^{2r}}{u^r(1+z)^r} \, du \, dz$$

$$= \frac{1}{(2\pi i)^2} \oint_{|z|=\beta_2} \mathbf{F}_k(z^2) \, z^{-2n-1}(1+z)^n \left(\oint_{|u|=\beta_3} (1+u)^n \frac{1}{u - \frac{z^2}{1+z}} \, du \right) dz.$$

Since $u = \frac{z^2}{1+z}$ is the only pole in the integral domain, Cauchy's integral formula implies

$$\oint_{|u|=\beta_3} (1+u)^n \frac{1}{u - \frac{z^2}{1+z}} \, du = 2\pi i \left(1 + \frac{z^2}{1+z} \right)^n.$$

Now we compute

$$t_k(n) = \frac{1}{2\pi i} \oint_{|z|=\beta_2} \mathbf{F}_k(z^2) \, z^{-1} z^{-2n}(1+z)^n \left(1 + \frac{z^2}{1+z} \right)^n dz$$

$$= \frac{1}{2\pi i} \oint_{|z|=\beta_2} \mathbf{F}_k(z^2) \, z^{-1} \left(\frac{1+z+z^2}{z^2} \right)^n dz$$

$$= \frac{1}{2\pi i} \oint_{|z|=\beta_2} \frac{1+z+z^2}{z+2} \mathbf{F}_k(z^2) \left(\frac{z^2}{1+z+z^2} \right)^{-n-1} d\left(\frac{z^2}{1+z+z^2} \right)$$

from which

$$\mathbf{T}_k \left(\frac{z^2}{1+z+z^2} \right) = \frac{1+z+z^2}{z+2} \mathbf{F}_k(z^2)$$

follows and the theorem is proved. $\quad\blacksquare$

Lemma 3.7, Theorem 2.8, and Proposition 2.24 imply for the asymptotics of tangles.

Theorem 3.8. *For $2 \leq k \leq 9$ the number of k-noncrossing tangles is asymptotically given by*

$$t_k(n) \sim c_k \, n^{-((k-1)^2 + \frac{k-1}{2})} \left(4(k-1)^2 + 2(k-1) + 1 \right)^n \quad \text{where } c_k > 0. \quad (3.6)$$

Proof. According to Lemma 3.7, we have the functional equation

$$\mathbf{T}_k \left(\frac{z^2}{z^2 + z + 1} \right) = \frac{z^2 + z + 1}{z+2} \mathbf{F}_k(z^2), \qquad (3.7)$$

where $|z| \leq \rho_k < 1$ and the function $\vartheta(z) = \frac{z^2}{z^2+z+1}$ is regular at $z = \pm\rho_k$ and $\rho_k = 1/2(k-1)$. Then

$$\vartheta(\rho_k) = \frac{\rho_k^2}{\rho_k^2 + \rho_k + 1} \quad \text{and} \quad \vartheta(-\rho_k) = \frac{\rho_k^2}{\rho_k^2 - \rho_k + 1}$$

are both singularities of $\mathbf{T}_k(z)$. We claim that $\vartheta(\rho_k)$ is the unique dominant positive real singularity of $\mathbf{T}_k(z)$. Indeed, $\vartheta(z)$ is strictly monotonously increasing and continuous for $0 < z \leq 1$, and $0 < \vartheta(z) \leq 1/3$. If there is a positive singularity γ of $\mathbf{T}_k(z)$

$$\gamma < \vartheta(\rho_k) \leq \vartheta\left(\frac{1}{2}\right) = \frac{1}{7},$$

there would exist $\vartheta^{-1}(\gamma) < \rho_k$ which is a contradiction to ρ_k being the dominant singularity of $\mathbf{T}_k(\vartheta(z))$. Next we show that $\vartheta(\rho_k)$ is unique. Suppose there exists a dominant singularity η different from $\vartheta(\rho_k)$, where $|\eta| = \vartheta(\rho_k)$. Then there exists $z_\eta \in \mathbb{C}$ such that $\vartheta(z_\eta) = \eta$ and $z_\eta \neq \rho_k$. Since $|\vartheta(z_\eta)| = \vartheta(\rho_k)$,

$$(\rho_k^2 + \rho_k + 1)|z_\eta|^2 = |z_\eta^2 + z_\eta + 1|\rho_k^2 \leq \left(|z_\eta^2| + |z_\eta| + 1\right)\rho_k^2,$$

whence $|z_\eta| \leq \rho_k$. Accordingly, z_η is a dominant singularity of $\mathbf{T}_k(\vartheta(z))$ which is a contradiction to eq. (3.7) which implies that $\mathbf{T}_k(\vartheta(z))$ has only the dominant singularities $\pm\rho_k$. Consequently, $\vartheta(\rho_k)$ is the unique dominant singularity of $\mathbf{T}_k(z)$.

According to Corollary 2.14, the generating function, $\mathbf{F}_k(z)$, is D-finite. Theorem 2.13 shows that the composition $F(G(z))$ of a D-finite function F and a rational function G, where $G(0) = 0$, is again D-finite, and the product of two D-finite functions is also D-finite, whence $\mathbf{T}_k(z)$ and $\mathbf{T}_k(\vartheta(z))$ are D-finite and accordingly have singular expansions. Let $S_{\mathbf{T}_k}(z - \vartheta(\rho_k))$ denote the singular expansion of $\mathbf{T}_k(z)$ at $z = \vartheta(\rho_k)$. Since $\vartheta(z)$ is regular at $z = \rho_k$ and $\vartheta'(\rho_k) \neq 0$, see Table 3.1, we are given the supercritical paradigm [42]. Indeed, we have $\vartheta'(\rho_k) \neq 0$, see Table 3.1 and derive

$$\mathbf{T}_k(\vartheta(z)) \sim S_{\mathbf{T}_k}(\vartheta(z) - \vartheta(\rho_k)) \qquad \text{as } \vartheta(z) \to \vartheta(\rho_k)$$
$$= \Theta(S_{\mathbf{T}_k}(z - \rho_k)) \qquad \text{as } z \to \rho_k.$$

Proposition 2.24 implies that for $z \to \rho_k^2$

$$\mathbf{F}_k(z) = \begin{cases} P_k(z - \rho_k^2) + c_k'(z - \rho_k^2)^{((k-1)^2 + (k-1)/2) - 1} \ln(z - \rho_k^2)(1 + o(1)) \\ P_k(z - \rho_k^2) + c_k'(z - \rho_k^2)^{((k-1)^2 + (k-1)/2) - 1}(1 + o(1)) \end{cases}$$

depending on k being odd and even. Here the terms $P_k(z)$ are polynomials of degree $\leq (k-1)^2 + (k-1)/2 - 1$ and c_k' is some constant. Let $S_{\mathbf{F}_k}(z - \rho_k^2)$

k	2	3	4	5	6	7	8	9
$\vartheta'(\rho_k)$	0.4082	0.3265	0.2531	0.2042	0.1704	0.1461	0.1277	0.1134

Table 3.1. The values of $\vartheta'(\rho_k)$ for $2 \leq k \leq 9$.

denote the singular expansion of $\mathbf{F}_k(z)$ at $z = \rho_k^2$. Equation (3.7) implies for $z \to \rho_k$

$$\mathbf{T}_k(\vartheta(z)) \sim \frac{\rho_k^2 + \rho_k + 1}{\rho_k + 2} S_{\mathbf{F}_k}(z^2 - \rho_k^2)$$

and thus

$$S_{\mathbf{T}_k}(z - \rho_k) = \Theta\left(S_{\mathbf{F}_k}(z - \rho_k)\right) \quad \text{as } z \to \rho_k.$$

Therefore, $\mathbf{T}_k(z)$ has at $v = \vartheta(\rho_k)$ exactly the same subexponential factors as $\mathbf{F}_k(z)$ at ρ_k^2, i.e., we have

$$[z^n]\,\mathbf{T}_k(z) \sim c_k\, n^{-\left((k-1)^2 + \frac{k-1}{2}\right)} \left(\frac{\rho_k^2}{\rho_k^2 + \rho_k + 1}\right)^{-n} \quad \text{for some } t_k > 0$$

and the theorem is proved.

4

Combinatorial analysis

In this chapter we develop the theory of k-noncrossing and k-noncrossing, σ-canonical structures. We derive their generating functions and obtain their singularity analysis, which produces simple, asymptotic formulas for the numbers of various types of k-noncrossing σ-canonical structures. This chapter is based on the results of [76, 77, 95, 107].

As introduced in Chapter 2, diagrams are labeled graphs over the vertex set $[n] = \{1, \ldots, n\}$ with vertex degrees ≤ 1, represented by drawing its vertices on a horizontal line and its arcs (i, j), where $i < j$, in the upper half plane. The length of an arc (i, j) is given by $j - i$. In a diagram two arcs (i_1, j_1) and (i_2, j_2) are called crossing if $i_1 < i_2 < j_1 < j_2$ holds. Accordingly, a k-crossing is a sequence of arcs $(i_1, j_1), \ldots, (i_k, j_k)$ such that

$$i_1 < i_2 < \cdots < i_k < j_1 < j_2 < \cdots < j_k,$$

see Fig. 4.1. Similarly, a k-nesting is a set of k distinct arcs such that

$$i_1 < i_2 < \cdots < i_k < j_k < \cdots < j_2 < j_1.$$

Let A, B be two sets of arcs, then A is nested in B if any element of A is nested in any element of B. Accordingly, k-noncrossing diagrams do not

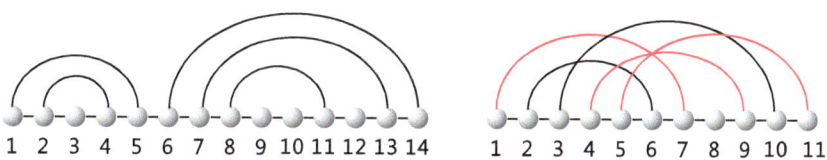

Fig. 4.1. k-noncrossing diagrams: a noncrossing (left) and a diagram exhibiting a 3-crossing (right) containing the three mutually crossing arcs $(1, 7), (4, 9), (5, 11)$.

contain any k-crossings. Denoting by $[i, j]$ an interval, i.e., the sequence of isolated vertices $(i, i + 1, \ldots, j - 1, j)$, we next specify further properties of k-noncrossing diagrams:

C. Reidys, *Combinatorial Computational Biology of RNA*,
DOI 10.1007/978-0-387-76731-4_4,
© Springer Science+Business Media, LLC 2011

■ A stack of length σ, $S_{i,j}^{\sigma}$, is a maximal sequence of "parallel" arcs,

$$((i,j),(i+1,j-1),\ldots,(i+(\sigma-1),j-(\sigma-1))).$$

We call a stack of length σ a σ-stack.

■ A stem of size s is a sequence

$$\left(S_{i_1,j_1}^{\sigma_1}, S_{i_2,j_2}^{\sigma_2}, \ldots, S_{i_s,j_s}^{\sigma_s}\right),$$

where $S_{i_m,j_m}^{\sigma_m}$ is nested in $S_{i_{m-1},j_{m-1}}^{\sigma_{m-1}}$ such that any arc nested in $S_{i_{m-1},j_{m-1}}^{\sigma_{m-1}}$ is either contained or nested in $S_{i_m,j_m}^{\sigma_m}$, for $2 \leq m \leq s$.

■ A hairpin loop is a pair $((i,j),[i+1,j-1])$, where (i,j) is an arc and $[i,j]$ is an interval, i.e., a sequence of consecutive, isolated vertices $(i, i+1, \ldots, j-1, j)$.

■ An interior loop is a quadruple $((i_1,j_1),[i_1+1,i_2-1],(i_2,j_2),[j_2+1,j_1-1])$, where (i_2,j_2) is nested in (i_1,j_1), i.e., $i_1 < i_2 < j_2 < j_1$.

For an illustration of the above structural features, see Fig. 4.2. Note that given a stem

$$\left(S_{i_1,j_1}^{\sigma_1}, S_{i_2,j_2}^{\sigma_2}, \ldots, S_{i_s,j_s}^{\sigma_s}\right)$$

the maximality of the stacks implies that any two nested stacks within a stem, $S_{i_m,j_m}^{\sigma_m}$ and $S_{i_{m-1},j_{m-1}}^{\sigma_{m-1}}$ are separated by a nonempty interval of isolated

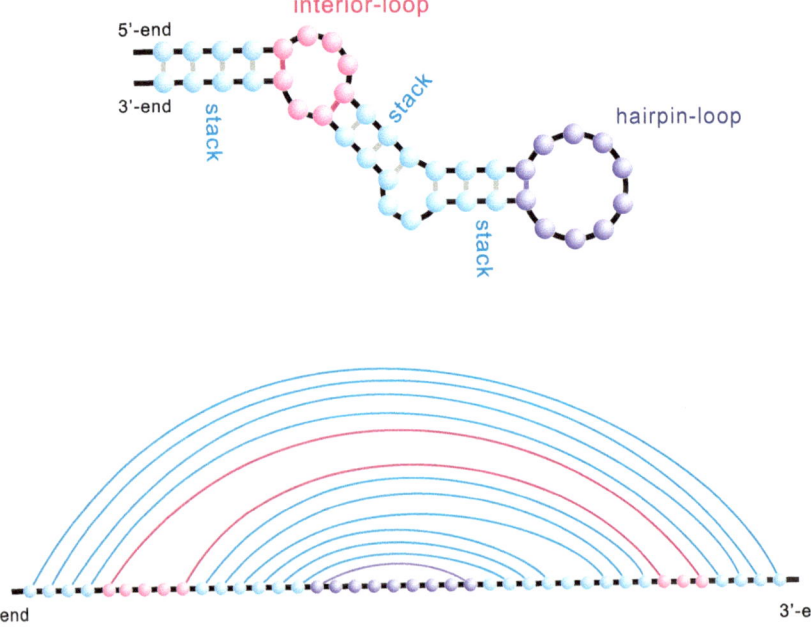

Fig. 4.2. Structural elements in RNA: we display three nested stacks forming a single stem, a hairpin loop, and an interior loop.

vertices between $i_{m-1} + (\sigma_{m-1} - 1)$ and i_m or $j_{m-1} - (\sigma_{m-1} - 1)$ and j_m, respectively.

Note that crossings of stems, see Fig. 4.3, are modular in the sense that *all* stacks of the stem have to be crossed simultaneously.

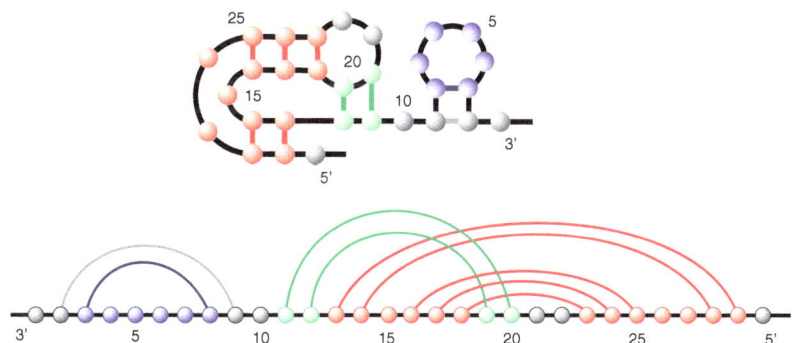

Fig. 4.3. Modular crossings: all stacks contained in a stem are crossed simultaneously.

Definition 4.1. *A k-noncrossing, σ-canonical structure is a k-noncrossing diagram with a minimum arc-length, $\lambda \geq 2$, and a minimum stack-length, σ; see Fig. 4.4.*

Let $\mathcal{T}_{k,\sigma}(n)$ denote the set of k-noncrossing, σ-canonical structures of length n and let $\mathsf{T}_{k,\sigma}(n)$ denote their cardinality. Similarly let $\mathcal{T}_{k,\sigma}(n,h)$ and $\mathsf{T}_{k,\sigma}(n,h)$ denote set and number of k-noncrossing, σ-canonical structures of length n having exactly h arcs. The set of k-noncrossing, σ-canonical structures that satisfy a minimum arc length condition of $\lambda > 2$ is denoted by $\mathcal{T}_{k,\sigma}^{[\lambda]}(n)$. If no arc length is specified, we always implicitly assume $\lambda = 2$.

k-Noncrossing, σ-canonical structures are obtained by the "folding" of RNA sequences. Their vertices and arcs correspond to the nucleotides **A**, **G**, **U**, **C** and Watson–Crick (**A-U**, **G-C**) and (**U-G**) base pairs, respectively. The relevance of requiring minimum stack length greater than 1 stems from the fact that RNA structures are formed by Watson–Crick (**A-U**, **G-C**) and (**U-G**) base pairs. Due to the biochemistry of these base pairs, parallel bonds are thermodynamically more stable. Therefore, the minimum stack-length, σ, is a parameter of central importance and these structures are called σ-canonical [124]. In particular, for σ = 2, we refer to the structure as a canonical structure; see Fig. 4.4. Note that canonical structures contain no isolated base pairs.

We shall begin in Section 4.1 by introducing the notions of cores and V_k-shapes. The latter are a generalization of Giegerich's shapes for RNA secondary structures [139], originally designed for a different purpose: V_k-shapes

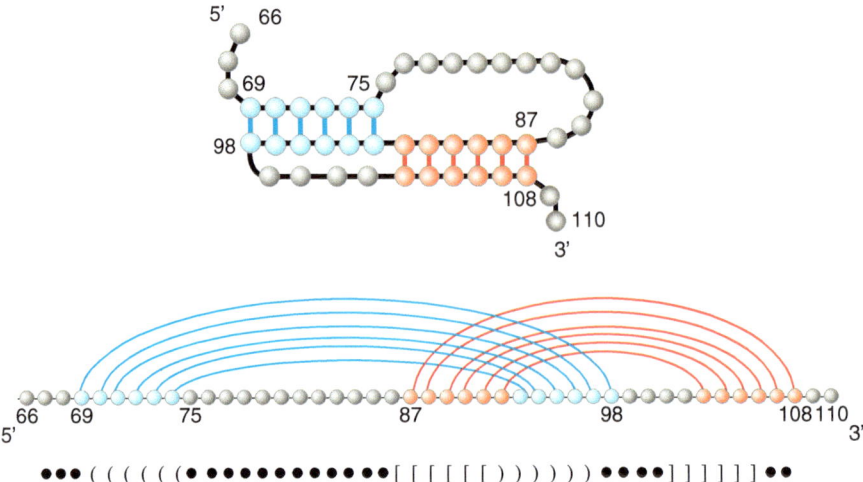

Fig. 4.4. An example of a 3-noncrossing, canonical structure: the pseudoknot structure of the PrP-encoding mRNA. Here "•" denotes an unpaired nucleotide and "(, ["
and "),]" denote origin and terminus of base pairs contained in the *blue* and *red*
stacks, respectively. Note that due to the crossing it is necessary to use distinct
labels for the base pairs of two respective stacks.

were developed in [107] in order to categorize k-noncrossing RNA pseudoknot structures. We later realized their central role for the computation of
the generating functions. In Section 4.2 we show how this inflation of V_k-
shapes via symbolic enumeration works. That is, via formal substitutions on
the level of generating functions, we derive the proofs for all relevant classes
of k-noncrossing RNA pseudoknot structures. In Section 4.3 we present exact and asymptotic enumeration results of k-noncrossing structures using the
previously derived generating functions. Finally, in Section 4.4 we give the
analysis of the remaining case by studying k-noncrossing, 2-canonical structures having minimum arc length 4.

4.1 Cores and Shapes

4.1.1 Cores

Definition 4.2. (Core) *A core structure is a k-noncrossing structure with
minimum arc length ≥ 2 in which each stack has length 1; see Fig. 4.5.*

We denote the set and number of core structures over $[n]$ by $\mathcal{C}_k(n)$ and $\mathsf{C}_k(n)$,
respectively. Analogously $\mathcal{C}_k(n, h)$ and $\mathsf{C}_k(n, h)$ denote the set and the number
of core structures having h arcs.

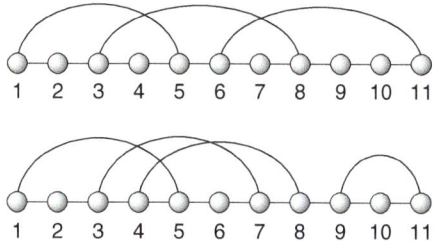

Fig. 4.5. Core structures: a 3-noncrossing core structure (*top*) and a 4-noncrossing core structure (*bottom*). All stacks in cores have length exactly 1.

In Lemma 4.3 we establish that the number of all k-noncrossing structures with stack-length $\geq \sigma$ is a sum of the number of k-noncrossing cores with positive integer coefficients.

Lemma 4.3. (Core lemma) *For $k, h, \sigma \in \mathbb{N}$, $k \geq 2$, $1 \leq h \leq n/2$ we have*

$$\mathsf{T}_{k,\sigma}(n,h) = \sum_{b=\sigma-1}^{h-1} \binom{b + (2-\sigma)(h-b) - 1}{h-b-1} \mathsf{C}_k(n-2b, h-b).$$

Proof. First, there exists a mapping from k-noncrossing structures with h arcs and minimum stack-length σ over $[n]$ into core structures:

$$c \colon \mathcal{T}_{k,\sigma}(n,h) \to \dot{\bigcup}_{0 \leq b \leq h-1} \mathcal{C}_k(n-2b, h-b), \quad \delta \mapsto c(\delta),$$

where the core structure $c(\delta)$ is obtained in two steps: first, we map arcs and isolated vertices as follows:

- $\forall \ell \geq \sigma - 1$; $((i-\ell, j+\ell), \ldots, (i,j)) \mapsto (i,j)$ and
- $q \mapsto q$ if q is isolated.

Second we relabel the vertices of the resulting diagram from left to right in increasing order, that is we replace each stack by a single arc and keep isolated vertices and then relabel; see Fig. 4.6. We have to prove that $c \colon \mathcal{T}_{k,\sigma}(n,h) \longrightarrow$

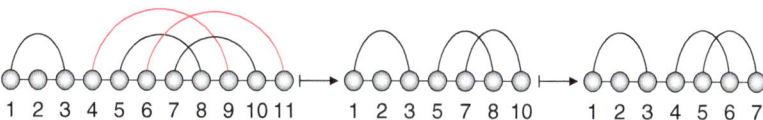

Fig. 4.6. The mapping $c \colon \mathcal{T}_{k,\sigma}(n,h) \longrightarrow \dot{\bigcup}_{0 \leq b \leq h-1} \mathcal{C}_k(n-2b, h-b)$ is obtained in two steps: first contraction of the stacks while keeping isolated points and second relabeling of the resulting diagram.

$\dot{\bigcup}_{0 \leq b \leq h-1} \mathcal{C}_k(n-2b, h-b)$ is well defined, i.e., that c cannot produce 1-arcs.

Indeed, since $\delta \in \mathcal{T}_{k,\sigma}(n, h)$, δ does not contain 1-arcs we can conclude that $c(\delta)$ has by construction arcs of length ≥ 2, c is by construction surjective. Keeping track of multiplicities gives rise to the map

$$f_{k,\sigma} \colon \mathcal{T}_{k,\sigma}(n, h) \to \dot{\bigcup}_{0 \leq b \leq h-1}$$

$$\left[\mathcal{C}_k(n - 2b, h - b) \times \left\{ (a_j)_{1 \leq j \leq h-b} \mid \sum_{j=1}^{h-b} a_j = b, \ a_j \geq \sigma - 1 \right\} \right],$$

$$(4.1)$$

given by $f_{k,\sigma}(\delta) = (c(\delta), (a_j)_{1 \leq j \leq h-b})$; see Fig. 4.7. We can conclude that $f_{k,\sigma}$ is well defined and a bijection. We proceed by computing the multiplicities of the resulting core structures:

Claim.

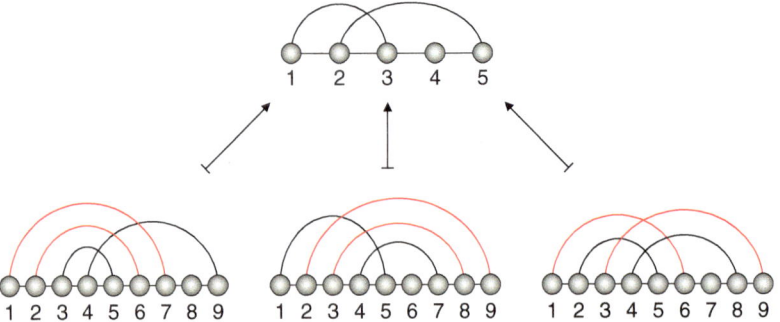

Fig. 4.7. A $\mathcal{C}_3(5, 2)$-core (*top*) and three structures contained in $\mathcal{T}_{3,1}(9, 4)$ (*bottom*). The bottom three $\mathcal{T}_{3,1}(9, 4)$-structures induce via $f_{k,\sigma}$ of eq. (4.1) the above $\mathcal{C}_3(5, 2)$-core.

$$\left| \left\{ (a_j)_{1 \leq j \leq h-b} \mid \sum_{j=1}^{h-b} a_j = b; \ a_j \geq \sigma - 1 \right\} \right| = \binom{b + (2 - \sigma)(h - b) - 1}{h - b - 1}.$$

Clearly, $a_j \geq \sigma - 1$ is equivalent to $\mu_j = a_j - \sigma + 2 \geq 1$ and we have

$$\sum_{j=1}^{h-b} \mu_j = \sum_{j=1}^{h-b} (a_j - \sigma + 2) = b + (2 - \sigma)(h - b).$$

We next show that

$$\left| \left\{ (\mu_j)_{1 \leq j \leq h-b} \mid \sum_{j=1}^{h-b} \mu_j = b + (2 - \sigma)(h - b); \ \mu_j \geq 1 \right\} \right|$$

is equal to the number of $(h-b-1)$-subsets in $\{1, 2, \ldots, b+(2-\sigma)(h-b)-1\}$. Consider the set

$$\{\mu_1, \mu_1 + \mu_2, \ldots, \mu_1 + \mu_2 + \cdots + \mu_{h-b-1}\}$$

consisting of $h-b-1$ distinct elements of $[b+(2-\sigma)(h-b)-1] = \{1, 2, \ldots, b+(2-\sigma)(h-b)-1\}$. Therefore $\{\mu_1, \mu_1 + \mu_2, \ldots, \mu_1 + \mu_2 + \cdots + \mu_{h-b-1}\}$ is an $(h-b-1)$-subset of $[b+(2-\sigma)(h-b)-1]$. Given any $(h-b-1)$-subset of $[b+(2-\sigma)(h-b)-1]$, we can arrange its elements in linear order and retrieve the sequence $\{\mu_i \mid 1 \le i \le h-b\}$ of positive integers with sum $b+(2-\sigma)(h-b)$. Therefore the above assignment is a bijection. Since the number of $(h-b-1)$-subsets of $[b+(2-\sigma)(h-b)-1]$ is given by $\binom{b+(2-\sigma)(h-b)-1}{h-b-1}$ the claim follows. We can conclude from the claim and eq. (4.1) that

$$\mathsf{T}_{k,\sigma}(n,h) = \sum_{b=\sigma-1}^{h-1} \binom{b+(2-\sigma)(h-b)-1}{h-b-1} \mathsf{C}_k(n-2b, h-b)$$

holds and the lemma follows.

We remark that Lemma 4.3 cannot be used in order to enumerate diagrams with arc length $\ge \lambda$, where $\lambda > 2$ and stack length σ. The key point here is that k-noncrossing structures with arc length $\ge \lambda$ have core structures with arc length 2; see Fig. 4.8. Instead of using Lemma 4.3 in order to establish

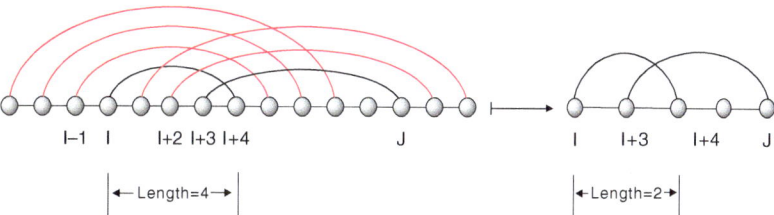

Fig. 4.8. Core structures in general have 2-arcs: the structure $\delta \in T_{3,3}^{[4]}(15)$ (lhs) is mapped into its core $c(\delta)$ (rhs). Clearly δ has arc length ≥ 4 and as a consequence of the contraction of the stack $((I+1, J+2), (I+2, J+1), (I+3, J))$ (the *red arcs* are being removed) into the arc $(I+3, J)$, $c(\delta)$ contains the arc $(I, I+4)$, which is, after relabeling, a 2-arc, i.e., an arc of the form $(i, i+2)$.

a functional relation between $\mathbf{T}_{k,\sigma}(z,u)$ and $\mathbf{C}_k(x,y)$, see Problem 4.7, we proceed by introducing first the combinatorial class of V_k-shapes. We show in Section 4.2 that these shapes allow the derivation of this functional relation in a natural way; see Proposition 4.17.

4.1.2 Shapes

Definition 4.4. (V_k-shape) *A V_k-shape is a k-noncrossing matching with stacks of length exactly 1.*

In other words, a V_k-shape is a core without any isolated vertices. Given a k-noncrossing, σ-canonical RNA structure δ, its V_k-shape, $V_k(\delta)$, is obtained by first removing all isolated vertices and second applying the core-map c (see eq. (4.1). By abuse of notation we refer to a V_k-shape simply as a shape.

Alternatively the V_k-shape can also be derived as follows: we first project into the core $c(\delta)$, second, we remove all isolated vertices, and third we apply the core-map c again; see Fig. 4.9. The second step is a projection from

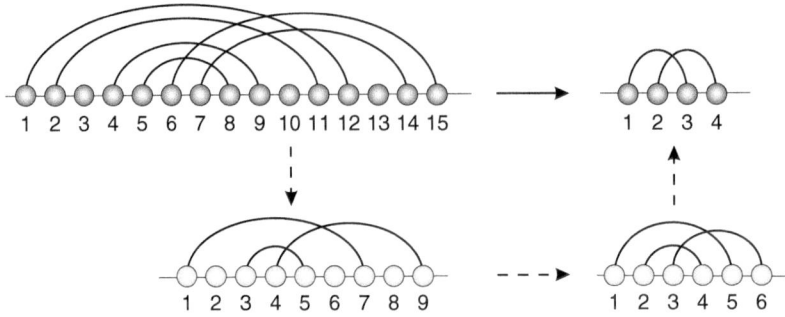

Fig. 4.9. Generation of V_k-shapes. A 3-noncrossing, 2-canonical RNA structure (*top left*) is mapped into its shape (*top right*).

k-noncrossing cores to k-noncrossing matchings, since for each k-noncrossing matching α, we can obtain a core structure by inserting isolated vertices between any two arcs contained in some stack. By construction, shapes do not preserve stack-lengths, isolated vertices, and interior loops, i.e., a sequence of the form

$$((i_1, j_1), [i_1 + 1, i_2 - 1], (i_2, j_2), [j_2 + 1, j_1 - 1]),$$

where (i_2, j_2) is an arc nested in (i_1, j_1) and $[i, j]$ is an interval. Let $\mathcal{I}_k(s, m)$ and $i_k(s, m)$ denote the set and number of shapes of length $2s$ having m 1-arcs and

$$\mathbf{I}_k(z, u) = \sum_{s \geq 0} \sum_{m=0}^{s} i_k(s, m) z^s u^m$$

be the bivariate generating function. Furthermore, let $i_k(s)$ denote the number of shapes of length $2s$ with generating function

$$\mathbf{I}_k(z) = \sum_{s \geq 0} i_k(s) z^s$$

and let $\mathcal{I}_k(m)$ and denote the set of shapes γ having m 1-arcs.

Before we study the generating functions $\mathbf{I}_k(z, u)$ and $\mathbf{I}_k(z)$ let us first study 1-arcs in shapes. For this purpose let $\mathcal{G}_k(s, m)$ denote the set of the k-noncrossing matchings of length $2s$ with m 1-arcs. Since a 1-arc cannot be involved in crossings, we obtain in Lemma 4.5 a linear recurrence between the cardinalities $g_k(s,m) = |\mathcal{G}_k(s,m)|$. We then derive from this recursion the bivariate generating function

$$\mathbf{G}_k(x,y) = \sum_{s \geq 0} \sum_{m=0}^{s} g_k(s,m) x^s y^m.$$

Lemma 4.5. (Reidys and Wang [107]) *Suppose $k,s,m \in \mathbb{N}$, $k \geq 2$, $0 \leq m \leq s$. Then $g_k(s,m)$ has the following properties:*

$$g_k(s,m) = 0 \quad for \ m > s, \tag{4.2}$$

$$\sum_{m=0}^{s} g_k(s,m) = f_k(2s,0) \tag{4.3}$$

and we have the recursion

$$(m+1)g_k(s+1,m+1) = (m+1)g_k(s,m+1) + (2s+1-m)g_k(s,m).$$

Furthermore, the generating function $\mathbf{G}_k(x,y)$ is given by

$$\mathbf{G}_k(x,y) = \frac{1}{x+1-yx} \mathbf{F}_k\left(\frac{x}{(x+1-yx)^2}\right). \tag{4.4}$$

Proof. By construction eq. (4.2) and $\sum_{m=0}^{s} g_k(s,m) = f_k(2s,0)$ hold, the latter being equivalent to

$$\mathbf{G}_k(x,1) = \mathbf{F}_k(x). \tag{4.5}$$

Choose a k-noncrossing matching $\delta \in \mathcal{G}_k(s+1,m+1)$ and label one 1-arc. We have $(m+1)g_k(s+1,m+1)$ different such labeled k-noncrossing matchings. On the other hand, in order to obtain such a labeled matching, we can also insert one labeled 1-arc in a k-noncrossing matching $\delta' \in \mathcal{G}_k(s,m+1)$. In this case, we can only put it inside one original 1-arc in δ' in order to preserve the number of 1-arcs. We may also insert a labeled 1-arc in a k-noncrossing matching $\delta'' \in \mathcal{G}_k(s,m)$. In this case, we can only insert the 1-arc between two vertices not forming a 1-arc; see Fig. 4.10. Therefore we have

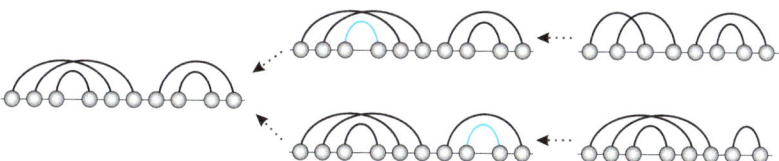

Fig. 4.10. Labeling the 1-arcs allows to trace how certain arc configurations arise.

$$(m+1)g_k(s, m+1) + (2s+1-m)g_k(s, m)$$

different such labeled matchings and

$$(m+1)g_k(s+1, m+1) = (m+1)g_k(s, m+1) + (2s+1-m)g_k(s, m).$$

The above recursion is equivalent to the partial differential equation

$$\frac{\partial \mathbf{G}_k(x, y)}{\partial y} = x \frac{\partial \mathbf{G}_k(x, y)}{\partial y} + 2x^2 \frac{\partial \mathbf{G}_k(x, y)}{\partial x} + x\mathbf{G}_k(x, y) - xy\frac{\partial \mathbf{G}_k(x, y)}{\partial y}. \quad (4.6)$$

We next claim

■

$$\mathbf{G}_k^*(x, y) = \frac{1}{x+1-yx}\mathbf{F}_k\left(\frac{x}{(x+1-yx)^2}\right)$$

is a solution of eq. (4.6),
- its coefficients, $g_k^*(s, m) = [x^s y^m]\mathbf{G}_k^*(x, y)$, satisfy $g_k^*(s, m) = 0$ for $m > s$,
- $\mathbf{G}_k^*(x, 1) = \mathbf{F}_k(x)$.

Indeed,

$$\frac{\partial \mathbf{G}_k^*(x, y)}{\partial y} = xu\,\mathbf{F}_k(xu) + 2xu\,\mathbf{F}_k'(xu), \quad (4.7)$$

$$\frac{\partial \mathbf{G}_k^*(x, y)}{\partial x} = (y-1)u\,\mathbf{F}_k(xu) + \frac{(1+yx-x)u}{x}\mathbf{F}_k'(xu), \quad (4.8)$$

where $u = (x+1-yx)^{-2}$ and $\mathbf{F}_k'(xu) = \sum_{s\geq 0} sf_k(2s, 0)(xu)^s$. Consequently,

$$(1+xy-x)\frac{\partial \mathbf{G}_k^*(x, y)}{\partial y} = 2x^2\frac{\partial \mathbf{G}_k^*(x, y)}{\partial x} + x\mathbf{G}_k^*(x, y) \quad (4.9)$$

which coincides with eq. (4.6). In order to prove $g_k^*(s, m) = 0$ for $m > s$ we first observe that $\mathbf{G}_k^*(x, y)$ is a power series, since it is analytic in $(0, 0)$. Note that the indeterminant y only appears in form of products xy, from which the assertion follows. The equality $\mathbf{G}_k^*(x, 1) = \mathbf{F}_k(x)$ is obvious. We next claim

$$\mathbf{G}_k^*(x, y) = \mathbf{G}_k(x, y). \quad (4.10)$$

By construction $g_k^*(s, m)$ satisfies

$$g_k^*(s, m) = 0 \quad \text{for } m > s$$

$$\sum_{m=0}^{s} g_k^*(s, m) = f_k(2s, 0),$$

$$(m+1)g_k^*(s+1, m+1) = (m+1)g_k^*(s, m+1) + (2s+1-m)g_k^*(s, m).$$

Using these properties we can prove by induction over s

$$\forall s, m \geq 0; \qquad g_k^*(s, m) = g_k(s, m),$$

whence eq. (4.4) and the lemma is proved.

Since any shape is in particular the core of some k-noncrossing matching, Lemma 4.5 allows us to establish a relation between the bivariate generating function of $i_k(s, m)$ and the generating function of $\mathbf{F}_k(z)$.

Theorem 4.6. (Shape theorem) (Reidys and Wang [107]) *Let k, s, m be natural numbers where $k \geq 2$, then the following assertions hold:*
(a) *The generating functions $\mathbf{I}_k(z, u)$ and $\mathbf{I}_k(z)$ satisfy*

$$\mathbf{I}_k(z, u) = \frac{1 + z}{1 + 2z - zu} \mathbf{F}_k \left(\frac{z(1 + z)}{(1 + 2z - zu)^2} \right), \tag{4.11}$$

$$\mathbf{I}_k(z) = \mathbf{F}_k \left(\frac{z}{1 + z} \right). \tag{4.12}$$

(b) *For $2 \leq k \leq 9$, the number of V_k-shapes of length $2s$ is asymptotically given by*

$$i_k(s) \sim c_k s^{-((k-1)^2 + (k-1)/2)} \left(\mu_k^{-1} \right)^s, \tag{4.13}$$

where μ_k is the unique minimum positive real solution of $\frac{z}{1+z} = \rho_k^2$ and c_k is some positive constant.

Proof. We first prove (a). For this purpose we consider the following map between k-noncrossing matchings with m 1-arcs and their V_k-shapes:

$$g \colon \mathcal{G}_k(s, m) \to \bigcup_{b=0}^{s-m} \left[\mathcal{I}_k(s - b, m) \times \left\{ (a_j)_{1 \leq j \leq s-b} \mid \sum_{j=1}^{s-b} a_j = b, \ a_j \geq 0 \right\} \right],$$

where $s \geq 1$. Here, for every $\delta \in \mathcal{G}_k(s, m)$, we have $g(\delta) = (c(\delta), (a_j)_{1 \leq j \leq s-b})$, where $c(\delta)$ is the core structure of δ and where $(a_j)_{1 \leq j \leq s-b}$ keeps track of the deleted arcs. It is straightforward to check that the map g is well defined, since all the 1-arcs of $c(\delta)$ are just the 1-arcs of δ. Furthermore, we observe that c is bijective. Since c is in particular surjective we obtain

$$\mathbf{G}_k(x, y) = \sum_{s,m} g_k(s, m) x^s y^m = \sum_m \sum_{\gamma \in \mathcal{I}_k(m)} \mathbf{G}_\gamma(x, y),$$

where $\mathcal{I}_k(m)$ is the set of V_k-shapes having m 1-arcs and $\mathbf{G}_\gamma(x, y)$ is the generating function of all k-noncrossing matchings having m 1-arcs that project into the shape γ. Suppose γ has s arcs. We consider the combinatorial classes of arcs $\mathcal{R}$ and 1-arcs $\mathcal{R}^*$ with generating functions $\mathbf{R}(x) = x$ and $\mathbf{R}^*(x, y) = yx$. Then

- each k-noncrossing matching having shape γ is obtained by inflating γ-arcs to stacks and the combinatorial class of stacks is given by $\mathcal{R} \times \text{SEQ}(\mathcal{R})$;
- the inflation of arcs does not affect the number of 1-arcs.

Therefore we derive

$$\mathbf{G}_\gamma(x, y) = \left(\frac{x}{1-x}\right)^s y^m.$$

For any $\gamma, \gamma_1 \in \mathcal{I}_k(m)$, having s arcs we have $\mathbf{G}_\gamma(x, y) = \mathbf{G}_{\gamma_1}(x, y)$, whence

$$\mathbf{G}_k(x, y) = \sum_m \sum_{\gamma \in \mathcal{I}_k(m)} \mathbf{G}_\gamma(x, y) = \sum_{s \geq 0} \sum_{m=0}^s i_k(s, m) \left(\frac{x}{1-x}\right)^s y^m. \quad (4.14)$$

According to Lemma 4.5, we have

$$\mathbf{G}_k(x, y) = \frac{1}{x+1-yx} \mathbf{F}_k\left(\frac{x}{(x+1-yx)^2}\right)$$

and setting $z = \frac{x}{1-x}$ and $u = y$, we arrive substituting for $\mathbf{G}_k(x, y)$ in eq. (4.14) at

$$\mathbf{I}_k(z, u) = \frac{1+z}{1+2z-zu} \mathbf{F}_k\left(\frac{z(1+z)}{(1+2z-zu)^2}\right).$$

In particular, setting $u = 1$, we derive

$$\mathbf{I}_k(z) = \mathbf{F}_k\left(\frac{z}{1+z}\right),$$

whence (a). Assertion (b) is a direct consequence of the supercritical paradigm; see Section 4.3.2. The ordinary generating function $\mathbf{F}_k(z) = \sum_{n \geq 0} f_k(2n, 0)z^n$ is D-finite (Corollary 4.14) and the inner function $\vartheta(z) = \frac{z}{1+z}$ is algebraic, satisfies $\vartheta(0) = 0$, and is analytic for $|z| < 1$. Using the fact that all singularities of $\mathbf{F}_k(z)$ are contained within the set of zeros of $q_{0,k}(z)$, see Proposition 2.22, we can then verify that $\mathbf{F}_k(\vartheta(z))$ has the unique dominant real singularity $\mu_k < 1$ satisfying $\vartheta(\mu_k) = \rho_k^2$ for $2 \leq k \leq 9$. In view of $\vartheta'(\mu_k) \neq 0$ Theorem 2.21 guarantees eq. (4.13)

$$i_k(s) \sim c_k s^{-((k-1)^2+(k-1)/2)} \left(\mu_k^{-1}\right)^s.$$

This proves (b) and completes the proof of the theorem.

We next study the number of $\mathbf{V}_k$-shapes induced by k-noncrossing, σ-canonical RNA structures of fixed length n, $\mathsf{u}_{k,\sigma}(n)$, and set

$$\mathbf{U}_{k,\sigma}(x) = \sum_{n \geq 0} \mathsf{u}_{k,\sigma}(n)x^n. \quad (4.15)$$

Theorem 4.7. (Reidys and Wang [107]) *Let $k, \sigma \in \mathbb{N}$, where $k \geq 2$. Then the following assertions hold:*
(a) *The generating function $\mathbf{U}_{k,\sigma}(x)$ is given by*

$$\mathbf{U}_{k,\sigma}(x) = \frac{(1+x^{2\sigma})}{(1-x)(1+2x^{2\sigma}-x^{2\sigma+1})} \mathbf{F}_k\left(\frac{x^{2\sigma}(1+x^{2\sigma})}{(1+2x^{2\sigma}-x^{2\sigma+1})^2}\right).$$

(b) *For $2 \le k \le 9$ and $1 \le \sigma \le 10$*

$$u_{k,\sigma}(n) \sim c_{k,\sigma} n^{-((k-1)^2+(k-1)/2)} \left(\zeta_{k,\sigma}^{-1}\right)^n, \tag{4.16}$$

where $c_{k,\sigma} > 0$ and $\zeta_{k,\sigma}$ is the unique minimum positive real solution of

$$\frac{x^{2\sigma}(1+x^{2\sigma})}{(1+2x^{2\sigma}-x^{2\sigma+1})^2} = \rho_k^2. \tag{4.17}$$

In Table 4.1 we list $\zeta_{k,\sigma}^{-1}$ for various k and σ.

σ/k	2	3	4	5	6	7	8
1	1.51243	3.67528	5.77291	7.82581	9.85873	11.88118	13.89746
2	1.26585	1.93496	2.41152	2.80275	3.14338	3.44943	3.72983
3	1.17928	1.55752	1.80082	1.98945	2.14693	2.28376	2.40567

Table 4.1. The exponential growth rates $\zeta_{k,\sigma}^{-1}$ of V_k-shapes induced by k-noncrossing, σ-canonical RNA structures of length n.

Proof. In order to prove (a) we construct for a given V_k-shape, γ, a unique structure of length n having γ as a shape. In fact, for any given V_k-shape, β, adding the minimal number of arcs to each stack such that every stack contains σ arcs, and inserting one isolated vertex in any 1-arc, we derive a k-noncrossing, σ-canonical structure having arc length≥ 2, of minimal length. We can then concatenate an interval of isolated vertices from the right, thereby arriving for fixed n at a unique k-noncrossing, σ-canonical structure, $\gamma^*(n)$, having arc length≥ 2 and length $n \ge 2\sigma s + m$. By construction, $\gamma_1 \ne \gamma_2$ implies $\gamma_1^*(n) \ne \gamma_1^*(n)$. In view of the injective map $\gamma \mapsto \gamma^*(n)$, we can express $U_{k,\sigma}(x)$, see eq.(4.15), via the bivariate generating function $I_k(z, u)$ as follows:

$$U_{k,\sigma}(x) = \sum_{m \ge 0} \sum_{\gamma \in \mathcal{I}_k(m)} U_\gamma(x),$$

where

$$U_\gamma(x) = \left(x^{2\sigma}\right)^{s-m} \left(x \, x^{2\sigma}\right)^m \frac{1}{1-x} = \left(\frac{x^{2\sigma s+m}}{1-x}\right)$$

is the generating function of inflated structures $\gamma^*(n)$. Since $U_\gamma(x)$ only depends on s and m we obtain

$$U_{k,\sigma}(x) = \sum_{s \ge 0} \sum_{m=0}^{s} i_k(s, m) U_\gamma(x).$$

Consequently,

$$\mathbf{U}_{k,\sigma}(x) = \frac{1}{1-x} \sum_{s \geq 0} \sum_{m=0}^{s} i_k(s,m) x^{2\sigma s} x^m$$

and in view of eq. (4.11), $\mathbf{I}_k(z,u) = \frac{1+z}{1+2z-zu} \mathbf{F}_k \left(\frac{z(1+z)}{(1+2z-zu)^2} \right)$, we derive

$$\mathbf{U}_{k,\sigma}(x) = \frac{(1+x^{2\sigma})}{(1-x)(1+2x^{2\sigma} - x^{2\sigma+1})} \mathbf{F}_k \left(\frac{x^{2\sigma}(1+x^{2\sigma})}{(1+2x^{2\sigma} - x^{2\sigma+1})^2} \right)$$

and (a) follows. As for (b), we observe that

$$\varphi_\sigma(x) = \frac{x^{2\sigma}(1+x^{2\sigma})}{(1+2x^{2\sigma} - x^{2\sigma+1})^2}$$

is algebraic and $\varphi_\sigma(0) = 0$. One verifies by explicit calculation that $\varphi_\sigma(x)$ is for $1 \leq \sigma \leq 10$ analytic for $|x| < r_\sigma$, where $r_\sigma < 1$. Furthermore, the factor

$$\phi_\sigma(x) = \frac{(1+x^{2\sigma})}{(1-x)(1+2x^{2\sigma} - x^{2\sigma+1})}$$

is analytic for $|x| < r_\sigma$. We distinguish the cases $k > 2$ and $k = 2$.
For $2 < k \leq 7$ and $1 \leq \sigma \leq 10$, the minimum positive real solution of eq. (4.17), $\zeta_{k,\sigma}$, is the unique dominant singularity of $\mathbf{U}_{k,\sigma}(x)$, $|\zeta_{k,\sigma}| < r_\sigma$, and $\varphi_\sigma'(\zeta_{k,\sigma}) \neq 0$. Therefore, Theorem 2.21 implies

$$\mathsf{u}_{k,\sigma}(n) \sim c_{k,\sigma} n^{-((k-1)^2+(k-1)/2)} \left(\zeta_{k,\sigma}^{-1} \right)^n,$$

where $c_{k,\sigma}$ is some positive constant.
In case of $k = 2$, we have

$$\mathbf{F}_2(z) = \sum_{n \geq 0} f_2(2n,0) z^n = \frac{2}{1 + \sqrt{1-4z}}. \tag{4.18}$$

Substituting $\varphi_\sigma(x)$ into eq. (4.18), we observe that the poles of $\varphi_\sigma(x)$ are not singularities of $\mathbf{U}_{2,\sigma}(x)$ whence the dominant singularity of $\mathbf{U}_{2,\sigma}(x)$ is the minimum positive solution of $\varphi_\sigma(x) = \rho_2^2$. Now Theorems 2.19 and 2.20 imply eq. (4.16) and the proof of the theorem is complete.

4.2 Generating functions

In this section we will compute various generating functions via symbolic enumeration; see Section 2.2 [42]. All generating functions are derived by inflating V_k-shapes. Symbolic enumeration has first been used in the context of RNA secondary structures in [94]. Let us illustrate in Fig. 4.11 the basic idea behind this section.

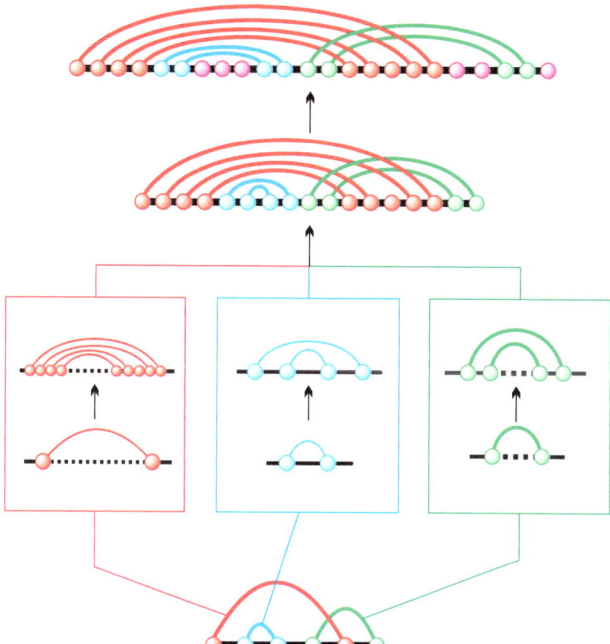

Fig. 4.11. From shapes to structures: we display the idea behind the inflation. A structure (*top*) is derived by inflating a V_k-shape (*bottom*) in two steps. First we individually inflate each arc in the shape into more complex configurations and second insert isolated vertices (*purple*).

4.2.1 The GF of cores

We begin by computing the generating function of core structures, via symbolic enumeration, see Chapter 2. Recall that given a k-noncrossing, τ-canonical RNA structure its shape is obtained by first removing all isolated vertices and second collapsing any stack into a single arc; see Fig. 4.12.

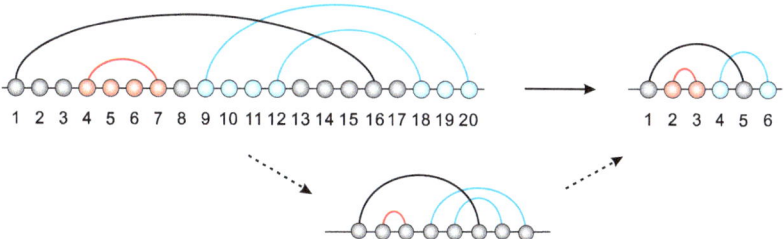

Fig. 4.12. A 3-noncrossing core structure (*top-left*) is mapped into its V_k-shape (*top-right*) in two steps. A stem (*blue*) is mapped into a single shape-arc (*blue*). A hairpin loop (*red*) is mapped into a shape-1-arc (*red*).

Theorem 4.8. (Core Structures) (Jin and Reidys [78]) *Suppose* $k \in \mathbb{N}$, $k \geq 2$, *let* z *be an indeterminant, and* $r(z) = \frac{1}{1+z^2}$. *Then*

$$\mathbf{C}_k(z) = \frac{1}{r(z)z^2 - z + 1} \mathbf{F}_k \left(\left(\frac{\sqrt{r(z)}z}{r(z)z^2 - z + 1} \right)^2 \right). \qquad (4.19)$$

Proof. $\mathcal{C}_k$ denotes the set of k-noncrossing cores, $\mathcal{I}_k$ denotes the set of all k-noncrossing shapes, and $\mathcal{I}_k(m)$ those having m 1-arcs; see Fig. 4.12. Then we have the surjective map,

$$\varphi: \mathcal{C}_k \to \mathcal{I}_k$$

inducing the partition $\mathcal{C}_k = \dot{\cup}_\gamma \varphi^{-1}(\gamma)$ where $\mathcal{C}_k(\gamma)$ is the set of k-noncrossing cores having shape γ. Then

$$\mathbf{C}_k(z) = \sum_{m \geq 0} \sum_{\gamma \in \mathcal{I}_k(m)} \mathbf{C}_\gamma(z),$$

where $\mathbf{C}_\gamma(z)$ is the generating function of the combinatorial class $\mathcal{C}_k(\gamma)$. We next compute $\mathbf{C}_\gamma(z)$ symbolically via inflation of shapes. Let $\mathcal{C}_\gamma$ denote the combinatorial class of cores derived by inflating the shape γ. To generate this class we consider the classes $\mathcal{M}$ (nested arc sequences), $\mathcal{L}$ (isolated vertices), $\mathcal{R}$ (arcs), $\mathcal{R}'$ (induced arcs), and $\mathcal{Z}$ (vertices), where $\mathbf{Z}(z) = z$ and $\mathbf{R}(z) = z^2$.

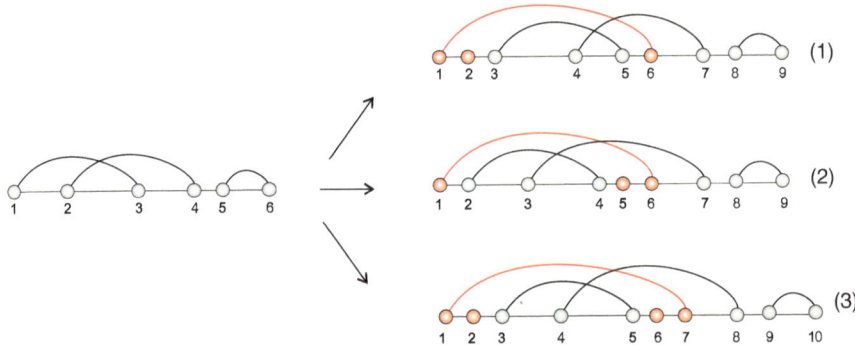

Fig. 4.13. Illustration of step I: shape arcs induce arc sequences, separated by intervals of isolated vertices.

The idea is to inflate a shape $\gamma \in \mathcal{I}_k(s)$ into two steps.

Step I: $\gamma \in \mathcal{I}_k(m)$ having s arcs, where $s \geq \max\{1, m\}$ is inflated to a core by inflating each arc in the shape to a stem of arcs; see Fig. 4.13. The nesting arcs are called induced and have to be separated by means of inserting isolated vertices: we either insert intervals of isolated vertices to the left, right, or on both sides of the arc. We generate the following:

- Isolated segments, i.e., sequences of isolated vertices. Plainly we have $\mathcal{L} = \text{SEQ}(\mathcal{Z})$, where

$$\mathbf{L}(z) = \frac{1}{1-z}.$$

- Induced arcs, that is, pairs consisting of arcs $\mathcal{R}$ and at least one nonempty interval of isolated vertices on either or both its sides. As the arc can be combined freely with these intervals we derive

$$\mathcal{R}' = \mathcal{R} \times \left(\mathcal{Z} \times \mathcal{L} + \mathcal{Z} \times \mathcal{L} + (\mathcal{Z} \times \mathcal{L})^2 \right),$$

having the generating function

$$\mathbf{R}'(z) = z^2 \cdot \left(\frac{z}{1-z} + \frac{z}{1-z} + \left(\frac{z}{1-z} \right)^2 \right).$$

- Stems, i.e., pairs consisting of the minimal arc $\mathcal{R}$ and an arbitrarily long sequence of induced arcs

$$\mathcal{M} = \mathcal{R} \times \text{SEQ}(\mathcal{R}'),$$

with generating function

$$\mathbf{M}(z) = z^2 \cdot \frac{1}{1 - \mathbf{R}'(z)}.$$

The resulting core has s nested sequences of arcs and $(2s+1)$ (possibly empty) intervals of isolated vertices.

Step II: we insert isolated vertices into the remaining $2s - 1 + 2$ positions; see Fig. 4.13. This second inflation is formally expressed by

- $\mathcal{J} = \mathcal{L}^{2s+1-m} \times (\mathcal{Z} \times \mathcal{L})^m$ where

$$\mathbf{J}(z) = \left(\frac{1}{1-z} \right)^{2s+1-m} \left(\frac{z}{1-z} \right)^m;$$

see Fig. 4.14. Combining steps I and II we arrive at

$$\mathcal{C}_\gamma = \mathcal{M}^s \times \mathcal{L}^{2s+1-m} \times (\mathcal{Z} \times \mathcal{L})^m$$

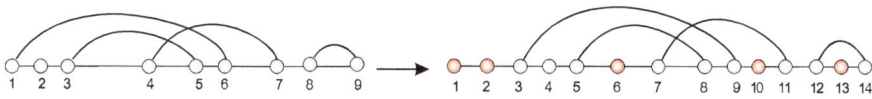

Fig. 4.14. Illustration of step II: $(2s + 1)$ intervals (possibly empty) consisting of isolated vertices are inserted into (1) of Fig. 4.13.

and we compute

$$\mathbf{C}_\gamma(z) = \left(\frac{z^2}{1 - z^2 \left(\frac{2z}{1-z} + \left(\frac{z}{1-z} \right)^2 \right)} \right)^s \cdot \left(\frac{1}{1-z} \right)^{2s+1-m} \left(\frac{z}{1-z} \right)^m$$

(4.20)

$$= (1-z)^{-1} \left(\frac{z^2}{1 - 2z + z^2 - 2z^3 + z^4} \right)^s z^m.$$

(4.21)

Since for any $\gamma, \gamma_1 \in \mathcal{I}_k(s, m)$ we have $\mathbf{C}_\gamma(z) = \mathbf{C}_{\gamma_1}(z)$

$$\mathbf{C}_k(z) = \sum_{m \geq 0} \sum_{\gamma \in \mathcal{I}_k(m)} \mathbf{C}_\gamma(z) = \sum_{s \geq 0} \sum_{m=0}^s i_k(s, m) \mathbf{C}_\gamma(z).$$

Therefore we obtain for the generating function of k-noncrossing cores

$$\mathbf{C}_k(z) = \sum_{s \geq 0} \sum_{m=0}^s i_k(s, m) \mathbf{C}_\gamma(z)$$

(4.22)

$$= (1-z)^{-1} \sum_{s \geq 0} \sum_{m=0}^s i_k(s, m) \left(\frac{z^2}{1 - 2z + z^2 - 2z^3 + z^4} \right)^s z^m.$$

(4.23)

According to Theorem 4.6

$$\sum_{s \geq 0} \sum_{m=0}^s i_k(s, m) x^s y^m = \frac{1+x}{1+2x-xy} \sum_{s \geq 0} f_k(2s, 0) \left(\frac{x(1+x)}{(1+2x-xy)^2} \right)^s.$$

(4.24)

Substituting $x = \frac{z^2}{1-2z+z^2-2z^3+z^4}$ and $y = z$ into eq. (4.24), we derive

$$\sum_{s \geq 0} \sum_{m=0}^s i_k(s, m) \left(\frac{z^2}{1 - 2z + z^2 - 2z^3 + z^4} \right)^s z^m =$$
$$\frac{(1-z)(1+z^2)}{1 - z + 2z^2 - z^3} \mathbf{F}_k \left(\frac{z^2(1+z^2)}{(1 - z + 2z^2 - z^3)^2} \right).$$

Substituting eq. (4.2.1) into eq. (4.23), we compute

$$\mathbf{C}_k(z) = \frac{1}{1-z} \cdot \frac{(1-z)(1+z^2)}{1 - z + 2z^2 - z^3} \mathbf{F}_k \left(\frac{z^2(1+z^2)}{(1 - z + 2z^2 - z^3)^2} \right)$$

$$= \frac{1}{1 - z + \frac{1}{1+z^2} z^2} \mathbf{F}_k \left(\left(\frac{\sqrt{\frac{1}{1+z^2}} z}{\frac{1}{1+z^2} z^2 - z + 1} \right)^2 \right),$$

whence eq. (4.19).

4.2.2 The GF of k-noncrossing, σ-canonical structures

We next use arguments analogous to those of Section 4.2.1 in order to compute the generating function of k-noncrossing, σ-canonical structures. Note that no result proved here applies to the case of k-noncrossing, 2-canonical structures with minimum arc length 4. The latter require a nontrivial refinement of V_k-shapes which allows us to deal with the then critical 2-arcs. The analysis of these structures is presented in Section 4.4.

Theorem 4.9. (Jin and Reidys [78]) *Suppose* $k, \sigma \in \mathbb{N}$, $k \geq 2, \sigma \geq 1$ *and* $u_\sigma(z) = \frac{(z^2)^{\sigma-1}}{z^{2\sigma}-z^2+1}$. *Then*

$$\mathbf{T}_{k,\sigma}(z) = \frac{1}{u_\sigma(z)z^2 - z + 1} \mathbf{F}_k \left(\left(\frac{\sqrt{u_\sigma(z)}z}{(u_\sigma(z)z^2 - z + 1)} \right)^2 \right).$$

In particular, setting $\sigma = 1$ *we have* $u_1(z) = 1$ *and*

$$\mathbf{T}_{k,1}(z) = \frac{1}{z^2 - z + 1} \mathbf{F}_k \left(\left(\frac{z}{z^2 - z + 1} \right)^2 \right).$$

Proof. Let $\mathcal{T}_{k,\sigma}$ denote the set of k-noncrossing, σ-canonical structures and $\mathcal{I}_k$ the set of all k-noncrossing shapes and $\mathcal{I}_k(m)$ those having m 1-arcs; see Fig. 4.15. Then we have the surjective map

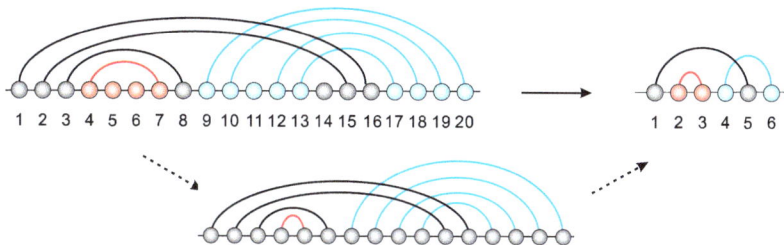

Fig. 4.15. A 3-noncrossing, 2-canonical RNA structure (*top left*) is mapped into its shape (*top right*). A stem (*blue*) is mapped into a single shape-arc (*blue*). A hairpin loop (*red*) is mapped into a shape-1-arc (*red*).

$$\varphi \colon \mathcal{T}_{k,\sigma} \to \mathcal{I}_k.$$

Indeed, for any shape γ in $\mathcal{I}_k(m)$, we can construct a k-noncrossing, σ-canonical structure with m hairpin loops, by adding at least $\sigma - 1$ arcs to each stack and inserting at least one isolated vertex in each 1-arc. $\varphi \colon \mathcal{T}_{k,\sigma} \to \mathcal{I}_k$ induces the partition $\mathcal{T}_{k,\sigma} = \dot{\cup}_\gamma \varphi^{-1}(\gamma)$. Then we have

$$\mathbf{T}_{k,\sigma}(z) = \sum_{m \geq 0} \sum_{\gamma \in \mathcal{I}_k(m)} \mathbf{T}_\gamma(z). \qquad (4.25)$$

We proceed by computing the generating function $\mathbf{T}_\gamma(z)$. We will construct $\mathbf{T}_\gamma(z)$ via simpler combinatorial classes as building blocks considering the classes $\mathcal{M}$ (stems), $\mathcal{K}^\sigma$ (stacks), $\mathcal{N}^\sigma$ (induced stacks), $\mathcal{L}$ (isolated vertices), $\mathcal{R}$ (arcs), and $\mathcal{Z}$ (vertices), where $\mathbf{Z}(z) = z$ and $\mathbf{R}(z) = z^2$. We inflate $\gamma \in \mathcal{I}_k(m)$ having s arcs, where $s \geq \max\{1, m\}$, to a structure in two steps.

Step I: we inflate any shape-arc to a stack of size at least σ and subsequently add additional stacks. The latter are called induced stacks and have to be separated by means of inserting isolated vertices; see Fig. 4.16. Note that during this first inflation step no intervals of isolated vertices, other than those necessary for separating the nested stacks, are inserted. We generate

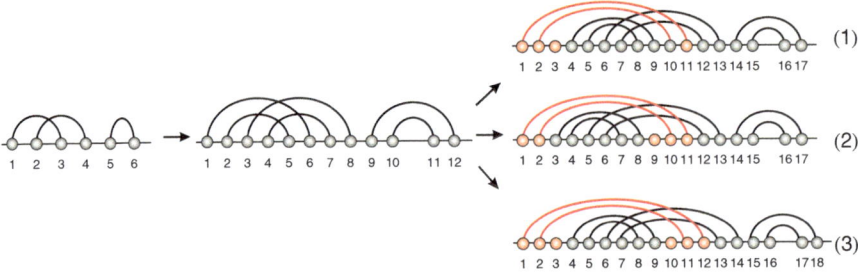

Fig. 4.16. Step I: a shape (*left*) is inflated to a 3-noncrossing, 2-canonical structure. First, every arc in the shape is inflated to a stack of size at least 2 (*middle*), and then the shape is inflated to a new 3-noncrossing, 2-canonical structure (*right*) by adding one stack of size 2. There are three ways to insert the interval isolated vertices.

- isolated segments, i.e., sequences of isolated vertices $\mathcal{L} = \text{SEQ}(\mathcal{Z})$, where

$$\mathbf{L}(z) = \frac{1}{1-z};$$

- stacks, i.e., pairs consisting of the minimal sequence of arcs $\mathcal{R}^\sigma$ and an arbitrary extension consisting of arcs of arbitrary finite length

$$\mathcal{K}^\sigma = \mathcal{R}^\sigma \times \text{SEQ}(\mathcal{R})$$

having the generating function

$$\mathbf{K}^\sigma(z) = z^{2\sigma} \cdot \frac{1}{1-z^2};$$

- induced stacks, i.e., stacks together with at least one nonempty interval of isolated vertices on either or both its sides

$$\mathcal{N}^\sigma = \mathcal{K}^\sigma \times \left(\mathcal{Z} \times \mathcal{L} + \mathcal{Z} \times \mathcal{L} + (\mathcal{Z} \times \mathcal{L})^2 \right),$$

with generating function

$$\mathbf{N}^\sigma(z) = \frac{z^{2\sigma}}{1 - z^2} \left(2 \frac{z}{1 - z} + \left(\frac{z}{1 - z} \right)^2 \right);$$

- stems, that is, pairs consisting of stacks $\mathcal{K}^\sigma$ and an arbitrarily long sequence of induced stacks

$$\mathcal{M}^\sigma = \mathcal{K}^\sigma \times \mathrm{SEQ}\left(\mathcal{N}^\sigma \right),$$

with generating function

$$\mathbf{M}^\sigma(z) = \frac{\mathbf{K}^\sigma(z)}{1 - \mathbf{N}^\sigma(z)} = \frac{\frac{z^{2\sigma}}{1 - z^2}}{1 - \frac{z^{2\sigma}}{1 - z^2} \left(2 \frac{z}{1 - z} + \left(\frac{z}{1 - z} \right)^2 \right)}.$$

Step II: here we insert additional isolated vertices at the remaining $(2s + 1)$ positions. For each 1-arc at least one such isolated vertex is necessarily inserted; see Fig. 4.17. Formally the second inflation is expressed via

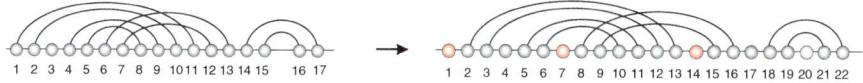

Fig. 4.17. Step II: the structure (*left*) obtained in (1) in Fig. 4.16 is inflated to a new 3-noncrossing, 2-canonical structures (*right*) by adding isolated vertices (*red*).

- $\mathcal{J} = \mathcal{L}^{2s+1-m} \times (\mathcal{Z} \times \mathcal{L})^m$, where

$$\mathbf{J}(z) = \left(\frac{1}{1 - z} \right)^{2s+1-m} \left(\frac{z}{1 - z} \right)^m.$$

Combining steps I and II we arrive at

$$\mathcal{T}_\gamma = (\mathcal{M}^\sigma)^s \times \mathcal{L}^{2s+1-m} \times (\mathcal{Z} \times \mathcal{L})^m$$

and accordingly

$$\mathbf{T}_\gamma(z) = \left(\frac{\frac{z^{2\sigma}}{1 - z^2}}{1 - \frac{z^{2\sigma}}{1 - z^2} \left(2 \frac{z}{1 - z} + \left(\frac{z}{1 - z} \right)^2 \right)} \right)^s \left(\frac{1}{1 - z} \right)^{2s+1-m} \left(\frac{z}{1 - z} \right)^m$$

$$= (1 - z)^{-1} \left(\frac{z^{2\sigma}}{(1 - z^2)(1 - z)^2 - (2z - z^2)z^{2\sigma}} \right)^s z^m.$$

Since for any $\gamma, \gamma_1 \in \mathcal{I}_k(s, m)$ we have $\mathbf{T}_\gamma(z) = \mathbf{T}_{\gamma_1}(z)$, we derive

$$\mathbf{T}_k(z) = \sum_{m \geq 0} \sum_{\gamma \in \mathcal{I}_k(m)} \mathbf{T}_\gamma(z) = \sum_{s \geq 0} \sum_{m=0}^{s} i_k(s, m) \mathbf{T}_\gamma(z).$$

Setting

$$\eta_\sigma(z) = \frac{z^{2\sigma}}{(1 - z^2)(1 - z)^2 - (2z - z^2)z^{2\sigma}},$$

we have according to eq. (4.25) and Theorem 4.6 the following situation:

$$\mathbf{T}_{k,\sigma}(z) = \sum_{s \geq 0} \sum_{m=0}^{s} i_k(s, m) \, \mathbf{T}_\gamma(z)$$

and Theorem 4.6 guarantees

$$\sum_{s \geq 0} \sum_{m=0}^{s} i_k(s, m) \, x^s \, y^m = \frac{1 + x}{1 + 2x - xy} \sum_{s \geq 0} f_k(2s, 0) \left(\frac{x(1 + x)}{(1 + 2x - xy)^2} \right)^s.$$

Therefore, setting $x = \eta_\sigma(z)$ and $y = z$ and $w_\sigma(z) = z^{2\sigma} - z^2 + 1$ we arrive at

$$\mathbf{T}_{k,\sigma}(z) = \frac{w_\sigma(z)}{(1 - z)w_\sigma(z) + z^{2\sigma}} \mathbf{F}_k \left(\frac{z^2 \, (z^2)^{\sigma-1} w_\sigma(z)}{((1 - z)w_\sigma(z) + z^{2\sigma})^2} \right)$$

$$= \frac{1}{(1 - z) + u_\sigma(z)z^2} \mathbf{F}_k \left(\frac{z^2 \, u_\sigma(z)}{((1 - z) + u_\sigma(z)z^2)^2} \right)$$

and the theorem follows.

We are now in position to establish a straightforward generalization of Theorem 4.9 that allows us to compute

- the generating function of k-noncrossing, canonical structures having minimum arc length 3 [79] as well as
- the generating function of k-noncrossing, 3-canonical structures having minimum arc length 4 [88].

The latter class is the target for the ab initio folding algorithm **cross**, discussed in Chapter 6.

Theorem 4.10. *Let $k, \sigma \in \mathbb{N}$, $k \geq 2$, z be an indeterminant, ρ_k^2 be the dominant, positive real singularity of $\mathbf{F}_k(z)$ and*

$$u_\sigma(z) = \frac{(z^2)^{\sigma-1}}{z^{2\sigma} - z^2 + 1},$$

$$v_\lambda(z) = 1 - z + u_\sigma(z) \sum_{h=2}^{\lambda} z^h.$$

Then, $\mathbf{T}_{k,\sigma}^{[\lambda]}(x)$, the generating function of k-noncrossing, σ-canonical structures with minimum arc length λ, $\lambda \leq \sigma + 1$ is given by

$$\mathbf{T}_{k,\sigma}^{[\lambda]}(z) = \frac{1}{v_\lambda(z)} \mathbf{F}_k\left(\left(\frac{\sqrt{u_\sigma(z)}\, z}{v_\lambda(z)}\right)^2\right). \tag{4.26}$$

Proof. Using the notation and approach of Theorem 4.9 one arrives at

$$
\begin{aligned}
T_\gamma^{[\lambda]} &= \mathcal{M}^s \times \mathcal{L}^{2s+1-m} \times (\mathcal{Z}^{\lambda-1} \times \mathcal{L})^m, \\
\mathcal{M} &= \mathcal{K}^\sigma \times \mathrm{SEQ}(\mathcal{N}^\sigma), \\
\mathcal{N}^\sigma &= \mathcal{K}^\sigma \times \left(\mathcal{Z} \times \mathcal{L} + \mathcal{Z} \times \mathcal{L} + (\mathcal{Z} \times \mathcal{L})^2\right), \\
\mathcal{K}^\sigma &= \mathcal{R} \times \mathrm{SEQ}(\mathcal{R}), \\
\mathcal{L} &= \mathrm{SEQ}(\mathcal{Z}).
\end{aligned}
$$

The only difference occurs during the second inflation step, where we have

- $\mathcal{J}^{[\lambda]} = \mathcal{L}^{2s+1-m} \times (\mathcal{Z}^{\lambda-1} \times \mathcal{L})^m$, where

$$\mathbf{J}^{[\lambda]}(z) = \left(\frac{1}{1-z}\right)^{2s+1-m} \left(\frac{z^{\lambda-1}}{1-z}\right)^m.$$

The key point here is that the condition $\lambda \leq \sigma + 1$ guarantees that any non 1-arc has after inflation a minimum arc length of $\sigma + 1$. The generating function of class $T_\gamma^{[\lambda]}$ is given by

$$\mathbf{T}_\gamma^{[\lambda]}(z) = \left(\frac{\frac{z^{2\sigma}}{1-z^2}}{1 - \frac{z^{2\sigma}}{1-z^2}\left(\frac{2z}{1-z} + \left(\frac{z}{1-z}\right)^2\right)}\right)^s \cdot \left(\frac{1}{1-z}\right)^{2s+1-m} \left(\frac{z^{\lambda-1}}{1-z}\right)^m$$

$$= (1-z)^{-1} \left(\frac{z^{2\sigma}}{(1-z)^2(1-z^2) - z^{2\sigma}(2z - z^2)}\right)^s (z^{\lambda-1})^m.$$

Since for any $\gamma, \gamma' \in \mathcal{I}_k(s,m)$ we have $\mathbf{T}_\gamma^{[\lambda]}(z) = \mathbf{T}_{\gamma'}^{[\lambda]}(z)$, we derive

$$\mathbf{T}_{k,\sigma}^{[\lambda]}(z) = \sum_{m\geq 0}\sum_{\gamma \in \mathcal{I}_k(m)} \mathbf{T}_\gamma^{[\lambda]}(z) = \sum_{s\geq 0}\sum_{m=0}^{s} i_k(s,m) \mathbf{T}_\gamma^{[\lambda]}(z),$$

whence

$$\mathbf{T}_{k,\sigma}^{[\lambda]}(z) = (1-z)^{-1} \sum_{s\geq 0}\sum_{m=0}^{s} i_k(s,m)\, \eta_\sigma(z)^s \left(z^{\lambda-1}\right)^m.$$

Substituting $x = \eta_\sigma(z)$ and $y = z^{\lambda-1}$ into eq. (4.11) we derive

$$\mathbf{T}_{k,\sigma}^{[\lambda]}(z) = (1-z)^{-1}\frac{1+\eta_\sigma(z)}{1+2\eta_\sigma(z)-\eta_\sigma(z)z^{\lambda-1}}\mathbf{F}_k\left(\frac{\eta_\sigma(z)(1+\eta_\sigma(z))}{(1+2\eta_\sigma(z)-\eta_\sigma(z)z^{\lambda-1})^2}\right)$$

$$= \frac{(1-z)(1-z^2+z^{2\sigma})}{1-2z+2z^3-z^4+2z^{2\sigma}-2z^{2\sigma+1}+z^{2\sigma+2}-z^{2\sigma+\lambda-1}} \times$$

$$\mathbf{F}_k\left(\frac{(1-z)^2z^{2\sigma}(1-z^2+z^{2\sigma})}{(1-2z+2z^3-z^4+2z^{2\sigma}-2z^{2\sigma+1}+z^{2\sigma+2}-z^{2\sigma+\lambda-1})^2}\right).$$

We compute

$$\frac{1}{v_\lambda(z)} = \frac{(1-z)(1-z^2+z^{2\sigma})}{1-2z+2z^3-z^4+2z^{2\sigma}-2z^{2\sigma+1}+z^{2\sigma+2}-z^{2\sigma+\lambda-1}}$$

$$\left(\frac{\sqrt{u_\sigma(z)}\,z}{v_\lambda(z)}\right)^2 = \frac{(1-z)^2z^{2\sigma}(1-z^2+z^{2\sigma})}{(1-2z+2z^3-z^4+2z^{2\sigma}-2z^{2\sigma+1}+z^{2\sigma+2}-z^{2\sigma+\lambda-1})^2},$$

where $u_\sigma(z) = \frac{(z^2)^{\sigma-1}}{1-z^2+z^{2\sigma}}$ and $v_\lambda(z) = 1-z+u_\sigma(z)\sum_{h=2}^{\lambda}z^h$. Thus eq. (4.26) follows and the proof of the theorem is complete.

We remark that in view of

$$\sum_{h=2}^{\lambda}z^h = \begin{cases} z^2 & \text{for } \lambda=2, \\ z^2+z^3 & \text{for } \lambda=3, \\ z^2+z^3+z^4 & \text{for } \lambda=4, \end{cases}$$

Theorem 4.10 immediately implies Theorem 4.9 and we furthermore have

Corollary 4.11. (Jin and Reidys [79]) *Let $k,\sigma \in \mathbb{N}$, $k,\sigma \geq 2$, z be an indeterminant. Then $\mathbf{T}_{k,\sigma}^{[3]}(z)$, the generating function of k-noncrossing, σ-canonical structures with $\lambda \geq 3$ is given by*

$$\mathbf{T}_{k,\sigma}^{[3]}(z) = \frac{1}{v_3(z)}\mathbf{F}_k\left(\left(\frac{\sqrt{u_\sigma(z)}\,z}{v_3(z)}\right)^2\right),$$

where $u_\sigma(z) = \frac{(z^2)^{\sigma-1}}{z^{2\sigma}-z^2+1}$ and furthermore $v_3(z) = u_\sigma(z)(z^3+z^2)-z+1$.

Corollary 4.12. (Ma and Reidys [88]) *Let $k,\sigma \in \mathbb{N}$, $k,\sigma \geq 3$, z be an indeterminant. Then $\mathbf{T}_{k,\sigma}^{[4]}(z)$, the generating function of k-noncrossing, σ-canonical structures with $\lambda \geq 4$ is given by*

$$\mathbf{T}_{k,\sigma}^{[4]}(z) = \frac{1}{v_4(z)}\mathbf{F}_k\left(\left(\frac{\sqrt{u_\sigma(z)}\,z}{v_4(z)}\right)^2\right),$$

where $u_\sigma(z) = \frac{(z^2)^{\sigma-1}}{z^{2\sigma}-z^2+1}$ and $v_4(z) = u_\sigma(z)(z^4+z^3+z^2)-z+1$.

Corollary 4.12 gives rise to ask whether it is possible to compute the generating function of k-noncrossing, canonical structures having minimum arc length 4. This class of structures can also be computed via symbolic enumeration, based on a refinement of V_k-shapes, see Section 4.4.

4.3 Asymptotics

In this section we compute directly the coefficients of the generating function $\mathbf{T}_{k,1}(z)$. This result (in combination with the cores of Section 4.1.1) can be used as the centerpiece for developing the theory of k-noncrossing structures [76, 78] with the exception of Section 4.4. It offers in particular a different proof of

$$\mathbf{T}_{k,1}(z) = \frac{1}{z^2 - z + 1} \, \mathbf{F}_k \left(\left(\frac{z}{z^2 - z + 1} \right)^2 \right).$$

In fact the proof via symbolic enumeration, given in Section 4.2.2, is based on the notion of V_k-shapes which appeared later [107]. We then present the singularity analysis of the various generating functions computed in Section 4.2.2. All of these computations are governed by the supercritical paradigm of Chapter 2 and are therefore connected to the results of Section 2.4. The key result of this section is the second assertion of Proposition 4.16.

4.3.1 k-Noncrossing structures

In this section we present the singularity analysis and further relations between the generating functions of Section 4.2. In order to motivate our results we begin by presenting an explicit formula for the numbers of k-noncrossing RNA pseudoknot structures. The result shows that even though explicit formulas can be derived they are not necessarily helpful in order to derive simple formulas for k-noncrossing RNA pseudoknot structures for large n.

Our construction uses k-noncrossing partial matchings as an intermediate in a procedure in which certain "bad" arcs are being placed over n vertices. We denote the number of RNA structures with exactly ℓ isolated vertices by $\mathsf{T}_{k,1}(n, \ell)$. Let $f_k(n, \ell)$ be the number of k-noncrossing diagrams over $[n]$ with exactly ℓ isolated vertices, and $\mathsf{M}_k(n) = \sum_{\ell > 0} f_k(n, \ell)$, i.e., the number of k-noncrossing partial matchings or the number of k-noncrossing diagrams over $[n]$. We next compute the coefficients of $\mathbf{T}_{k,1}(z)$.

Theorem 4.13. (Jin et al. [76]) *Let $k \in \mathbb{N}$ and $k \geq 2$. Then the number of RNA structures with ℓ isolated vertices, $\mathsf{T}_{k,1}(n, \ell)$, is given by*

$$\forall \, k \geq 2; \qquad \mathsf{T}_{k,1}(n, \ell) = \sum_{b \leq \lfloor \frac{n}{2} \rfloor} (-1)^b \binom{n - b}{b} f_k(n - 2b, \ell). \qquad (4.27)$$

Furthermore, the number of k-noncrossing RNA structures, $\mathsf{T}_{k,1}(n)$ is given by

$$\forall \, k \geq 2; \qquad \mathsf{T}_{k,1}(n) = \sum_{b \leq \lfloor \frac{n}{2} \rfloor} (-1)^b \binom{n - b}{b} \mathsf{M}_k(n - 2b).$$

n	1	2	3	4	5	6	7	8	9	10	11	12	13	14	15
$T_{3,1}(n)$	1	1	2	5	13	36	105	321	1018	3334	11,216	38,635	135,835	486,337	1,769,500

Table 4.2. The first 15 numbers of 3-noncrossing RNA structures.

In Table 4.2 we list the first 15 numbers of 3-noncrossing RNA structures.

Proof. Suppose $k \geq 2$ and let $\mathcal{G}_{n,k}(\ell, j)$ be the set of all k-noncrossing diagrams having exactly ℓ isolated points and exactly j 1-arcs. Setting $G_k(n, \ell, j) = |\mathcal{G}_{n,k}(\ell, j)|$, we have in particular

$$T_{k,1}(n, \ell) = G_k(n, \ell, 0). \tag{4.28}$$

We first prove

$$\sum_{j \geq b} \binom{j}{b} G_k(n, \ell, j) = \binom{n-b}{b} f_k(n - 2b, \ell). \tag{4.29}$$

For this purpose we construct a family $\mathcal{F}$ of $\mathcal{G}_{n,k}$-diagrams, having exactly ℓ

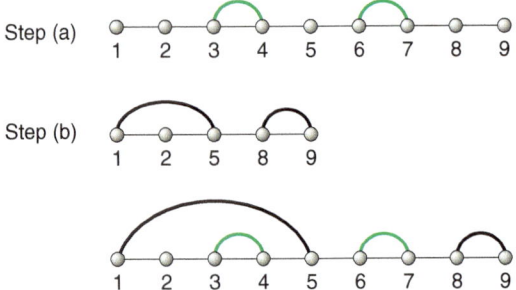

Fig. 4.18. Constructing an element of the family $\mathcal{F}$ of $\mathcal{G}_{9,2}$-diagrams for $b = 2$ and $\ell = 1$.

isolated points and having at least b 1-arcs as follows: select **(a)** b 1-arcs and **(b)** an arbitrary k-noncrossing diagram with exactly ℓ isolated points over the remaining $n - 2b$ vertices. Let $\mathcal{F}$ be the resulting family of diagrams; see Fig. 4.18.

Claim 1. Each element $\theta \in \mathcal{F}$ is contained in $\mathcal{G}_{n,k}(\ell, j)$ for some $j \geq b$.
To prove this we observe that an 1-arc cannot cross any other arc, i.e., cannot be contained in a set of mutually crossing arcs. As a result for $k \geq 2$ our construction generates diagrams that are k-noncrossing. Clearly, θ has exactly ℓ isolated vertices and in step **(b)** we potentially derive additional 1-arcs, whence $j \geq b$.
Claim 2.

$$|\mathcal{F}| = \binom{n-b}{b} f_k(n - 2b, \ell).$$

Let $\lambda(n,b)$ denote the number of ways to select b 1-arcs over $\{1,\ldots,n\}$. We observe that $\lambda(n,b) = \binom{n-b}{b}$. Identifying the two incident vertices of a 1-arc we conclude that we can choose the b 1-arcs in $\binom{n-b}{b}$ ways. Obviously, ℓ isolated vertices can be obtained in $\binom{n-2b}{\ell}$ different ways and it remains to select an arbitrary k-noncrossing diagram with exactly ℓ isolated points over $n - 2b$ vertices, whence Claim 2 is proved; see Fig. 4.19.

In view of the fact that any of the k-noncrossing diagrams can introduce

Fig. 4.19. All 12 elements in $\mathcal{F}$ for $n = 5$, $k = 3$, $b = 1$, and $\ell = 1$. The *blue arcs* are the "bad" arcs selected in step (a) while the *black arcs* are those selected in step (b).

additional 1-arcs we set

$$\mathcal{F}(j) = \{\theta \in \mathcal{F} \mid \theta \text{ has exactly } j \text{ 1-arcs}\}.$$

Obviously, $\mathcal{F} = \dot{\bigcup}_{j \geq b} \mathcal{F}(j)$. Suppose $\theta \in \mathcal{F}(j)$. According to Claim 1, $\theta \in \mathcal{G}_{n,k}(\ell, j)$ and furthermore θ occurs with multiplicity $\binom{j}{b}$ in $\mathcal{F}$ since by construction any b-element subset of the j 1-arcs is counted, respectively, in $\mathcal{F}$. Therefore we have

$$|\mathcal{F}(j)| = \binom{j}{b} G_k(n, \ell, j)$$

and

$$\sum_{j \geq b} \binom{j}{b} G_k(n, \ell, j) = \sum_{j \geq b} |\mathcal{F}(j)| = \binom{n-b}{b} f_k(n - 2b, \ell),$$

whence eq. (4.29). We next set $F_k(x) = \sum_{j \geq 0} G_k(n, \ell, j) x^j$, taking the bth derivative and letting $x = 1$ we obtain

$$\frac{1}{b!} F_k^{(b)}(1) = \sum_{j \geq b} \binom{j}{b} G_k(n, \ell, j) 1^{j-b}. \tag{4.30}$$

Claim 2 provides an interpretation of the right-hand side of eq. (4.30)

$$\sum_{j \geq b} \binom{j}{b} G_k(n, \ell, j) 1^{j-b} = \binom{n-b}{b} f_k(n - 2b, \ell).$$

In order to connect $F_k(x)$ and $\frac{1}{b!} F^{(b)}(1)$ we consider the Taylor expansion of $F_k(x)$ at $x = 1$ and compute

$$F_k(x) = \sum_{b \geq 0} \frac{1}{b!} F^{(b)}(1)(x-1)^b = \sum_{b=0}^{(n-\ell)/2} \binom{n-b}{b} f_k(n - 2b, \ell)(x-1)^b.$$

According to eq. (4.28) we have $\mathsf{T}_{k,1}(n, \ell) = G_k(n, \ell, 0)$ and the latter is the constant term of $F_k(x)$, whence

$$\mathsf{T}_{k,1}(n, \ell) = \sum_{b=0}^{(n-\ell)/2} (-1)^b \binom{n-b}{b} f_k(n - 2b, \ell).$$

It remains to prove eq. (4.27). Summing over all possible values of isolated vertices, we derive

$$\mathsf{T}_{k,1}(n) = \sum_{b=0}^{\lfloor n/2 \rfloor} (-1)^b \binom{n-b}{b} \left\{ \sum_{\ell=0}^{n-2b} f_k(n - 2b, \ell) \right\},$$

where $\mathsf{M}(n - 2b) = \sum_{\ell=0}^{n-2b} f_k(n - 2b, \ell)$ is given by eq. (2.8) and the proof of the theorem is complete.

We next study the asymptotics of the coefficients of various generating functions. Theorem 4.13 shows that $\mathsf{T}_{k,1}(n)$ is an alternating sum. Therefore even the knowledge about the exact coefficients does not directly imply asymptotic formulas. In the following we use the fact that $\mathbf{T}_{k,\sigma}(z)$ is of the type $\mathbf{F}_k(\vartheta(z))$, where $\vartheta(z)$ is algebraic and satisfies $\vartheta(0) = 0$. Therefore, $\mathbf{T}_{k,\sigma}(z)$ is D-finite and has a solution of an ODE generic asymptotic form [42]. We then apply Theorem 2.21 and derive simply asymptotic expressions for the coefficients.

Proposition 4.14. (Jin and Reidys [77]) *Suppose $2 \leq k \leq 9$ and $\rho_k = \frac{1}{2(k-1)}$. Then the number of k-noncrossing structures is asymptotically given by*

$$\mathsf{T}_{k,1}(n) \sim c_{k,1}\, n^{-((k-1)^2+(k-1)/2)}\, (\gamma_{k,1}^{-1})^n,\ \text{ for some } c_{k,1} > 0, \quad (4.31)$$

where $\gamma_{k,1}$ is the minimal, positive real solution of $\vartheta(z) = \rho_k^2$ and

$$\vartheta(z) = \left(\frac{z}{z^2 - z + 1}\right)^2;$$

see Table 4.3.

k	2	3	4	5	6	7	8	9
$\theta(n)$	$n^{-\frac{3}{2}}$	n^{-5}	$n^{-\frac{21}{2}}$	n^{-18}	$n^{-\frac{55}{2}}$	n^{-39}	$n^{-\frac{105}{2}}$	n^{-68}
$\gamma_{k,1}^{-1}$	2.6180	4.7913	6.8541	8.8875	10.9083	12.9226	14.9330	16.9410

Table 4.3. Exponential growth rates $\gamma_{k,1}^{-1}$ and subexponential factors $\theta(n)$, for k-noncrossing RNA structures with minimum arc length ≥ 2.

Proof. According to Theorem 4.9 we have

$$\mathbf{T}_{k,1}(z) = \frac{1}{z^2 - z + 1}\, \mathbf{F}_k\left(\left(\frac{z}{z^2 - z + 1}\right)^2\right)$$

and according to Corollary 2.14, $\mathbf{F}_k(z)$ is D-finite. Equations (2.20), (2.21), (2.22), (2.23), (2.24), (2.25), (2.26), (2.27) and (2.9) show for $2 \leq k \leq 9$ that the dominant singularities are given by ρ_k^2.

Claim. Suppose $\vartheta(z)$ is algebraic over $\mathbb{C}(z)$, analytic for $|z| < \delta$, and satisfies $\vartheta(0) = 0$. Suppose further $\gamma_{k,1}$ is the unique dominant singularity of $\mathbf{F}_k(\vartheta(z))$ with modulus $< \delta$ and satisfies the equation $\vartheta(z) = \rho_k^2$ and $\vartheta'(\gamma_{k,1}) \neq 0$. Then

$$[z^n]\, \mathbf{F}_k(\vartheta(z)) \sim a_k\, n^{-((k-1)^2+(k-1)/2)}\, \left(\gamma_{k,1}^{-1}\right)^n,$$

where $a_k > 0$ is some constant. Since $\vartheta(z)$ is algebraic over $\mathbb{C}(z)$ and satisfies $\vartheta(0) = 0$ we can conclude that the composition $\mathbf{F}_k(\vartheta(z))$ is D-finite [125]. According to Theorem 2.8 we have

$$f_k(2n, 0) \sim c_k\, n^{-((k-1)^2+(k-1)/2)}\, (2(k-1))^{2n}$$

for some $c_k > 0$ and Theorem 2.24 implies

$$\mathbf{F}_k(z) = \begin{cases} P_k(z - \rho_k^2) + c_k'(z - \rho_k^2)^{((k-1)^2+(k-1)/2)-1} \log(z - \rho_k^2)\,(1 + o(1)) \\ P_k(z - \rho_k^2) + c_k'(z - \rho_k^2)^{((k-1)^2+(k-1)/2)-1}\,(1 + o(1)) \end{cases}$$

depending on k being odd or even and where $P_k(z)$ are polynomials of degree not larger than $(k-1)^2+(k-1)/2-1$, c'_k is some constant, and $\rho_k = 1/2(k-1)$. By assumption, $\vartheta(z)$ is regular at $\gamma_{k,1}$ and $\vartheta'(\gamma_{k,1}) \neq 0$, whence we are given the supercritical case of singularity analysis; see Section 2.3.2. Consequently we have

$$[z^n]\, \mathbf{F}_k(\vartheta(z)) \sim a_k\, n^{-((k-1)^2+(k-1)/2)} \left(\gamma_{k,1}^{-1}\right)^n,$$

for some constant a_k and the claim is proved.

We proceed proving eq. (2.31). In view of

$$\mathbf{T}_{k,1}(z) = \frac{1}{z^2 - z + 1}\, \mathbf{F}_k\left(\left(\frac{z}{z^2 - z + 1}\right)^2\right),$$

we observe $\vartheta(z) = \left(\frac{z}{z^2-z+1}\right)^2$. Clearly, $\vartheta(z)$ is algebraic, analytic for $|z| < 1$, and satisfies $\vartheta(0) = 0$. By construction, the factor $\frac{z}{z^2-z+1}$ does not induce any singularities of modulus strictly smaller than those of $\vartheta(z)$. Hence

$$\mathbf{T}_{k,1}(z) = \Theta\left(\mathbf{F}_k\left(\left(\frac{z}{z^2 - z + 1}\right)^2\right)\right).$$

Since $\vartheta(z)\colon [0,1/2] \longrightarrow [0,4/9]$ is strictly monotonously increasing whence the positive real solution of $\vartheta(z) = \rho_k^2$, denoted by $\gamma_{k,1}$, is the minimum positive singularity of $\mathbf{F}_k(\vartheta(z))$. Table 4.4 shows that $\gamma_{k,1}$ is the unique solution of $\vartheta(z) = \rho_k^2$ of minimum modulus. Thus $\gamma_{k,1}$ is the unique dominant singularity of $\sum_{n\geq 0} \mathbf{T}_{k,1}(n)\, z^n$. According to Table 4.5, $\vartheta'(\gamma_{k,1}) \neq 0$, whence the supercritical paradigm applies and we derive

$$\mathbf{T}_{k,1}(n) \sim c_{k,1}\, n^{-((k-1)^2+(k-1)/2)} \left(\gamma_{k,1}^{-1}\right)^n$$

for some $c_{k,1} > 0$ and the proposition follows.

4.3.2 Canonical structures

Proposition 4.15. (Jin and Reidys [78]) *Suppose $k \in \mathbb{N}$, $k \geq 2$ and z is an indeterminant. Then*
(a) *the number of k-noncrossing core structures with h arcs, $\mathsf{C}_k(n,h)$ is given by*

$$\mathsf{C}_k(n,h) = \sum_{b=0}^{h-1} (-1)^{h-b-1} \binom{h-1}{b} \mathsf{T}_{k,1}(n - 2h + 2b + 2, b + 1). \tag{4.32}$$

(b) *Furthermore*

k	$\vartheta(z) = \rho_k^2$	$\|z\|$	k	$\vartheta(z) = \rho_k^2$	$\|z\|$
2	0.3820	0.3820	6	0.0917	0.0917
	2.6180	2.6180		10.9083	10.9083
	$-0.5000 \pm 0.8660i$	1		-0.1125	0.1125
				-8.8875	8.8875
3	0.2087	0.2087	7	0.0774	0.0774
	4.7913	4.7913		12.9226	12.9226
	-0.3820	0.3820		-0.0917	0.0917
	-2.6180	2.6180		-10.9083	10.9083
4	0.14590	0.14590	8	0.0670	0.0670
	6.8541	6.8541		14.9330	14.9330
	-0.2087	0.2087		-0.0774	0.0774
	-4.7913	4.7913		-12.9226	12.9226
5	0.1125	0.1125	9	0.0590	0.0590
	8.8875	8.8875		16.9410	16.9410
	-0.1459	0.1459		-0.0670	0.0670
	-6.8541	6.8541		-14.9330	14.9330

Table 4.4. The solutions of $\vartheta(z) = \rho_k^2$ for $2 \le k \le 9$ and their respective modulus.

k	2	3	4	5	6	7	8	9
$\vartheta'(\gamma_{k,1})$	1.4635	0.6861	0.4257	0.3046	0.2360	0.1921	0.1618	0.1396

Table 4.5. $\vartheta'(\gamma_{k,1})$ for $2 \le k \le 9$ obtained by MAPLE.

$$C_k(n) \sim c_k n^{-((k-1)^2 + (k-1)/2)} \left(\frac{1}{\kappa_k}\right)^n, \quad k = 3, 4, \ldots, 9,$$

where κ_k is the unique dominant positive real singularity of $\mathbf{C}_k(z)$ and the minimal positive real solution of the equation $\left(\frac{\sqrt{r(x)}\,x}{r(x)x^2 - x + 1}\right)^2 = \rho_k^2$ for $k = 3, 4, \ldots, 9$.

Proof. To prove (a) we set

$$\forall 0 \le i \le h - 1; \quad a(i) = C_k(n - 2(h - 1 - i), i + 1),$$
$$\forall 0 \le i \le h - 1; \quad b(i) = T_{k,1}(n - 2(h - 1 - i), i + 1).$$

We first employ Lemma 4.3 for $\sigma = 1$:

$$T_k^{[2]}(n, h) = \sum_{b=0}^{h-1} \binom{h-1}{b} C_k(n - 2b, h - b) \iff b(h-1) = \sum_{i=0}^{h-1} \binom{h-1}{i} a(i).$$

Via Möbius inversion we arrive at $a(h-1) = \sum_{i=0}^{h-1} (-1)^{h-1-i} \binom{h-1}{i} b(i)$, which is equivalent to

$$\mathsf{C}_k(n,h) = \sum_{b=0}^{h-1} (-1)^{h-b-1} \binom{h-1}{b} \mathsf{T}_{k,1}^{[2]}(n - 2h + 2b + 2, b + 1),$$

whence eq. (4.32). The proof of assertion (b) follows the logic of Proposition 4.14. According to Theorem 4.8 we have

$$\mathbf{C}_k(z) = \frac{1}{r(z)z^2 - z + 1} \mathbf{F}_k \left(\left(\frac{\sqrt{r(z)}z}{r(z)z^2 - z + 1} \right)^2 \right)$$

and Pringsheim's theorem [134] guarantees that $\mathbf{C}_k(z)$ has a dominant real positive singularity κ_k. We verify that there exists a unique solution of minimal modulus of

$$w(x) = \left(\frac{\sqrt{r(x)}\, x}{r(x)x^2 - x + 1} \right)^2 = \rho_k^2$$

for $3 \leq k \leq 9$. This solution necessarily equals κ_k, the therefore unique, dominant real singularity of $\mathbf{C}_k(z)$. Furthermore, since κ_k is strictly smaller than the singularity of $w(x)$ and $w'(\kappa_k) \neq 0$, the composite function $\mathbf{F}_k(w(x))$ belongs to the supercritical paradigm. Since κ_k is strictly smaller than the singularities of the factor $\frac{1}{r(x)x^2-x+1}$, and $w(x)$ is algebraic, using eqs. (2.21), (2.22), (2.23), (2.24), (2.25), (2.26), and (2.27) it is straightforward to verify that Theorem 2.21 applies for $k = 3, 4, \ldots, 9$; see the SM. Thus we have

$$\mathsf{C}_k(n) \sim c_k n^{-((k-1)^2 + (k-1)/2)} (\kappa_k^{-1})^n \quad \text{for some } c_k > 0,$$

whence Proposition 4.15.

We proceed by studying the generating function of k-noncrossing canonical RNA pseudoknot structures with minimum stack-length σ.

Proposition 4.16. (Jin and Reidys [78]) *Let $k, \sigma \in \mathbb{N}$, $k \geq 2$, let x be an indeterminant, $u_\sigma(x) = \frac{(x^2)^{\sigma-1}}{(x^2)^\sigma - x^2 + 1}$, and ρ_k^2 the dominant, positive real singularity of $\mathbf{F}_k(z)$. Then*

$$\mathsf{T}_{k,\sigma}(n,h) = \sum_{b=\sigma-1}^{h-1} \sum_{j=0}^{(h-b)-1} \binom{b + (2-\sigma)(h-b) - 1}{h - b - 1} (-1)^{(h-b)-j-1}$$

$$\times \binom{(h-b)-1}{j} \mathsf{T}_{k,1}(n - 2h + 2j + 2, j + 1).$$

Furthermore

$$T_{k,\sigma}(n) \sim c_{k,\sigma} n^{-((k-1)^2+(k-1)/2)} \left(\frac{1}{\gamma_{k,\sigma}}\right)^n \quad 2 \le k \le 9 \text{ and } 1 \le \sigma \le 9,$$

where $\gamma_{k,\sigma}$ is the dominant real singularity of $\mathbf{T}_{k,\sigma}(x)$ and the minimal positive real solution of the equation

$$q_\sigma(x) = \left(\frac{\sqrt{u_\sigma(x)}x}{u_\sigma(x)x^2 - x + 1}\right)^2 = \rho_k^2;$$

see Table 4.6.

k	2	3	4	5	6	7	8	9
$\gamma_{k,1}$	2.6180	4.7913	6.8541	8.8875	10.9083	12.9226	14.9330	16.9410
$\gamma_{k,2}$	1.9680	2.5881	3.0382	3.4138	3.7438	4.0420	4.3162	4.5715
$\gamma_{k,3}$	1.7160	2.0477	2.2704	2.4466	2.5955	2.7259	2.8427	2.9490
$\gamma_{k,4}$	1.5782	1.7984	1.9410	2.0511	2.1423	2.2209	2.2904	2.3529
$\gamma_{k,5}$	1.4899	1.6528	1.7561	1.8347	1.8991	1.9540	2.0022	2.0454
$\gamma_{k,6}$	1.4278	1.5563	1.6368	1.6973	1.7466	1.7883	1.8248	1.8573
$\gamma_{k,7}$	1.3815	1.4872	1.5528	1.6019	1.6415	1.6750	1.7041	1.7300
$\gamma_{k,8}$	1.3454	1.4351	1.4903	1.5314	1.5645	1.5923	1.6165	1.6378
$\gamma_{k,9}$	1.3164	1.3941	1.4417	1.4770	1.5054	1.5291	1.5497	1.5679

Table 4.6. The exponential growth rates for various classes of k-noncrossing structures computed via Proposition 4.16. $\sigma = 1$ corresponds to structures with isolated arcs and $\sigma = 2$ corresponds to canonical structures.

Proof. The first assertion follows from Lemma 4.3 and eq. (4.32), which allows us to express the terms $C_k(n - 2b, h - b)$ via $T_{k,1}(n', h')$.
As for the second assertion we use Theorem 4.9

$$\mathbf{T}_{k,\sigma}(z) = \frac{1}{u_\sigma(z)z^2 - z + 1} \mathbf{F}_k\left(\left(\frac{\sqrt{u_\sigma(z)}z}{(u_\sigma(z)z^2 - z + 1)}\right)^2\right)$$

and verify that all dominant singularities of $\mathbf{T}_{k,\sigma}(x)$ are singularities of $\mathbf{F}_k\left(\frac{\sqrt{u_\sigma(x)}x}{u_\sigma(x)x^2-x+1}\right)$ and that $\gamma_{k,\sigma}$ is the unique dominant singularity for both functions. In analogy to Proposition 4.14 we can eventually conclude via Theorem 2.21

$$\mathbf{T}_{k,\sigma}(n) \sim c_{k,\sigma} n^{-((k-1)^2+(k-1)/2)} (\gamma_{k,\sigma}^{-1})^n,$$

whence Proposition 4.16.

Next, we compute the bivariate generating function of $\mathsf{T}_{k,\sigma}(n,h)$,

$$\mathbf{T}_{k,\sigma}(x,u) = \sum_{n\geq 0}\sum_{0\leq h\leq\frac{n}{2}} \mathsf{T}_{k,\sigma}(n,h)u^h x^n.$$

Proposition 4.17. *Let $k,\sigma \in \mathbb{N}$, $k \geq 2$ and let u,x be indeterminants. Then*

$$\mathbf{T}_{k,\sigma}(x,u) = \sum_{n\geq 0}\sum_{0\leq h\leq\frac{n}{2}} \mathsf{C}_k(n,h)\left(\frac{u\cdot(ux^2)^{\sigma-1}}{1-ux^2}\right)^h x^n \qquad (4.33)$$

which in particular implies setting $u=1$:

$$\mathbf{T}_{k,\sigma}(x,1) = \sum_{n\geq 0}\sum_{0\leq h\leq\frac{n}{2}} \mathsf{C}_k(n,h)\left(\frac{(x^2)^{\sigma-1}}{1-x^2}\right)^h x^n. \qquad (4.34)$$

Proof. Let $n,h \in \mathbb{N}$ where $n \geq 2h$ and let $\mathcal{C}_k$ denote the set of k-noncrossing cores. Let $\mathcal{T}_{k,\sigma}$ denote the set of all k-noncrossing, σ-canonical structures. Lemma 4.3 implies the existence of the surjective map $\varphi'\colon \mathcal{T}_{k,\sigma} \longrightarrow \mathcal{C}_k$. For a given core, contained in $\mathcal{C}_k(n,h)$, such a preimage is obtained by inflating each core-arc to a stack of length at least σ; see Fig. 4.20.

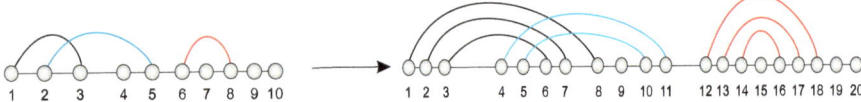

Fig. 4.20. From cores to structures: inflating each core-arc to a stack of length at least σ, we obtain a 3-noncrossing 2-canonical structure.

We consider the combinatorial classes $\mathcal{T}_\gamma$ consisting of k-noncrossing, σ-canonical structures obtained by inflating the core γ, $\mathcal{K}^\sigma$ (stacks of length at least σ), $\mathcal{R}$ (arcs), and $\mathcal{Z}$ (vertices). In view of

$$\mathbf{C}(x,u) = \sum_{n\geq 0}\sum_{h\leq n/2} \mathsf{C}_k(n,h)\, x^n u^h = \sum_{n\geq 0}\sum_{h\leq n/2}\sum_{\gamma\in\mathcal{C}_k(n,h)} \left((ux^2)^h \cdot x^{n-2h}\right)$$

we observe $\mathbf{Z}(x) = x$ and $\mathbf{R}(x,u) = ux^2$. The inflation process can be expressed symbolically via

$$\mathcal{T}_\gamma = (\mathcal{K}^\sigma)^h \times \mathcal{Z}^{n-2h},$$
$$\mathcal{K}^\sigma = (\mathcal{R})^\sigma \times \mathrm{SEQ}(\mathcal{R}).$$

Since the number of isolated vertices remains invariant under inflation we derive

$$\mathbf{T}_\gamma(x,u) = \left(\frac{(ux^2)^\sigma}{1-ux^2} \right)^h x^{n-2h} = \left(\frac{u \cdot (ux^2)^{\sigma-1}}{1-ux^2} \right)^h x^n.$$

Now we can easily compute

$$\mathbf{T}_{k,\sigma}(x,u) = \sum_{n\geq 0} \sum_{h\leq n/2} \sum_{\gamma\in\mathcal{C}_k(n,h)} \mathbf{T}_\gamma(x,u)$$

$$= \sum_{n\geq 0} \sum_{h\leq n/2} \mathsf{C}_k(n,h) \mathbf{T}_\gamma(x,u),$$

whence eq. (4.33). Setting $u = 1$, we derive eq. (4.34) and Proposition 4.17 follows.

The generating functions of k-noncrossing, σ-canonical structures $\mathbf{T}_{k,\sigma}^{[4]}(x)$ are of particular importance since it represents the folding target of the algorithm cross discussed in Chapter 6. According to Corollaries 4.11 and 4.12 we have

$$\mathbf{T}_{k,\sigma}^{[4]}(x) = \frac{1}{v_4(x)} \mathbf{F}_k\left(\left(\frac{\sqrt{u_\sigma(x)}\,x}{v_4(x)} \right)^2 \right),$$

$$\mathbf{T}_{k,\sigma}^{[3]}(z) = \frac{1}{v_3(z)} \mathbf{F}_k\left(\left(\frac{\sqrt{u_\sigma(z)}\,z}{v_3(z)} \right)^2 \right),$$

where $u_\sigma(x) = \frac{(x^2)^{\sigma-1}}{x^{2\sigma}-x^2+1}$ and $v_\lambda(x) = u_\sigma(x)\sum_{h=2}^\lambda x^h - x + 1$. We next present the asymptotic formulas for the coefficients $[z^n]\mathbf{T}_{k,\sigma}^{[4]}(x)$ and $\mathbf{T}_{k,\sigma}^{[3]}(n)$ that can be computed in analogy to Proposition 4.14.

Proposition 4.18. (Ma and Reidys [88]) *Let* $k,\sigma \in \mathbb{N}$, $k \geq 2$ *and* $\sigma \geq 3$. *Then*

$$\mathsf{T}_{k,\sigma}^{[4]}(n) \sim c_{k,\sigma}^{[4]}\, n^{-(k-1)^2-\frac{k-1}{2}} \left(\frac{1}{\gamma_{k,\sigma}^{[4]}} \right)^n \qquad for \ \ 2 \leq k \leq 9, \ 3 \leq \sigma \leq 10$$

holds, where $c_{k,\sigma}^{[4]}$ *is some positive constant and* $\gamma_{k,\sigma}^{[4]}$ *is the unique positive real dominant singularity of* $\mathbf{T}_{k,\sigma}^{[4]}(x)$; *see Table 4.7.*

Proposition 4.19. (Jin and Reidys [79]) *Let* $k,\sigma \in \mathbb{N}$ *where* $k,\sigma \geq 2$. *Then*

$$\mathsf{T}_{k,\sigma}^{[3]}(n) \sim c_{k,\sigma}^{[3]}\, n^{-(k-1)^2-\frac{k-1}{2}} \left(\frac{1}{\gamma_{k,\sigma}^{[3]}} \right)^n \qquad for \ \ 2 \leq k \leq 9, \ 2 \leq \sigma \leq 9$$

holds, where $c_{k,\sigma}^{[3]}$ *is some positive constant and* $\gamma_{k,\sigma}^{[3]}$ *is the unique positive real dominant singularity of* $\mathbf{T}_{k,\sigma}^{[3]}(x)$; *see Table 4.8.*

k	2	3	4	5	6	7	8	9
$\sigma = 3$	1.6521	2.0348	2.2644	2.4432	2.5932	2.7243	2.8414	2.9480
$\sigma = 4$	1.5375	1.7898	1.9370	2.0488	2.1407	2.2198	2.2896	2.3523
$\sigma = 5$	1.4613	1.6465	1.7532	1.8330	1.8979	1.9532	2.0016	2.0449
$\sigma = 6$	1.4065	1.5515	1.6345	1.6960	1.7457	1.7877	1.8243	1.8569
$\sigma = 7$	1.3649	1.4834	1.5510	1.6008	1.6408	1.6745	1.7038	1.7297
$\sigma = 8$	1.3320	1.4319	1.4888	1.5305	1.5639	1.5919	1.6162	1.6376
$\sigma = 9$	1.3053	1.3915	1.4405	1.4763	1.5049	1.5288	1.5494	1.5677

Table 4.7. Exponential growth rates of k-noncrossing, σ-canonical structures with stacks of length at least 3 and minimum arc length 4.

k	2	3	4	5	6	7	8	9
$\sigma = 2$	1.89900	2.57207	3.03057	3.40923	3.74072	4.03973	4.31449	4.57020
$\sigma = 3$	1.68016	2.03917	2.26625	2.44418	2.59389	2.72470	2.84176	2.94826
$\sigma = 4$	1.55580	1.79299	1.93841	2.04952	2.14123	2.22012	2.28981	2.35249
$\sigma = 5$	1.47437	1.64895	1.75428	1.83360	1.89832	1.95349	2.00184	2.04504
$\sigma = 6$	1.41635	1.55344	1.63538	1.69651	1.74601	1.78794	1.82451	1.85703
$\sigma = 7$	1.37262	1.48498	1.55175	1.60122	1.64108	1.67470	1.70390	1.72979
$\sigma = 8$	1.33831	1.43323	1.48943	1.53086	1.56411	1.59206	1.61627	1.63769
$\sigma = 9$	1.31057	1.39259	1.44102	1.47660	1.50506	1.52893	1.54955	1.56777

Table 4.8. Exponential growth rates of k-noncrossing, canonical structures having minimum arc length 3.

4.4 Modular k-noncrossing structures

In the context of sequence to structure maps in RNA, canonical structures with minimum arc length ≥ 4 are considered to be the most relevant class. In case of noncrossing base pairs these structures are well understood [69] and important properties are tied to their combinatorics. Point in case is that the existence of neutral networks of these structures [69] is connected to the asymptotics of the coefficients of their generating function; see Section 1.3 for details. For noncrossing arcs the key is the formula given in Section 1.3, eq. (1.2), which is a result of a straightforward substitution; see Proposition 4.20. However, in the presence of cross-serial interactions considerably more effort has to be made.

Let us begin by revisiting V_k-shapes. In Section 4.1.2 we relate these shapes to the bivariate generating function $\mathbf{G}_k(x, y)$; see Theorem 4.6. There we specifically accounted for 1-arcs since these require a particular insertion of isolated vertices. The derivation of $\mathbf{G}_k(x, y)$ is based on the recurrence of Lemma 4.5, whose existence is not entirely trivial as k-noncrossing structures, despite having D-finite generation functions for all k, cannot be inductively constructed. Of course, this recurrence exists since 1-arcs are special: They are not involved in crossings at all.

It is straightforward to see that the key to compute the generating function of k-noncrossing canonical structures with minimum arc length 4 are precisely the 2-arcs of V_k-shapes. While 2-arcs exhibit crossings and are therefore more complicated than the 1-arcs discussed above, their crossing can be classified easily. As a result, we can establish via Lemmas 4.21 and 4.23 the relevant recurrences. The latter eventually facilitate the proof of Theorem 4.25.

In the following we refer to k-noncrossing, canonical RNA structures with minimum arc length 4 as modular, k-noncrossing structures. Let $\mathbf{Q}_k(n)$ denote their number and

$$\mathbf{Q}_k(z) = \sum_{n=0}^{\infty} \mathbf{Q}_k(n)z^n \tag{4.35}$$

their generating function.

We begin our analysis of modular structures by studying first the noncrossing case [69] and present a new proof based on the framework developed in Section 4.2.

Proposition 4.20. *The generating function of modular noncrossing structures is given by*

$$\mathbf{Q}_2(z) = \frac{1 - z^2 + z^4}{1 - z - z^2 + z^3 + 2z^4 + z^6} \cdot \mathbf{F}_2 \left(\frac{z^4 - z^6 + z^8}{(1 - z - z^2 + z^3 + 2z^4 + z^6)^2} \right) \tag{4.36}$$

and $\mathbf{Q}_2(n)$ satisfies

$$\mathbf{Q}_2(n) \sim c_2 n^{-3/2} \gamma_2^{-n},$$

where γ_2 is the minimal, positive real solution of $\vartheta(z) = 1/4$, and

$$\vartheta(z) = \frac{z^4 - z^6 + z^8}{(1 - z - Sz^2 + z^3 + 2z^4 + z^6)^2}. \tag{4.37}$$

Furthermore we have $\gamma_2 \approx 1.8489$ and $c_2 \approx 1.4848$.

Proof. Using the notation of Theorem 4.9 one expresses $\mathbf{Q}_2(z)$ via V_2-shapes, γ, having s arcs, m of which are 1-arcs. This gives rise to the combinatorial classes:

$$\begin{aligned}
\mathcal{Q}_\gamma &= \mathcal{M}^s \times \mathcal{L}^{2s+1-m} \times (\mathcal{Z}^3 \times \mathcal{L})^m, \\
\mathcal{M} &= \mathcal{K}^\sigma \times \text{SEQ}(\mathcal{N}^\sigma), \\
\mathcal{N}^\sigma &= \mathcal{K}^\sigma \times (\mathcal{Z} \times \mathcal{L} + \mathcal{Z} \times \mathcal{L} + (\mathcal{Z} \times \mathcal{L})^2), \\
\mathcal{K}^\sigma &= \mathcal{R} \times \text{SEQ}(\mathcal{R}), \\
\mathcal{L} &= \text{SEQ}(\mathcal{Z}).
\end{aligned}$$

We first insert a segment containing at least three isolated vertices into any 1-arc $\mathcal{J}^{[3]} = \mathcal{L}^{2s+1-m} \times (\mathcal{Z}^3 \times \mathcal{L})^m$, i.e.,

$$\mathbf{J}^{[3]}(z) = \left(\frac{1}{1-z}\right)^{2s+1-m} \left(\frac{z^3}{1-z}\right)^m.$$

Since we have only nested arcs in V_2-shapes, any non-1-arc can, after the above insertion of isolated vertices, be arbitrarily inflated. Therefore

$$\mathbf{Q}_\gamma(z) = \left(\frac{\frac{z^{2\sigma}}{1-z^2}}{1 - \frac{z^{2\sigma}}{1-z^2}\left(\frac{2z}{1-z} + \left(\frac{z}{1-z}\right)^2\right)}\right)^s \cdot \left(\frac{1}{1-z}\right)^{2s+1-m} \left(\frac{z^3}{1-z}\right)^m$$

$$= (1-z)^{-1}\left(\frac{z^{2\sigma}}{(1-z)^2(1-z^2) - z^{2\sigma}(2z-z^2)}\right)^s (z^3)^m.$$

Since for any $\gamma, \gamma' \in \mathcal{I}_2(s,m)$ we have $\mathbf{Q}_\gamma(z) = \mathbf{Q}_{\gamma'}(z)$,

$$\mathbf{Q}_2(z) = \sum_{m,s\geq 0}\sum_{\gamma\in\mathcal{I}_2(s,m)} \mathbf{Q}_\gamma(z) = \sum_{s\geq 0}\sum_{m=0}^s i_2(s,m)\,\mathbf{Q}_\gamma(z),$$

whence

$$\mathbf{Q}_2(z) = (1-z)^{-1}\sum_{s\geq 0}\sum_{m=0}^s i_2(s,m)\,\eta_\sigma(z)^s \left(z^3\right)^m.$$

Substituting $x = \eta_\sigma(z)$ and $y = z^3$ into eq. (4.11) we derive

$$\mathbf{Q}_2(z) = (1-z)^{-1}\frac{1+\eta_\sigma(z)}{1+2\eta_\sigma(z)-\eta_\sigma(z)z^3}\mathbf{F}_2\left(\frac{\eta_\sigma(z)(1+\eta_\sigma(z))}{(1+2\eta_\sigma(z)-\eta_\sigma(z)z^3)^2}\right)$$

$$= \frac{(1-z)(1-z^2+z^{2\sigma})}{1-2z+2z^3-z^4+2z^{2\sigma}-2z^{2\sigma+1}+z^{2\sigma+2}-z^{2\sigma+3}} \times$$

$$\mathbf{F}_2\left(\frac{(1-z)^2z^{2\sigma}(1-z^2+z^{2\sigma})}{(1-2z+2z^3-z^4+2z^{2\sigma}-2z^{2\sigma+1}+z^{2\sigma+2}-z^{2\sigma+3})^2}\right).$$

Setting $\sigma = 2$ we obtain eq. (4.36). The asymptotic formula follows immediately from the supercritical paradigm; see Theorem 2.21.

We remark that the proof of Proposition 4.20 works for any σ and λ. Thus, noncrossing canonical structures of any minimal arc length are straightforwardly derived.

The situation is quite different in case of $k > 2$. In order to understand modular, k-noncrossing structures, we have to distinguish a variety of 2-arcs, i.e., arcs of the form $(i, i + 2)$. Each such class requires its specific inflation procedure; see Theorem 4.25. Let us next have a closer look at these classes:

- C_1 the class of 1-arcs,
- C_2 the class of arc pairs consisting of mutually crossing 2-arcs,

- $\mathbf{C}_3$ the class of arc pairs (α, β) where α is the unique 2-arc crossing β and β has length at least 3.
- $\mathbf{C}_4$ the class of arc triples $(\alpha_1, \beta, \alpha_2)$, where α_1 and α_2 are 2-arcs that cross β.

In Fig. 4.21 we illustrate how these classes are induced by modular, k-noncrossing structures.

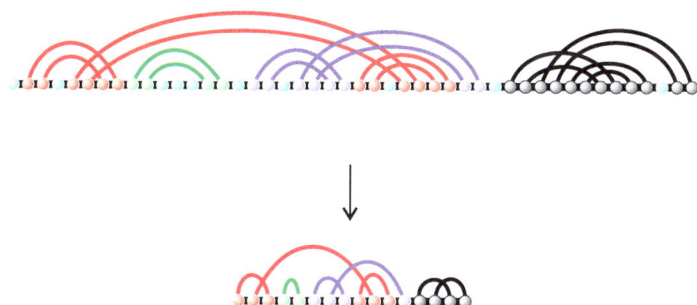

Fig. 4.21. Colored V_k-shapes: a modular, 2-noncrossing structures (*top*) and its colored V_k-shape (*bottom*). In the resulting V_k-shape we color the four classes as follows: $\mathbf{C}_1$(*green*), $\mathbf{C}_2$(*black*), $\mathbf{C}_3$(*blue*), and $\mathbf{C}_4$(*red*).

4.4.1 Colored shapes

In this section we refine V_k-shapes into two stages. For this purpose let $\mathcal{I}_k(s, u_1, u_2)$ and $i_k(s, u_1, u_2)$ denote the set and cardinality of V_k-shapes having s-arcs, u_1 1-arcs, and u_2 pairs of mutually crossing 2-arcs. Our first objective consists in computing the generating function

$$\mathbf{W}_k(x, y, w) = \sum_{s \geq 0} \sum_{u_1=0}^{s} \sum_{u_2=0}^{\lfloor \frac{s-u_1}{2} \rfloor} i_k(s, u_1, u_2) \, x^s y^{u_1} w^{u_2},$$

that is, we first take the classes $\mathbf{C}_1$ and $\mathbf{C}_2$ into account.

Lemma 4.21. *For $k > 2$, the coefficients $i_k(s, u_1, u_2)$ satisfy*

$$i_k(s, u_1, u_2) = 0 \quad for \ u_1 + 2u_2 > s, \tag{4.38}$$

$$\sum_{u_2=0}^{\lfloor \frac{s-u_1}{2} \rfloor} i_k(s, u_1, u_2) = i_k(s, u_1), \tag{4.39}$$

where $i_k(s, u_1)$ denotes the number of V_k-shapes having s arcs and u_1 1-arcs. Furthermore, we have the recursion

$$(u_2 + 1)\, i_k(s + 1, u_1, u_2 + 1) = (u_1 + 1)\, i_k(s, u_1 + 1, u_2)$$
$$+ (u_1 + 1)\, i_k(s - 1, u_1 + 1, u_2) \quad (4.40)$$

and the solution of eqs. (4.38), (4.39), and (4.40) is unique.

Proof. By construction, eqs. (4.38) and (4.39) hold. In order to prove eq. (4.40) we choose a shape $\delta \in \mathcal{I}_k(s + 1, u_1, u_2 + 1)$ and label exactly one of the $(u_2 + 1)$ $\mathbf{C}_2$-elements. We denote the leftmost $\mathbf{C}_2$-arc by α. Let $\mathcal{L}$ be the set of these labeled shapes, λ, then

$$|\mathcal{L}| = (u_2 + 1)\, i_k(s + 1, u_1, u_2 + 1).$$

We next observe that the removal of α results in either a shape or a matching. Let the elements of the former set be $\mathcal{L}_1$ and those of the latter $\mathcal{L}_2$. By construction,

$$\mathcal{L} = \mathcal{L}_1 \dot\cup \mathcal{L}_2.$$

Claim 1.

$$|\mathcal{L}_1| = (u_1 + 1)\, i_k(s, u_1 + 1, u_2).$$

To prove Claim 1, we consider the labeled $\mathbf{C}_2$-element (α, β). Let $\mathcal{L}_1^\alpha$ be the set of shapes induced by removing α. It is straightforward to verify that the removal of α can lead to only one additional $\mathbf{C}_1$-element, β. Therefore $\mathcal{L}_1$-shapes induce unique $\mathcal{I}_k(s, u_1 + 1, u_2)$-shapes, having a labeled 1-arc, β, see Fig. 4.22. This proves Claim 1.

Fig. 4.22. The term $(u_1 + 1)\, i_k(s, u_1 + 1, u_2)$.

Claim 2.

$$|\mathcal{L}_2| = (u_1 + 1)\, i_k(s - 1, u_1 + 1, u_2).$$

To prove Claim 2, we consider $\mathcal{M}_2^\alpha$, the set of matchings, μ_2^α, obtained by removing α. Such a matching contains exactly one stack of length 2, (β_1, β_2), where β_2 is nested in β_1. Let $\mathcal{L}_2^\alpha$ be the set of shapes induced by collapsing (β_1, β_2) into β_2. We observe that α crosses β_2 and that β_2 becomes a 1-arc. Therefore, $\mathcal{L}_2$ is the set of labeled shapes that induce unique $\mathcal{I}_k(s - 1, u_1 + 1, u_2)$-shapes having a labeled 1-arc, β_2; see Fig. 4.23. This proves Claim 2.

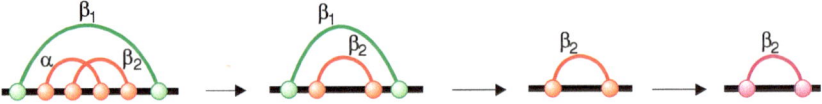

Fig. 4.23. The term $(u_1 + 1)\, i_k(s - 1, u_1 + 1, u_2)$.

Combining Claims 1 and 2 we derive eq. (4.40).

It remains to show by induction on s that the numbers $i_k(s, u_1, u_2)$ can be uniquely derived from eqs. (4.38), (4.39), and (4.40), whence the lemma.

We next proceed by computing $\mathbf{W}_k(x, y, w)$.

Proposition 4.22. *For $k > 2$, we have*

$$\mathbf{W}_k(x, y, w) = (1 + x)v\, \mathbf{F}_k\left(x(1+x)v^2\right), \tag{4.41}$$

where $v = \left((1 - w)x^3 + (1 - w)x^2 + (2 - y)x + 1\right)^{-1}$.

Proof. According to Theorem 4.6, we have

$$\mathbf{I}_k(z, u) = \frac{1 + z}{1 + 2z - zu}\mathbf{F}_k\left(\frac{z(1 + z)}{(1 + 2z - zu)^2}\right).$$

This generating function is connected to $\mathbf{W}_k(x, y, z)$ via eq. (4.39) as follows: setting $w = 1$, we have $\mathbf{W}_k(x, y, 1) = \mathbf{I}_k(x, y)$. The recursion of eq. (4.40) gives rise to the partial differential equation

$$\frac{\partial \mathbf{W}_k(x, y, w)}{\partial w} = x\frac{\partial \mathbf{W}_k(x, y, w)}{\partial y} + x^2\frac{\partial \mathbf{W}_k(x, y, w)}{\partial y}. \tag{4.42}$$

We next show

- the function

$$\mathbf{W}_k^*(x, y, w) = \frac{(1 + x)}{(1 - w)x^3 + (1 - w)x^2 + (2 - y)x + 1} \times$$

$$\mathbf{F}_k\left(\frac{(1 + x)x}{((1 - w)x^3 + (1 - w)x^2 + (2 - y)x + 1)^2}\right) \tag{4.43}$$

  is a solution of eq. (4.42);
- its coefficients, $i_k^*(s, u_1, u_2) = [x^s y^{u_1} w^{u_2}]\mathbf{W}_k^*(x, y, w)$, satisfy

$$i_k^*(s, u_1, u_2) = 0 \quad \text{for} \quad u_1 + 2u_2 > s;$$

- $\mathbf{W}_k^*(x, y, 1) = \mathbf{I}_k(x, y)$.

First,

$$\frac{\partial \mathbf{W}_k^*(x, y, w)}{\partial y} = u\, \mathbf{F}_k\left(u\right) + 2u\, \mathbf{F}_k'\left(u\right), \tag{4.44}$$

$$\frac{\partial \mathbf{W}_k^*(x, y, w)}{\partial w} = x(1 + x)u\, \mathbf{F}_k\left(u\right) + 2x(1 + x)u\mathbf{F}_k'\left(u\right), \tag{4.45}$$

where

$$u = \frac{x(1 + x)}{\left((1 - w)x^3 + (1 - w)x^2 + (2 - y)x + 1\right)^2}$$

and $\mathbf{F}'_k(u) = \sum_{n \geq 0} n f_k(2n, 0)(u)^n$. Consequently, we derive

$$\frac{\partial \mathbf{W}^*_k(x, y, w)}{\partial w} = x \frac{\partial \mathbf{W}^*_k(x, y, w)}{\partial y} + x^2 \frac{\partial \mathbf{W}^*_k(x, y, w)}{\partial y}. \qquad (4.46)$$

Second we prove $i^*_k(s, u_1, u_2) = 0$ for $u_1 + 2u_2 > s$. To this end we observe that $\mathbf{W}^*_k(x, y, w)$ is a power series, since it is analytic in $(0, 0, 0)$. It now suffices to note that the indeterminants y and w only appear in form of products xy and $x^2 w$ or $x^3 w$. Third, the equality $\mathbf{W}^*_k(x, y, 1) = \mathbf{I}_k(x, y)$ is obvious.
Claim.

$$\mathbf{W}^*_k(x, y, w) = \mathbf{W}_k(x, y, w). \qquad (4.47)$$

By construction the coefficients $i^*_k(s, u_1, u_2)$ satisfy eq. (4.40) and we just proved $i^*_k(s, u_1, u_2) = 0$ for $u_1 + 2u_2 > s$. In view of $\mathbf{W}^*_k(x, y, 1) = \mathbf{I}_k(x, y)$ we have

$$\forall s, u_1; \qquad \sum_{u_2=0}^{\lfloor \frac{s-u_1}{2} \rfloor} i^*_k(s, u_1, u_2) = i_k(s, u_1).$$

Using these three properties it follows via induction over s

$$\forall s, u_1, u_2 \geq 0; \qquad i^*_k(s, u_1, u_2) = i_k(s, u_1, u_2),$$

whence the claim and the proposition is proved.

In addition to $\mathbf{C}_1$ and $\mathbf{C}_2$, we consider next the classes $\mathbf{C}_3$ and $\mathbf{C}_4$. For this purpose we have to identify two new recursions; see Lemma 4.23. Setting $\mathbf{u} = (u_1, \ldots, u_4)$ we denote by $\mathcal{I}_k(s, \mathbf{u})$ and $i_k(s, \mathbf{u})$, the set and number of colored V_k-shapes over s arcs, containing u_i elements of class $\mathbf{C}_i$, where $1 \leq i \leq 4$. The key result is

Lemma 4.23. *For $k > 2$, the coefficients $i_k(s, \mathbf{u})$ satisfy*

$$i_k(s, u_1, u_2, u_3, u_4) = 0 \quad for\ u_1 + 2u_2 + 2u_3 + 3u_4 > s, \quad (4.48)$$

$$\sum_{u_3, u_4 \geq 0} i_k(s, u_1, u_2, u_3, u_4) = i_k(s, u_1, u_2). \qquad (4.49)$$

Furthermore we have the recursions

$$(u_3 + 1)i_k(s + 1, u_1, u_2, u_3 + 1, u_4) =$$
$$2u_1 i_k(s - 1, u_1, u_2, u_3, u_4)$$
$$+ 4(u_2 + 1)i_k(s - 1, u_1, u_2 + 1, u_3, u_4)$$
$$+ 4(u_2 + 1)i_k(s - 1, u_1, u_2 + 1, u_3 - 1, u_4)$$
$$+ 4(u_2 + 1)i_k(s - 2, u_1, u_2 + 1, u_3 - 1, u_4)$$
$$+ 2u_3 i_k(s - 1, u_1, u_2, u_3, u_4)$$
$$+ 6(u_3 + 1)i_k(s - 1, u_1, u_2, u_3 + 1, u_4)$$
$$+ 2(u_3 + 1)i_k(s - 2, u_1, u_2, u_3 + 1, u_4)$$

$$+ 2u_3 i_k(s - 2, u_1, u_2, u_3, u_4)$$
$$+ 4(u_4 + 1)i_k(s - 1, u_1, u_2, u_3 - 1, u_4 + 1)$$
$$+ 4u_4 i_k(s - 1, u_1, u_2, u_3, u_4)$$
$$+ 4(u_4 + 1)i_k(s - 1, u_1, u_2, u_3, u_4 + 1)$$
$$+ 4u_4 i_k(s - 2, u_1, u_2, u_3, u_4)$$
$$+ 2(u_4 + 1)i_k(s - 2, u_1, u_2, u_3, u_4 + 1)$$
$$+ 2(2(s - 1) - 2u_1 - 4u_2 - 4u_3 - 6u_4)i_k(s - 1, u_1, u_2, u_3, u_4)$$
$$+ (2(s - 2) - 4u_2 - 4u_3 - 6u_4)i_k(s - 2, u_1, u_2, u_3, u_4)$$
$$+ 2(u_3 + 1)i_k(s, u_1, u_2, u_3 + 1, u_4)$$
$$+ 4(u_4 + 1)i_k(s, u_1, u_2, u_3 - 1, u_4 + 1)$$
$$+ (2s - 2u_1 - 4u_2 - 4u_3 - 6u_4)i_k(s, u_1, u_2, u_3, u_4) \tag{4.50}$$

and

$$2(u_4 + 1)i_k(s + 1, u_1, u_2, u_3, u_4 + 1) = (u_3 + 1)i_k(s, u_1, u_2, u_3 + 1, u_4)$$
$$+ 2(u_2 + 1)_k(s, u_1, u_2 + 1, u_3, u_4). \tag{4.51}$$

The sequence satisfying eqs. (4.48), (4.49), (4.50), and (4.51) is unique.

The proof of Lemma 4.23 is outlined in Problem 4.8 and all details are given in the SM.

Proposition 4.22 and Lemma 4.23 put us in position to compute the generating function of colored V_k-shapes

$$\mathbf{I}_k(x, y, z, w, t) = \sum_{s, u_1, u_2, u_3, u_4} i_k(s, \mathbf{u})\, x^s y^{u_1} z^{u_2} w^{u_3} t^{u_4}. \tag{4.52}$$

Proposition 4.24. *For $k > 2$, the generating function of colored V_k-shapes is given by*

$$\mathbf{I}_k(x, y, z, w, t) = \frac{1 + x}{\theta} \mathbf{F}_k\left(\frac{x(1 + (2w - 1)x + (t - 1)x^2)}{\theta^2}\right), \tag{4.53}$$

where $\theta = 1 - (y - 2)x + (2w - z - 1)x^2 + (2w - z - 1)x^3$.

Proof. The first recursion of Lemma 4.23 implies the partial differential equation

$$\frac{\partial \mathbf{I}_k}{\partial w} = \frac{\partial \mathbf{I}_k}{\partial x}(2x^2 + 4x^3 + 2x^4) - \frac{\partial \mathbf{I}_k}{\partial y}(2xy + 2x^2 y)$$

$$+ \frac{\partial \mathbf{I}_k}{\partial z}(-4xz + 4x^2 w + 4x^2 - 4x^3 z - 8x^2 z + 4x^3 w)$$

$$+ \frac{\partial \mathbf{I}_k}{\partial w}(-4xw + 2x - 6x^2 w + 6x^2 - 2x^3 w + 2x^3)$$

$$+ \frac{\partial \mathbf{I}_k}{\partial t}(-6xt + 4xw - 8x^2 t + 4x^2 w + 4x^2 - 2x^3 t + 2x^3). \tag{4.54}$$

Analogously, the second recursion of Lemma 4.23 gives rise to the partial differential equation

$$2\frac{\partial \mathbf{I}_k}{\partial t} = \frac{\partial \mathbf{I}_k}{\partial w}x + \frac{\partial \mathbf{I}_k}{\partial z}2x. \tag{4.55}$$

Aside from being a solution of eqs. (4.54) and (4.55), we take note of the fact that eq. (4.49) is equivalent to

$$\mathbf{I}_k(x, y, z, 1, 1) = \mathbf{W}_k(x, y, z). \tag{4.56}$$

We next show

■

$$\mathbf{I}_k^*(x, y, z, w, t) = \frac{1+x}{1 - (y-2)x + (2w - z - 1)x^2 + (2w - z - 1)x^3} \times$$
$$\mathbf{F}_k\left(\frac{x(1 + (2w-1)x + (t-1)x^2)}{(1 - (y-2)x + (2w - z - 1)x^2 + (2w - z - 1)x^3)^2}\right)$$

is a solution of eqs. (4.54) and (4.55);
- its coefficients, $i_k^*(s, u_1, u_2, u_3, u_4) = [x^s y^{u_1} z^{u_2} w^{u_3} t^{u_4}] \mathbf{I}_k^*(x, y, z, w, t)$, satisfy $i_k^*(s, u_1, u_2, u_3, u_4) = 0$ for $u_1 + 2u_2 + 2u_3 + 3u_4 > s$;
- $\mathbf{I}_k^*(x, y, z, 1, 1) = \mathbf{W}_k(x, y, z)$.

We verify by direct computation that $\mathbf{I}_k^*(x, y, z, w, t)$ satisfies eq. (4.54) as well as eq. (4.55). Next we prove $i_k^*(s, u_1, u_2, u_3, u_4) = 0$ for $u_1 + 2u_2 + 2u_3 + 3u_4 > s$. Since $\mathbf{I}_k^*(x, y, z, w, t)$ is analytic in $(0, 0, 0, 0, 0)$, it is a power series. As the indeterminants y, z, w, and t appear only in form of products xy, x^2z, or x^3z; x^2w, or x^3w; and x^3t, respectively, the assertion follows.

Claim.
$$\mathbf{I}_k^*(x, y, z, w, t) = \mathbf{I}_k(x, y, z, w, t).$$

By construction, $i_k^*(s, \mathbf{u})$ satisfies the recursions (4.50) and (4.51) as well as $i_k^*(s, u_1, u_2, u_3, u_4) = 0$ for $u_1 + 2u_2 + 2u_3 + 3u_4 > s$. Equation (4.56) implies

$$\sum_{u_3, u_4 \geq 0} i_k^*(s, u_1, u_2, u_3, u_4) = i_k(s, u_1, u_2).$$

Using these properties we can show via induction over s

$$\forall s, u_1, u_2, u_3, u_4 \geq 0; \quad i_k^*(s, u_1, u_2, u_3, u_4) = i_k(s, u_1, u_2, u_3, u_4)$$

and the proposition is proved.

4.4.2 The main theorem

Now we compute $\mathbf{Q}_k(z)$, given in eq. (4.35).

Theorem 4.25. *Suppose $k > 2$, then*

$$\mathbf{Q}_k(z) = \frac{1 - z^2 + z^4}{q(z)} \mathbf{F}_k\left(\vartheta(z)\right), \tag{4.57}$$

where

$$q(z) = 1 - z - z^2 + z^3 + 2z^4 + z^6 - z^8 + z^{10} - z^{12},$$
$$\vartheta(z) = \frac{z^4(1 - z^2 - z^4 + 2z^6 - z^8)}{q(z)^2}. \tag{4.58}$$

Furthermore, for $3 \le k \le 9$, $\mathbf{Q}_k(n)$ satisfies

$$\mathbf{Q}_k(n) \sim c_k\, n^{-((k-1)^2 + (k-1)/2)}\, \gamma_k^{-n} \quad \text{for some } c_k > 0, \tag{4.59}$$

where γ_k is the minimal, positive real solution of $\vartheta(z) = \rho_k^2$; see Table 4.9.

k	3	4	5	6	7	8	9
$\theta(n)$	n^{-5}	$n^{-\frac{21}{2}}$	n^{-18}	$n^{-\frac{55}{2}}$	n^{-39}	$n^{-\frac{105}{2}}$	n^{-68}
γ_k^{-1}	2.5410	3.0132	3.3974	3.7319	4.0327	4.3087	4.5654

Table 4.9. Exponential growth rates γ_k^{-1} and subexponential factors $\theta(n)$, for modular, k-noncrossing structures.

Proof. Let $\mathcal{Q}_k$ denote the set of modular, k-noncrossing structures and let $\mathcal{I}_k$ and $\mathcal{I}_k(s, \mathbf{u})$ denote the set of all k-noncrossing V_k-shapes and those having

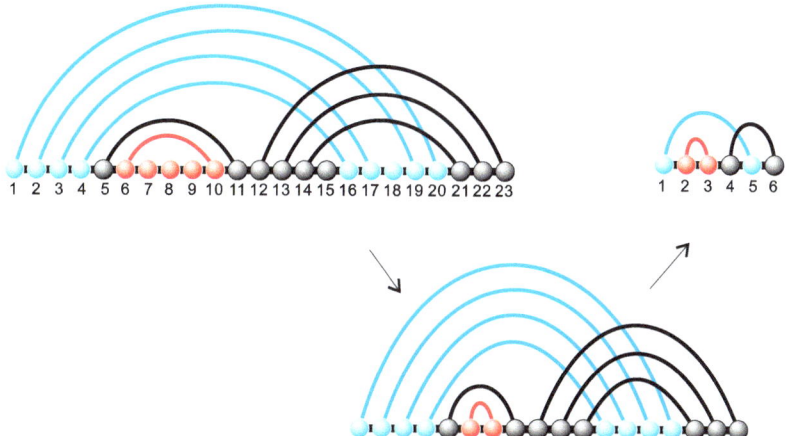

Fig. 4.24. A modular, 3-noncrossing structure (*top left*) is mapped into its V_k-shape (*top right*). A stem (*blue*) is mapped into a single shape-arc (*blue*). A hairpin loop (*red*) is mapped into a 1-arc of the shape (*red*).

s-arcs and u_i elements belonging to class $\mathbf{C}_i$, where $1 \leq i \leq 4$; see Fig. 4.24. Then we have the surjective map,

$$\varphi: \ \mathcal{Q}_k \to \mathcal{I}_k,$$

inducing the partition $\mathcal{Q}_k = \dot{\cup}_\gamma \varphi^{-1}(\gamma)$. This partition allows us to organize $\mathbf{Q}_k(z)$ with respect to colored V_k-shapes, γ, as follows:

$$\mathbf{Q}_k(z) = \sum_{s,\mathbf{u}} \ \sum_{\gamma \in \mathcal{I}_k(s,\mathbf{u})} \mathbf{Q}_\gamma(z). \tag{4.60}$$

We proceed by computing the generating function $\mathbf{Q}_\gamma(z)$ following the strategy of Theorem 4.9, also using the notation therein. The key point is that the inflation procedures are specific to the $\mathbf{C}_i$-classes. In the following we will inflate all "critical" arcs, i.e., arcs that require the insertion of additional isolated vertices in order to satisfy the minimum arc length condition. In the following we refer to a stem different from a 2-stack as a †-stem. Accordingly, the combinatorial class of †-stems is given by $(\mathcal{M} - \mathcal{R}^2)$.

- **$\mathbf{C}_1$-class:** here we insert isolated vertices, see Fig. 4.25, and obtain immediately

$$\mathbf{C}_1(z) = \frac{z^3}{1-z}. \tag{4.61}$$

Fig. 4.25. $\mathbf{C}_1$-class: insertion of at least three vertices (*red*).

- **$\mathbf{C}_2$-class:** any such element is a pair $((i, i+2), (i+1, i+3))$ and we shall distinguish the following scenarios:
  - Both arcs are inflated to stacks of length 2; see Fig. 4.26. Ruling out the cases where no isolated vertex is inserted and the two scenarios, where there is no insertion into the interval $[i+1, i+2]$ and only in either $[i, i+1]$ or $[i+2, i+3]$, see Fig. 4.26, we arrive at

$$\mathcal{C}_2^{(a)} = \mathcal{R}^4 \times [(\mathrm{SEQ}(\mathcal{Z}))^3 - \mathcal{E} - 2(\mathcal{Z} \times \mathrm{SEQ}(\mathcal{Z}))].$$

This combinatorial class has the generating function

$$\mathbf{C}_2^{(a)}(z) = z^8 \left(\left(\frac{1}{1-z} \right)^3 - 1 - \frac{2z}{1-z} \right).$$

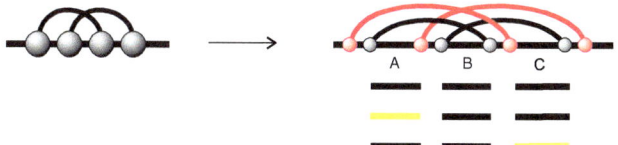

Fig. 4.26. C_2-class: inflation of both arcs to 2-stacks. Inflated arcs are colored *red* while the original arcs of the shape are colored *black*. We set $A = [i+1, i+2]$, $B = [i+2, i+3]$, and $C = [i+2, i+3]$ and illustrate the "bad" insertion scenarios as follows: an insertion of some isolated vertices is represented by an *yellow segment* and no insertion by a *black segment*. See text for details.

Fig. 4.27. C_2-class: inflation of only one arc to a 2-stack. Arc coloring and labels as in Fig. 4.26.

- One arc, $(i+1, i+3)$ or $(i, i+2)$, is inflated to a 2-stack, while its counterpart is inflated to an arbitrary †-stem; see Fig. 4.27. Ruling out the cases where no vertex is inserted in $[i+1, i+2]$ and $[i+2, i+3]$ or $[i, i+1]$ and $[i+2, i+3]$, we obtain

$$\mathcal{C}_2^{(b)} = 2\mathcal{R}^2 \times (\mathcal{M} - \mathcal{R}^2) \times ((\text{Seq}(\mathcal{Z}))^2 - \mathcal{E}) \times \text{Seq}(\mathcal{Z}),$$

having the generating function

$$\mathbf{C}_2^{(b)}(z) = 2z^4 \left(\frac{\frac{z^4}{1-z^2}}{1 - \frac{z^4}{1-z^2}\left(\frac{2z}{1-z} + \left(\frac{z}{1-z}\right)^2\right)} - z^4 \right)$$
$$\times \left(\left(\frac{1}{1-z}\right)^2 - 1 \right) \cdot \left(\frac{1}{1-z}\right).$$

- Both $\mathcal{C}_2$-arcs are inflated to an arbitrary †-stem, respectively; see Fig. 4.28. In this case the insertion of isolated vertices is arbitrary, whence

$$\mathcal{C}_2^{(c)} = (\mathcal{M} - \mathcal{R}^2)^2 \times (\text{Seq}(\mathcal{Z}))^3,$$

with generating function

$$\mathbf{C}_2^{(c)}(z) = \left(\frac{\frac{z^4}{1-z^2}}{1 - \frac{z^4}{1-z^2}\left(\frac{2z}{1-z} + \left(\frac{z}{1-z}\right)^2\right)} - z^4 \right)^2 \left(\frac{1}{1-z}\right)^3.$$

Fig. 4.28. $\mathbf{C}_2$**-class:** inflation of both arcs to an arbitrary †-stem. Arc coloring and labels as in Fig. 4.26.

As the above scenarios are mutually exclusive, the generating function of the $\mathcal{C}_2$-class is given by

$$\mathbf{C}_2(z) = \mathbf{C}_2^{(\mathrm{a})} + 2\mathbf{C}_2^{(\mathrm{b})} + \mathbf{C}_2^{(\mathrm{c})}. \tag{4.62}$$

Furthermore note that *both* arcs of the $\mathcal{C}_2$-class are inflated in cases (a), (b), and (c).

- $\mathbf{C}_3$**-class:** this class consists of arc pairs (α, β) where α is the unique 2-arc crossing β and β has length at least 3. Without loss of generality we can restrict our analysis to the case $((i, i+2), (i+1, j))$, $(j > i+3)$:
  - The arc $(i+1, j)$ is inflated to a 2-stack. Then we have to insert at least one isolated vertex in either $[i, i+1]$ or $[i+1, i+2]$; see Fig. 4.29. Therefore, we have

    $$\mathcal{C}_3^{(\mathrm{a})} = \mathcal{R}^2 \times (\mathrm{SEQ}(\mathcal{Z})^2 - \mathcal{E}),$$

    with generating function

    $$\mathbf{C}_3^{(\mathrm{a})}(z) = z^4 \left(\left(\frac{1}{1-z} \right)^2 - 1 \right).$$

    Note that the arc $(i, i+2)$ is not considered here, it can be inflated without any restrictions.
  - The arc $(i+1, j)$ is inflated to an arbitrary †-stem; see Fig. 4.29. Then

    $$\mathcal{C}_3^{(\mathrm{b})} = (\mathcal{M} - \mathcal{R}^2) \times \mathrm{SEQ}(\mathcal{Z})^2,$$

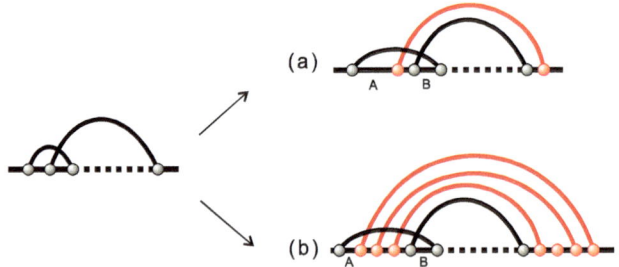

Fig. 4.29. $\mathbf{C}_3$**-class:** only one arc is inflated here and its inflation distinguishes two subcases. Arc coloring as in Fig. 4.26.

with generating function

$$\mathbf{C}_3^{(b)}(z) = \left(\frac{\frac{z^4}{1-z^2}}{1 - \frac{z^4}{1-z^2}\left(\frac{2z}{1-z} + \left(\frac{z}{1-z}\right)^2\right)} - z^4 \right) \cdot \left(\frac{1}{1-z}\right)^2.$$

Consequently, this inflation process leads to a generating function

$$\mathbf{C}_3(z) = \mathbf{C}_3^{(a)}(z) + \mathbf{C}_3^{(b)}(z). \tag{4.63}$$

Note that during inflation (a) and (b) only *one* of the two arcs of a $\mathbf{C}_3$-class element is being inflated.

- $\mathbf{C}_4$-**class:** this class consists of arc triples $(\alpha_1, \beta, \alpha_2)$, where α_1 and α_2 are 2-arcs, respectively, that cross β.
  - β is inflated to a 2-stack; see Fig. 4.30. Using similar arguments as in the case of $\mathbf{C}_3$-class, we arrive at

$$\mathcal{C}_4^{(a)} = \mathcal{R}^2 \times (\mathrm{SEQ}(\mathcal{Z})^2 - \mathcal{E}) \times (\mathrm{SEQ}(\mathcal{Z})^2 - \mathcal{E}),$$

with generating function

$$\mathbf{C}_4^{(a)}(z) = z^4 \left(\left(\frac{1}{1-z}\right)^2 - 1 \right)^2.$$

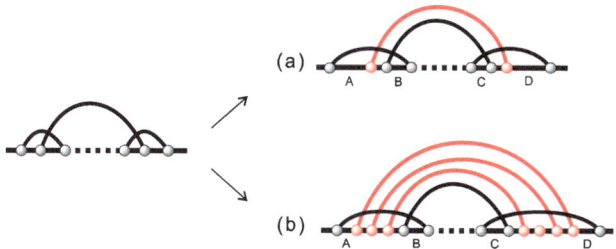

Fig. 4.30. $\mathbf{C}_4$-class: as for the inflation of $\mathbf{C}_3$ only the non-2-arc is inflated, distinguishing two subcases. Arc coloring as in Fig. 4.26.

- The arc β is inflated to an arbitrary †-stem; see Fig. 4.30,

$$\mathcal{C}_4^{(b)} = (\mathcal{M} - \mathcal{R}^2) \times \mathrm{SEQ}(\mathcal{Z})^4,$$

with generating function

$$\mathbf{C}_4^{(b)}(z) = \left(\frac{\frac{z^4}{1-z^2}}{1 - \frac{z^4}{1-z^2}\left(\frac{2z}{1-z} + \left(\frac{z}{1-z}\right)^2\right)} - z^4 \right) \cdot \left(\frac{1}{1-z}\right)^4.$$

Accordingly we arrive at

$$\mathbf{C}_4(z) = \mathbf{C}_4^{(\mathrm{a})}(z) + \mathbf{C}_4^{(\mathrm{b})}(z). \qquad (4.64)$$

The inflation of any γ-arc not considered in the previous steps follows the logic of Theorem 4.9. We observe that

$$s - 2u_2 - u_3 - u_4$$

arcs of the shape γ have not been considered. Furthermore,

$$2s + 1 - u_1 - 3u_2 - 2u_3 - 4u_4$$

intervals were not considered for the insertion of isolated vertices. The inflation of these along the lines of Theorem 4.9 gives rise to the class

$$\mathcal{S} = \mathcal{M}^{s-2u_2-u_3-u_4} \times (\mathrm{SEQ}(\mathcal{Z}))^{2s+1-u_1-3u_2-2u_3-4u_4},$$

having the generating function

$$\mathbf{S}(z) = \left(\frac{\frac{z^4}{1-z^2}}{1 - \frac{z^4}{1-z^2}\left(\frac{2z}{1-z} + \left(\frac{z}{1-z}\right)^2\right)} \right)^{s-2u_2-u_3-u_4} \times \left(\frac{1}{1-z} \right)^{2s+1-u_1-3u_2-2u_3-4u_4}.$$

Since all these inflations can freely be combined, we have

$$\mathcal{Q}_\gamma = \mathcal{C}_1^{u_1} \times \mathcal{C}_2^{u_2} \times \mathcal{C}_3^{u_3} \times \mathcal{C}_4^{u_4} \times \mathcal{S},$$

whence

$$\mathbf{Q}_\gamma(z) = \mathbf{C}_1(z)^{u_1} \cdot \mathbf{C}_2(z)^{u_2} \cdot \mathbf{C}_3(z)^{u_3} \cdot \mathbf{C}_4(z)^{u_4} \cdot \mathbf{S}(z)$$
$$= \frac{1}{1-z} \varsigma_0(z)^s \varsigma_1(z)^{u_1} \varsigma_2(z)^{u_2} \varsigma_3(z)^{u_3} \varsigma_4(z)^{u_4},$$

where

$$\varsigma_0(z) = \frac{z^4}{1 - 2z + 2z^3 - z^4 - 2z^5 + z^6},$$
$$\varsigma_1(z) = z^3,$$
$$\varsigma_2(z) = \frac{z(1 - 4z^3 + 2z^4 + 8z^5 - 6z^6 - 7z^7 + 8z^8 + 2z^9 - 4z^{10} + z^{11})}{1 - z},$$
$$\varsigma_3(z) = z(2 - 2z^2 + z^3 + 2z^4 - z^5),$$
$$\varsigma_4(z) = z^2(5 - 4z - 3z^2 + 6z^3 + 2z^4 - 4z^5 + z^6).$$

Observing that $\mathbf{Q}_{\gamma_1}(z) = \mathbf{Q}_{\gamma_2}(z)$ for any $\gamma_1, \gamma_2 \in \mathcal{I}_k(s, \mathbf{u})$, we have according to eq. (4.60)

$$\mathbf{Q}_k(z) = \sum_{s,\mathbf{u}\geq 0} i_k(s, \mathbf{u})\, \mathbf{Q}_\gamma(z),$$

where $\mathbf{u} \geq 0$ denotes $u_i \geq 0$ for $1 \leq i \leq 4$. Proposition 4.24 guarantees

$$\sum_{s,\mathbf{u}\geq 0} i_k(s, \mathbf{u})\, x^n y^{u_1} z^{u_2} w^{u_3} t^{u_4}$$

$$= \frac{1 + x}{1 - (y - 2)x + (2w - z - 1)x^2 + (2w - z - 1)x^3} \times$$

$$\mathbf{F}_k\left(\frac{x(1 + (2w - 1)x + (t - 1)x^2)}{(1 - (y - 2)x + (2w - z - 1)x^2 + (2w - z - 1)x^3)^2}\right).$$

Setting $x = \varsigma_0(z)$, $y = \varsigma_1(z)$, $r = \varsigma_2(z)$, $w = \varsigma_3(z)$, $t = \varsigma_4(z)$, we arrive at

$$\mathbf{Q}_k(z) = \frac{1 - z^2 + z^4}{1 - z - z^2 + z^3 + 2z^4 + z^6 - z^8 + z^{10} - z^{12}} \times$$

$$\mathbf{F}_k\left(\frac{z^4(1 - z^2 - z^4 + 2z^6 - z^8)}{(1 - z - z^2 + z^3 + 2z^4 + z^6 - z^8 + z^{10} - z^{12})^2}\right).$$

By Corollary 2.14, $\mathbf{Q}_k(z)$ is D-finite. Pringsheim's theorem [134] guarantees that $\mathbf{Q}_k(z)$ has a dominant real positive singularity γ_k. We verify that for $3 \leq k \leq 9$, γ_k is the unique solution of minimum modulus of the equation $\vartheta(z) = \rho_k^2$. According to Table 4.10, this solution is strictly smaller than

k	$\vartheta(z) = \rho_k^2$	$|z|$	k	$\vartheta(z) = \rho_k^2$	$|z|$
3	0.3935	0.3935	7	0.2480	0.2480
	$0.1979 \pm 0.4983i$	0.5361		$0.0468 \pm 0.2928i$	0.2965
	$0.1979 \pm 0.4986i$	0.7309		-0.3274	0.3274
	Other solutions	≥ 0.8762		Other solutions	≥ 0.8684
4	0.3319	0.3319	8	0.2680	0.2680
	$0.1116 \pm 0.4181i$	0.4327		$0.0393 \pm 0.2705i$	0.3691
	-0.4984	0.4984		$--0.3003$	-0.3003
	Other solutions	≥ 0.8734		Other solutions	≥ 0.8684
5	0.2943	0.2943	9	0.2190	0.2190
	$0.0763 \pm 0.3611i$	0.3691		$0.0339 \pm 0.2526i$	0.2548
	-0.4144	0.4144		-0.2789	0.2789
	Other solutions	≥ 0.8693		Other solutions	≥ 0.8685
6	0.2680	0.2680			
	$0.0580 \pm 0.3218i$	0.3269			
	-0.3633	0.3633			
	Other solutions	≥ 0.8685			

Table 4.10. The solutions of $\vartheta(z) = \rho_k^2$ for $3 \leq k \leq 9$ and their respective modulus.

k	3	4	5	6	7	8	9
$\vartheta'(\gamma_k)$	0.739849	0.402041	0.253552	0.176398	0.131015	0.101895	0.081989

Table 4.11. $\vartheta'(\gamma_k)$ for $3 \leq k \leq 9$.

the singularity of $\vartheta(z)$ and Table 4.11 shows that $\vartheta'(z) \neq 0$. Therefore, the composite function

$$\mathbf{F}_k \left(\frac{z^4(1 - z^2 - z^4 + 2z^6 - z^8)}{q(z)^2} \right)$$

is governed by the supercritical paradigm of Theorem 2.21 for $k = 3, \ldots, 9$ and follows the logic of Proposition 4.14; see the SM. According to Theorem 2.21 we therefore have

$$Q_k(n) \sim c_k \, n^{-((k-1)^2 + (k-1)/2)} \, (\gamma_k^{-1})^n \quad \text{for some } c_k > 0$$

and the proof of Theorem 4.25 is complete.

We remark that Theorem 4.25 does not hold for $k = 2$, i.e., we cannot compute the generating function $\mathbf{Q}_2(z)$ via eq. (4.57). The reason is that Lemma 4.23 only holds for $k > 2$ and indeed we find

$$\mathbf{Q}_2(z) \neq \frac{1 - z^2 + z^4}{q(z)} \mathbf{F}_2 \left(\frac{z^4(1 - z^2 - z^4 + 2z^6 - z^8)}{q(z)^2} \right). \tag{4.65}$$

However, the computation of the generating function $\mathbf{Q}_2(z)$ in Proposition 4.20 is based on Theorem 4.6, which does hold for $k = 2$.

4.5 Exercises

4.1. (lv_k^1-shapes) (Reidys and Wang [107]) An lv_k^1-shape is a k-noncrossing structure in which each stack and each segment of isolated vertices have length exactly 1.

That is, given a k-noncrossing, σ-canonical RNA structure its lv_k^1-shape is derived as follows: first, we apply the core map, second, we replace a segment of isolated vertices by a single isolated vertex, and third relabel the vertices of the resulting diagram; see Fig. 4.31. lv_k^1-shapes do not preserve stack-lengths and project intervals of isolated vertices into singletons. Let $\mathcal{J}_k$ and $\mathcal{I}_k$ denote the set of lv_k^1-shapes and lv_k^5-shapes, respectively. There is a map between lv_k^1-shapes and lv_k^5-shapes

$$\phi: \quad \mathcal{J}_k \rightarrow \mathcal{I}_k,$$

obtained by removing all isolated vertices from lv_k^1-shapes. By construction, ϕ is surjective (for any lv_k^5-shape, we can, inserting one isolated vertex in any

1-arc, obtain an lv_k^1-shape). Let $\mathcal{J}_k(n, h)$ $(j_k(n, h))$ denote the set (number) of lv_k^1-shapes of length n having h-arcs and let $j_k(n)$ be the number of all lv_k^1-shapes of length n and

$$\mathbf{J}_k(z, u) = \sum_{h \geq 0} \sum_{n=2h}^{4h+1} j_k(n, h) z^n u^h \quad \text{and} \quad \mathbf{J}_k(z) = \sum_{n \geq 0} j_k(n) z^n.$$

Prove the following theorem.

Theorem 4.26. *For $k, n, h \in \mathbb{N}$, $k \geq 2$, the following assertions hold:*
(a) *The generating functions $\mathbf{J}_k(z, u)$ and $\mathbf{J}_k(z)$ are given by*

$$\mathbf{J}_k(z, u) = \frac{(1+z)(1+uz^2)}{uz^3 + 2uz^2 + 1} \mathbf{F}_k \left(\frac{(1+z)^2(1+uz^2)uz^2}{(uz^3 + 2uz^2 + 1)^2} \right),$$

$$\mathbf{J}_k(z) = \frac{(1+z)(1+z^2)}{z^3 + 2z^2 + 1} \mathbf{F}_k \left(\frac{(1+z)^2(1+z^2)z^2}{(z^3 + 2z^2 + 1)^2} \right).$$

(b) *For $2 \leq k \leq 7$, the number of lv_k^1-shapes of length n satisfies*

$$j_k(n) \sim c_k' n^{-((k-1)^2 + (k-1)/2)} \left(\mu_k'^{-1} \right)^n,$$

where $c_k' > 0$ and μ_k' is the unique minimum positive real solution of

$$\frac{(1+z)^2(1+z^2)z^2}{(z^3 + 2z^2 + 1)^2} = \rho_k^2.$$

4.2. (lv_k^1-shapes of k-noncrossing structures of length n) Consider the lv_k^1-shapes introduced in Problem 4.1. Here we will compute the number of lv_k^1-shapes induced by k-noncrossing, σ-canonical RNA structures of fixed length n, $\mathsf{lv}_{k,\sigma}^1(n)$. Let

$$\mathbf{Lv}_{k,\sigma}^1(x) = \sum_{n \geq 0} \mathsf{lv}_{k,\sigma}^1(n) x^n.$$

Prove the following proposition.

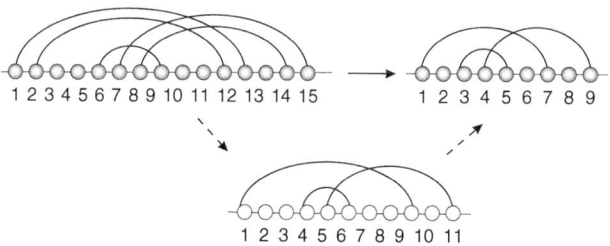

Fig. 4.31. lv_k^1-shapes via the core map and subsequent identification of unpaired nucleotides: A 3-noncrossing, 1-canonical RNA structure (*top left*) is mapped into its lv_3^1-shape (*top right*).

Proposition 4.27. *Let $k, \sigma \in \mathbb{N}$, where $k \geq 2$. Then the following assertions hold:*

(a) *The generating function $\mathbf{Lv}_{k,\sigma}^1(x)$ is given by*

$$\mathbf{Lv}_{k,\sigma}^1(x) = \frac{(1+x)(1+x^{2\sigma})}{(1-x)(x^{2\sigma+1}+2x^{2\sigma}+1)} \mathbf{F}_k\left(\frac{(1+x)^2 x^{2\sigma}(1+x^{2\sigma})}{(x^{2\sigma+1}+2x^{2\sigma}+1)^2}\right).$$

(b) *For $2 \leq k \leq 7$ and $1 \leq \sigma \leq 10$, we have*

$$\mathsf{lv}_{k,\sigma}^1(n) \sim c'_{k,\sigma} n^{-((k-1)^2+(k-1)/2)}\left(\chi_{k,\sigma}^{-1}\right)^n,$$

where $c'_{k,\sigma} > 0$ and $\chi_{k,\sigma}$ is the unique minimum positive real solution of (Table 4.12)

$$\frac{(1+x)^2 x^{2\sigma}(1+x^{2\sigma})}{(x^{2\sigma+1}+2x^{2\sigma}+1)^2} = \rho_k^2. \tag{4.66}$$

σ/k	2	3	4	5	6	7	8
1	2.09188	4.51263	6.65586	8.73227	10.7804	12.8137	14.8381
2	1.56947	2.31767	2.81092	3.21184	3.55939	3.87079	4.15552
3	1.38475	1.80408	2.05600	2.24968	2.41081	2.55050	2.67477

Table 4.12. The exponential growth rates $\chi_{k,\sigma}^{-1}$ of lv_k^1-shapes induced by k-noncrossing, σ-canonical RNA structures of length n.

4.3. Prove

$$\mathbf{T}_{k,1}(z) = \frac{1}{z^2 - z + 1} \mathbf{F}_k\left(\left(\frac{z}{z^2 - z + 1}\right)^2\right).$$

directly, using Theorem 4.13 [77].

4.4. Prove [116], Waterman's formula for the number of RNA secondary structures with exactly ℓ isolated vertices, $\mathsf{T}_{2,1}(n, \ell)$.

Proposition 4.28. (Schmitt and Waterman [116])

$$\mathsf{T}_{2,1}(n, \ell) = \frac{2}{n - \ell}\binom{\frac{n+\ell}{2}}{\frac{n-\ell}{2}+1}\binom{\frac{n+\ell}{2}-1}{\frac{n-\ell}{2}-1}. \tag{4.67}$$

Furthermore, show using MAPLE: $\mathsf{T}_{2,1}(n, \ell)$ satisfies the recursion

$$(n - \ell)(n - \ell + 2) \cdot \mathsf{T}_{2,1}(n, \ell) - (n + \ell)(n + \ell - 2) \cdot \mathsf{T}_{2,1}(n - 2, \ell) = 0. \tag{4.68}$$

4.5. We compute here the generating function $\mathbf{T}_{k,1}^{[4]}(z)$, the number of k-noncrossing RNA structures with arc length ≥ 4. These structures are more realistic since they respect the fact that bio-molecular configurations satisfy a minimum arc length 4. In contrast to the situation for RNA secondary structures increasing the minimum arc length imposes (technical) difficulties. However, when passing to the level of generating functions, the minimum arc length 4 leads to "just" a more complicated inner rational function. We set

$$u(z) = \sqrt{1 + 4z - 4z^2 - 6z^3 + 4z^4 + z^6}, \tag{4.69}$$

$$h_j(z) = -\frac{-2z^2 + z^3 - 1 + (-1)^j\, u(z)}{2(1 - 2z - z^2 + z^4)}. \tag{4.70}$$

Note that $h_j(z)$ is an algebraic function over the function field $\mathbb{C}(z)$, i.e., there exists a polynomial with coefficients being polynomials in z for which $h_j(z)$ is a root. This fact will be important when computing the subexponential factors of the asymptotic formula for $\mathbf{T}_{k,1}^{[4]}(n)$. We can now compute the generating function $\mathbf{T}_{k,1}^{[4]}(z)$ in analogy to Theorem 4.13.

Prove the following statement [59].

Proposition 4.29. *let $k > 3$ be a positive integer, $H_1(z) = \frac{h_2(z)-1}{h_2(z)-h_1(z)}$ and $H_2(z) = \frac{h_1(z)-1}{h_1(z)-h_2(z)}$, where $h_1(z)$ and $h_2(z)$ be given by eq. (4.70). Then we have*

$$\mathsf{T}_{k,1}^{[4]}(n) = \sum_{b \leq \lfloor \frac{n}{2} \rfloor} (-1)^b\, \lambda(n,b)\, \mathsf{M}_k(n - 2b) \tag{4.71}$$

and $\lambda(n,b)$ satisfies the recurrence formula

$$\lambda(n,b) =$$
$$\lambda(n-1,b) + \lambda(n-4,b-2) + \lambda(n-5,b-2) + \lambda(n-6,b-3)$$
$$+ \sum_{i=1}^{b} [\lambda(n-2i,b-i) + 2\lambda(n-2i-1,b-i) + \lambda(n-2i-2,b-i)] \tag{4.72}$$
$$- \lambda(n-3,b-1),$$

where $\lambda(n,0) = 1$, $\lambda(n,1) = 3n - 6$, and $n \geq 2b$. Furthermore we have the functional equation

$$\mathbf{T}_{k,1}^{[4]}(z) = \frac{H_1(-z^2)}{1 - zh_1(-z^2)}\mathbf{F}_k\left(\left(\frac{z\,h_1(-z^2)}{1 - zh_1(-z^2)}\right)^2\right) +$$
$$\frac{H_2(-z^2)}{1 - zh_2(-z^2)}\mathbf{F}_k\left(\left(\frac{z\,h_2(-z^2)}{1 - zh_2(-z^2)}\right)^2\right).$$

4.6. Let

$$\vartheta_1(z) = \left(\frac{z\, h_1(-z^2)}{1 - zh_1(-z^2)}\right)^2, \tag{4.73}$$

$$\vartheta_2(z) = \left(\frac{z\, h_2(-z^2)}{1 - zh_2(-z^2)}\right)^2. \tag{4.74}$$

Note that $\vartheta_1(z)$ and $\vartheta_2(z)$ are algebraic functions over the function field $\mathbb{C}(z)$. Prove

Proposition 4.30. *Let $3 < k \leq 9$ be a positive integer. Then the number of k-noncrossing RNA structures with arc length ≥ 4 is for $3 < k \leq 9$ asymptotically given by*

$$\mathsf{T}_{k,1}^{[4]}(n) \sim c_{k,1}^{[4]}\, n^{-((k-1)^2+(k-1)/2)}\left(\gamma_{\vartheta_1,k}^{-1}\right)^n,$$

where $\gamma_{\vartheta_1,k}$ is the unique minimum positive real solution of the equation $\vartheta_1(z) = \rho_k^2$ and $c_{k,1}^{[4]}$ is a positive constant (Table 4.13).

k	4	5	6	7	8	9
$\theta(n)$	$n^{-\frac{21}{2}}$	n^{-18}	$n^{-\frac{55}{2}}$	n^{-39}	$n^{-\frac{105}{2}}$	n^{-68}
$\gamma_{\vartheta_1,k}^{-1}$	6.5290	8.6483	10.7176	12.7635	14.7963	16.8210

Table 4.13. The exponential growth rates $\gamma_{\vartheta_1,k}^{-1}$ and subexponential factors $\theta(n)$, for k-noncrossing RNA structures with minimum arc length ≥ 4.

4.7. Prove Proposition 4.17 using the core lemma, Lemma 4.3.

4.8. Prove Lemma 4.23.
Hint: By construction, eqs. (4.48) and (4.49) hold. In order to prove eq. (4.50) choose a shape $\delta \in \mathcal{I}_k(s+1, u_1, u_2, u_3 + 1, u_4)$ and label exactly one of the $(u_3 + 1)$ $\mathbf{C}_3$-elements containing a unique 2-arc, α. We denote the set of these labeled shapes, λ, by $\mathcal{L}$. Clearly

$$|\mathcal{L}| = (u_3 + 1)i_k(s+1, u_1, u_2, u_3 + 1, u_4).$$

We observe that the removal of α results in either a shape ($\mathcal{L}_1$) or a matching ($\mathcal{L}_2$), i.e., we have

$$\mathcal{L} = \mathcal{L}_1 \,\dot\cup\, \mathcal{L}_2.$$

Prove:

Claim 1.

$$
\begin{aligned}
|\mathcal{L}_1| = {}& 2(u_3 + 1)\, i_k(s, u_1, u_2, u_3 + 1, u_4) + \\
& 4(u_4 + 1)\, i_k(s, u_1, u_2, u_3 - 1, u_4 + 1) + \\
& (2(s - u_1 - 2u_2 - 2u_3 - 3u_4))\, i_k(s, u_1, u_2, u_3, u_4).
\end{aligned}
$$

Prove:

Claim 2. Let $(\beta_1, \ldots, \beta_\ell)$ denote a μ_2^α-stack $((\beta_1, \ldots, \beta_\ell) \prec \mu_2^\alpha)$. Then we have

$$
\mathcal{L}_2 = \mathcal{L}_{2,1} \dot\cup \mathcal{L}_{2,2} \dot\cup \mathcal{L}_{2,3},
$$

where

$$
\begin{aligned}
\mathcal{L}_{2,1} =&\{\lambda \in \mathcal{L}_2 \mid \alpha, \beta_i \in \lambda, i = 1, 2;\ (\beta_1, \beta_2) \prec \mu_2^\alpha;\ \alpha \text{ crosses } \beta_2\}, \\
\mathcal{L}_{2,2} =&\{\lambda \in \mathcal{L}_2 \mid \alpha, \beta_i \in \lambda, i = 1, 2;\ (\beta_1, \beta_2) \prec \mu_2^\alpha;\ \alpha \text{ crosses } \beta_1\}, \\
\mathcal{L}_{2,3} =&\{\lambda \in \mathcal{L}_2 \mid \alpha, \beta_i \in \lambda, i = 1, 2, 3;\ (\beta_1, \beta_2, \beta_3) \prec \mu_2^\alpha;\ \alpha \text{ crosses } \beta_2\}.
\end{aligned}
$$

Prove:

Claim 2.1

$$
\begin{aligned}
|\mathcal{L}_{2,1}| = {}& 4(u_2 + 1)\, i_k(s - 1, u_1, u_2 + 1, u_3, u_4) \\
& + 4(u_3 + 1)\, i_k(s - 1, u_1, u_2, u_3 + 1, u_4) \\
& + [4(u_4 + 1)\, i_k(s - 1, u_1, u_2, u_3 - 1, u_4 + 1) \\
& + 2(u_4 + 1)\, i_k(s - 1, u_1, u_2, u_3, u_4 + 1)] \\
& + 2((s - 1) - u_1 - 2u_2 - 2u_3 - 3u_4))\, i_k(s - 1, u_1, u_2, u_3, u_4).
\end{aligned}
$$

Prove:

Claim 2.2.

$$
\begin{aligned}
|\mathcal{L}_{2,2}| = {}& 2u_1\, i_k(s - 1, u_1, u_2, u_3, u_4) \\
& + 4(u_2 + 1)\, i_k(s - 1, u_1, u_2 + 1, u_3 - 1, u_4) \\
& + [2u_3\, i_k(s - 1, u_1, u_2, u_3, u_4) \\
& + 2(u_3 + 1)\, i_k(s - 1, u_1, u_2, u_3 + 1, u_4)] \\
& + [4u_4\, i_k(s - 1, u_1, u_2, u_3, u_4) \\
& + 2(u_4 + 1)\, i_k(s - 1, u_1, u_2, u_3, u_4 + 1)] \\
& + 2((s - 1) - u_1 - 2u_2 - 2u_3 - 3u_4))i_k(s - 1, u_1, u_2, u_3, u_4).
\end{aligned}
$$

Prove:

Claim 2.3

$$
\begin{aligned}
|\mathcal{L}_{2,3}| = {}& 2u_1\, i_k(s - 2, u_1, u_2, u_3, u_4) \\
& + 4(u_2 + 1)\, i_k(s - 2, u_1, u_2 + 1, u_3 - 1, u_4)
\end{aligned}
$$

$$+ \; [2u_3 \, i_k(s - 2, u_1, u_2, u_3, u_4)$$
$$+ \; 2(u_3 + 1)i_k(s - 2, u_1, u_2, u_3 + 1, u_4)]$$
$$+ \; [4u_4 i_k(s - 2, u_1, u_2, u_3, u_4)$$
$$+ \; 2(u_4 + 1)i_k(s - 2, u_1, u_2, u_3, u_4 + 1)]$$
$$+ \; 2((s - 2) - u_1 - 2u_2 - 2u_3 - 3u_4))i_k(s - 2, u_1, u_2, u_3, u_4).$$

Equation (4.50) now follows from Claims 1, 2.1, 2.2, and 2.3.

Next we prove eq. (4.51). We choose some $\eta \in \mathcal{I}_k(s + 1, u_1, u_2, u_3, u_4 + 1)$ and label one $\mathbf{C}_4$-element denoting one of its two 2-arcs by α. We denote the set of these labeled shapes, λ, by $\mathcal{L}_*$. Clearly,

$$|\mathcal{L}_*| = 2(u_4 + 1) \, i_k(s + 1, u_1, u_2, u_3, u_4 + 1).$$

Let γ be the arc crossing α. The removal of α can lead to either an additional $\mathbf{C}_2$- or an additional $\mathbf{C}_3$-element in a shape, whence

$$\mathcal{L}_* = \mathcal{L}_*^{\mathbf{C}_2} \, \dot{\cup} \, \mathcal{L}_*^{\mathbf{C}_3},$$

where $\mathcal{L}_*^{\mathbf{C}_i}$ denotes the set of labeled shapes, $\lambda \in \mathcal{L}_*$, that induce shapes having a labeled $\mathbf{C}_i$-element containing γ.

Prove:

$$|\mathcal{L}_*^{\mathbf{C}_2}| = 2(u_2 + 1) \, i_k(s, u_1, u_2 + 1, u_3, u_4),$$
$$|\mathcal{L}_*^{\mathbf{C}_3}| = (u_3 + 1) \, i_k(s, u_1, u_2, u_3 + 1, u_4).$$

5

Probabilistic Analysis

5.1 Uniform generation

In this section we prove that k-noncrossing RNA structures can be generated efficiently with uniform probability. The results presented here are derived from [26] and are based on Section 2.1. For RNA secondary structures ($k = 2$), the uniform generation is well known [67] and can be derived in linear time, using the framework of Flajolet et al. [33]. The situation is, however, for pseudoknotted structures ($k > 2$) more complicated. Due to the cross-serial interactions, the numbers of pseudoknot structures do not satisfy a recursion of the type of eq. (1.1), rendering the ab initio folding into minimum free energy configurations [87] as well as the derivation of detailed statistical properties, a nontrivial task. Indeed, in order to derive statistical properties, the entire space of structures has to be exhaustively generated, which is only possible for small sequence lengths.

In the following we will show that after polynomial preprocessing time, k-noncrossing RNA pseudoknot structures can be generated uniformly, in linear time. Our approach is based on the interpretation of k-noncrossing structures as ∗-tableaux, see Fig. 5.1, which in turn are viewed as sampling paths of a stochastic process. Biophysical realism can be added by modifying the transition rates of this process.

There exists no general framework for the uniform generation of elements of a non-inductive combinatorial class. However, in the context of graphs, the subject of uniform generation via Markov processes has been studied. Most notably here is the paper of Wilf [146] as well as the book [147].

The main idea is to translate k-noncrossing structures into lattice walks, see Theorem 2.2, and view the latter as sampling paths of a stochastic process, see Fig. 5.2. The key observation is that the generating function of these walks is D-finite or equivalently, P-recursive, see Section 2.1.5. As a result, the *numbers* of these walks can be derived in linear time which allows us to compute the transition probabilities of the process displayed in Fig. 5.2.

C. Reidys, *Combinatorial Computational Biology of RNA*,   143
DOI 10.1007/978-0-387-76731-4_5,
© Springer Science+Business Media, LLC 2011

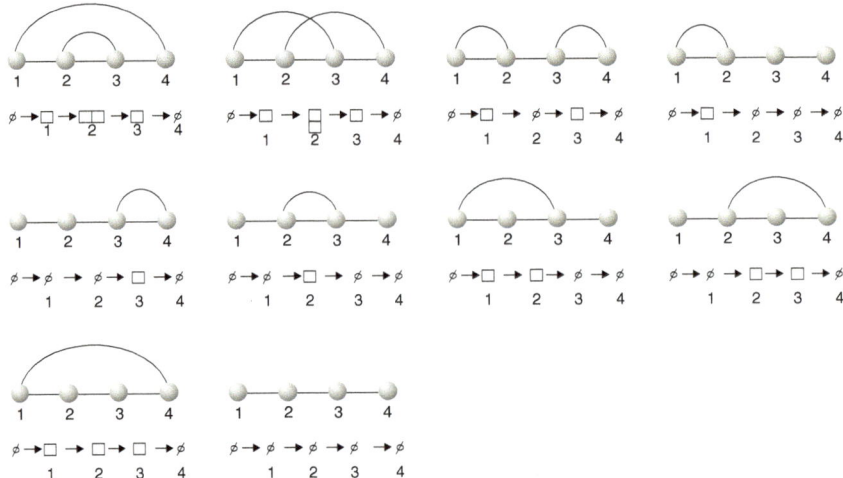

Fig. 5.1. A diagram corresponds uniquely to a sequence of "shapes," i.e., ∗-tableaux (Theorem 2.2), and the latter is viewed as a sampling path of a stochastic process. We display all 3-noncrossing diagrams over four vertices and draw their corresponding sequences of shapes underneath.

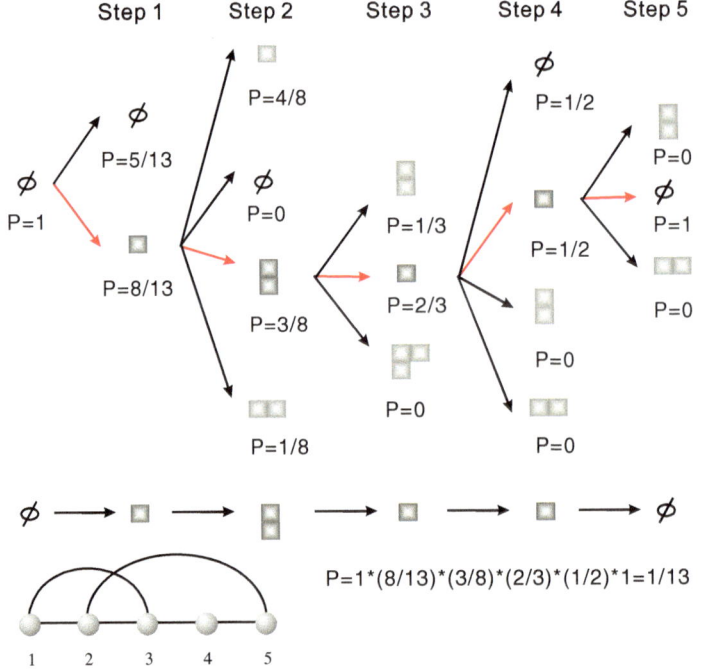

Fig. 5.2. Uniform generation: the stochastic process over shapes (*top*), a sampling path (*middle*), and its pseudoknot structure (*bottom*). The transition probabilities are computed in Theorem 5.4 as a pre-processing step.

Therefore, each $*$-tableaux of length n, containing shapes with at most $(k-1)$-rows, corresponds uniquely to a k-noncrossing partial matching on $[n]$ [25]. We denote the numbers of $*$-tableaux and those without hesitating steps (oscillating tableaux) of shape λ^i and length $(n-i)$, by $O_k^*(\lambda^i, n-i)$ and $O_k^0(\lambda^i, n-i)$, respectively.

5.1.1 Partial matchings

In Theorem 2.5 we derived the following relation between the exponential generating function of oscillating tableaux and a determinant of Bessel functions

$$\sum_{n \geq 0} \Gamma_n'^+(a, b) \frac{x^n}{n!} = \det[I_{b_j - a_i}(2x) - I_{a_i + b_j}(2x)]|_{i,j=1}^{k-1}. \tag{5.1}$$

According to Theorem 4.13, for any $k \geq 2$, the numbers of k-noncrossing RNA pseudoknot structures with minimum arc length 2 are P-recursive and given by

$$\mathsf{T}_{k,1}(n) = \sum_{b \leq \lfloor \frac{n}{2} \rfloor} (-1)^b \binom{n-b}{b} O_k^*(\varnothing^0, n - 2b), \tag{5.2}$$

where $O_k^*(\lambda^i, n-i)$ satisfies

$$O_k^*(\lambda^i, n-i) = \begin{cases} \sum_{l=0}^{\frac{n-i}{2}} \binom{n-i}{2l} O_k^0(\lambda^i, n-i-2l), \\ \qquad\qquad\text{for } (n-i) \text{ even} \\ \sum_{l=0}^{\frac{n-i}{2}} \binom{n-i}{2l+1} O_k^0(\lambda^i, n-i-2l-1), \\ \qquad\qquad\text{for } (n-i) \text{ odd.} \end{cases} \tag{5.3}$$

As a result, the number of k-noncrossing RNA pseudoknot structures can be derived from the quantities $O_k^0(\lambda^i, n)$, given by eq. (5.1).

Equation (5.1) combined with the fact that D-finite functions form an algebra [125] implies that the ordinary generating function $\sum_{n \geq 0} \Gamma_n'^+(a, b) x^n$ is D-finite. Since D-finiteness is equivalent to P-recursiveness, see Lemma 2.11, we derive

Corollary 5.1. *For fixed shape λ with at most $(k-1)$ rows and $n \in \mathbb{N}$, there exists some $m \in \mathbb{N}$ and polynomials $p_0(n), \ldots, p_m(n)$ such that*

$$p_m(n+m) O_k^0(\lambda, n+m) + \cdots + p_0(n) O_k^0(\lambda, n) = 0. \tag{5.4}$$

In particular, given the coefficients $p_0(n), \ldots, p_m(n+m)$, the numbers $O_k^0(\lambda, n)$ can be computed in $O(n)$ time.

We remark that for fixed n and λ, the derivation of eq. (5.4) is a pre-processing step. In special cases we can employ Zeilberger's algorithm [114, 149].

We next generate k-noncrossing partial matchings with uniform probability. The construction is as follows: First, we compute for any shape λ, having at most $(k-1)$ rows, the recursion relation of Corollary 5.1. Second, we compute the array $(O_k^*(\lambda^i, n-i))_{\lambda,(n-i)}$, indexed by λ and $(n-i)$. Then we specify a Markov process that constructs a k-noncrossing partial matching with uniform probability with linear time and space complexity.

Theorem 5.2. *Random k-noncrossing partial matchings can be generated with uniform probability in polynomial time. The algorithmic implementation, see Algorithm 5.3, has $O(n^{k+1})$ preprocessing time and $O(n^k)$ space complexity. Each k-noncrossing partial matching is generated with $O(n)$ time and space complexity.*

Algorithm 5.3.

1: *Pascal* $\leftarrow$ Binomial(n) (computation of all binomial coefficients, $B(n,h)$)
2: *PShape* $\leftarrow$ ArrayP(n,k) (computation of $O_k^*(\lambda^i, n-i)$, $i = 0, 1, \ldots, n-1$, λ^i, stored in the $k \times n$ array, *PShape*)
3: **while** $i < n$ **do**
4:    **for** j from 0 to $k-1$ **do**
5:       X[j]$\leftarrow$ $O_k^*(\lambda^{i+1}, n-(i+1))$
6:       $sum \leftarrow sum + $X[j]
7:    **end for**
8:    *Shape* $\leftarrow$ Random(sum) (Random generates the random shape λ^{i+1})
9:    $i \leftarrow i + 1$
10:   Insert *Shape* into *Tableaux* (generates the sequence of shapes).
11: **end while**
12: Map(*Tableaux*) (maps *Tableaux* into its corresponding partial matching)

Fig. 5.3 illustrates that Algorithm 5.3 indeed generates each k-noncrossing partial matching with uniform probability.

Proof. Suppose $(\lambda^i)_{i=0}^n$ is an $*$-tableaux of shape λ having at most $(k-1)$ rows. By definition, a shape λ^{i+1} does only depend on its predecessor, λ^i. Accordingly, we can interpret any given $*$-tableaux of shape λ as a path of a Markov process $(X^i)_{i=0}^n$ over shapes, given as follows:

- $X^0 = X^n = \varnothing$ and X^i is a shape having at most $(k-1)$ rows,
- for $0 \le i \le n-1$, X^i and X^{i+1} differ by at most one square,
- the transition probabilities are given by

$$\mathbb{P}_n(X^{i+1} = \lambda^{i+1} \mid X^i = \lambda^i) = \frac{O_k^*(\lambda^{i+1}, n-(i+1))}{O_k^*(\lambda^i, n-i)}.$$

We next observe

$$\prod_{i=0}^n \mathbb{P}_n(X^{i+1} = \lambda^{i+1} \mid X^i = \lambda^i) = \frac{1}{O_k^*(\varnothing, n)} = \frac{1}{f_k^*(n)},$$

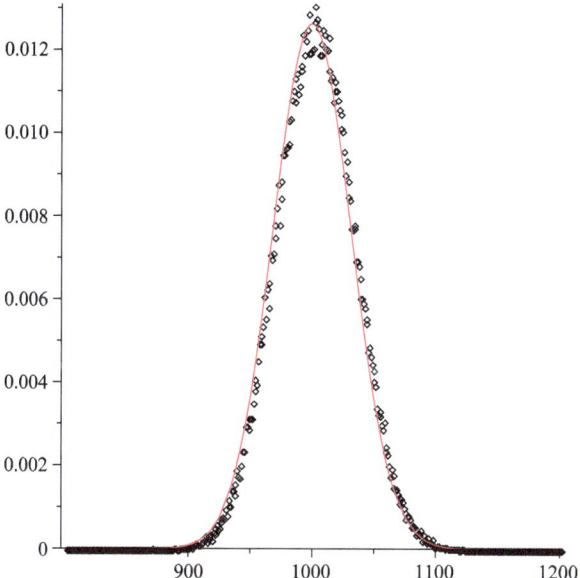

Fig. 5.3. Uniform generation of k-noncrossing partial matchings: for $n = 12$ we have $m = 99,991$ distinct 3-noncrossing partial matchings. We generate via Algorithm 5.3 $N = 10^8$ and display the frequency distribution of their multiplicities (*black dots*) versus the distribution $\binom{N}{\ell}(1/m)^{\ell}(1 - 1/m)^{N-\ell}$, resulting from uniform sampling (*red curve*).

where $f_k^*(n)$ denotes the number of *-tableaux of length n and $\mathsf{O}_k^*(\lambda^i, n-i)$ is given by eq. (5.3).

Accordingly, the Markov process, $(X^i)_{i=0}^n$, generates k-noncrossing partial matchings with uniform probability. Clearly, the Pascal triangle of binomial coefficients can be generated in $O(n^2)$ time and space and for any fixed λ^i, having at most $(k-1)$ rows, we can via Corollary 5.1 compute $\mathsf{O}_k^*(\lambda^i, n-i)$ in $O(n)$ time. Consequently, we can generate the array of numbers $\mathsf{O}_k^0(\lambda^i, n-i)$ as well as $\mathsf{O}_k^*(\lambda^i, n-i)$ for all shapes λ in $O(n^2) + O(n)\,O(n)\,O(n^{k-1})$ time and $O(n^k)$ space. The first factor $O(n)$ represents the time complexity for deriving the recursion and the second comes from the computation of all numbers $\mathsf{O}_k^0(\lambda^i, n-i)$ for fixed $\lambda = \lambda^i$ for all $(n-i)$. As for the generation of a random k-noncrossing partial matching, for each shape λ^i, the transition probabilities can be derived in $O(1)$ time. Therefore, a k-noncrossing partial matching can be computed with $O(n)$ time and space complexity, whence the theorem.

5.1.2 k-Noncrossing structures

Theorem 5.4. *A random k-noncrossing structure can be generated, after polynomial pre-processing time, with uniform probability in linear time. The* *algorithmic implementation, see Algorithm 5.5, has $O(n^{k+1})$ pre-processing time and $O(n^k)$ space complexity. Each k-noncrossing structure is generated with $O(n)$ space and time complexity.*

Let $W_k^*(\lambda^i, n - i)$ denote the number of $*$-tableaux of shape λ^i with at most $(k - 1)$ rows of length $(n - i)$ that do not contain any $(+\square_1, -\square_1)$-steps, then we have

Algorithm 5.5.
1: $Pascal \leftarrow$ Binomial(n) (computation of all binomial coefficients, $B(n, h)$)
2: $PShape \leftarrow$ ArrayP(n,k) (computation of $O_k^*(\lambda^i, n - i)$, $i = 0, 1, \ldots,$
    $n - 1, \lambda^i$)
3: $SShape \leftarrow$ ArrayS(n,k) (computation of $W_k^*(\lambda_j^i, n - i)$, $j = 0, 1^+, 1^-, \ldots,$
    $(k - 1)^+, (k - 1)^-$; $i = 0, 1, \ldots, n - 1$, stored in the $k \times n$ array $SShape$)
4: $flag \leftarrow 1$
5: **while** $i < n$ **do**
6:     $X[0] \leftarrow W_k^*(\lambda_0^{i+1}, n - (i + 1))$
7:     $X[1] \leftarrow W_k^*(\lambda_{1^+}^{i+1}, n - (i + 1)) - W_k^*(\lambda_{1^-}^{i+2}, n - (i + 2))$
8:     **if** $flag$=0 **then**
9:         $X[2] \leftarrow 0$
10:    **else**
11:        $X[2] \leftarrow W_k^*(\lambda_{1^-}^{i+1}, n - (i + 1))$
12:    **end if**
13:    $sum \leftarrow X[0]+X[1]+X[2]$
14:    **for** j from 2 to $k - 1$ **do**
15:        $X[2j\text{-}1] \leftarrow W_k^*(\lambda_{j^+}^{i+1}, n - (i + 1))$
16:        $X[2j] \leftarrow W_k^*(\lambda_{j^-}^{i+1}, n - (i + 1))$
17:        $sum \leftarrow sum+X[2j\text{-}1]+X[2j]$
18:    **end for**
19:    $Shape \leftarrow$ Random(sum) (Random generates the random shape λ_j^{i+1} with
        probability $X[j]/sum$)
20:    **if** $Shape = \lambda_{1^+}^{i+1}$ **then**
21:        $flag \leftarrow 0$
22:    **else**
23:        $flag \leftarrow 1$
24:    **end if**
25:    Insert λ_j^{i+1} into $Tableaux$
26:    $i \leftarrow i + 1$
27: **end while**
28: Map($Tableaux$)

Fig. 5.4 illustrates that Algorithm 5.5 generates k-noncrossing RNA structures with uniform probability. Before we come to the proof of Theorem 5.4, we observe the following: a 1-arc corresponds to a subsequence of shapes $(\lambda^i, \lambda^{i+1}, \lambda^{i+2} = \lambda^i)$, obtained by first adding and then removing a square in the

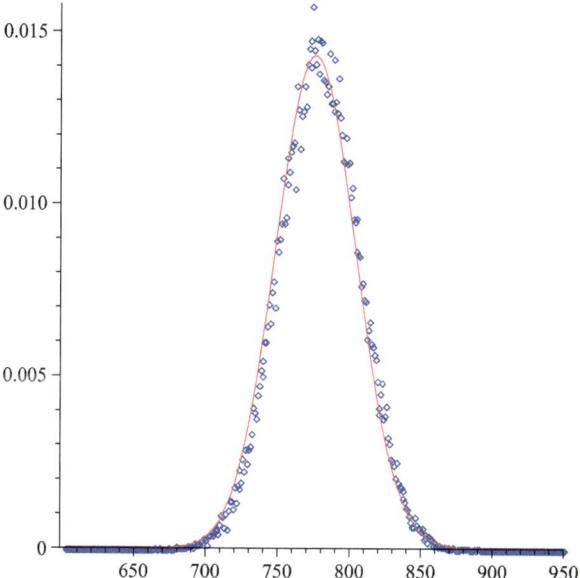

Fig. 5.4. Uniform generation of k-noncrossing structures. For $n = 12$ we have $m = 38,635$ distinct 3-noncrossing RNA structures; see Theorem 4.13. We generate via Algorithm 5.5 $N = 3 \times 10^7$ of these structures and display the frequency distribution of their multiplicities (*blue dots*) and the distribution induced by uniform sampling, $\binom{N}{\ell}(1/m)^\ell(1 - 1/m)^{N-\ell}$ (*red curve*).

first row. This sequence corresponds to a pair of steps $(+\square_1, -\square_1)$, where $+\square_1$ and $-\square_1$ indicate that a square is added and subtracted in the first row, respectively. In terms of $*$-tableaux having at most $(k-1)$ rows, eq. (5.2) can be rewritten as follows:

$$W_k^*(\varnothing, n) = \sum_{b=0}^{\frac{n}{2}}(-1)^b\binom{n - b}{b}O_k^*(\varnothing, n - 2b).$$

In order to prove Theorem 5.4 we have to generalize this relation from the empty shape, $\varnothing$, to arbitrary shapes, λ.

Lemma 5.6. *Let λ^i be an arbitrary shape with at most $(k-1)$ rows, then*

$$W_k^*(\lambda^i, n - i) = \sum_{b=0}^{\frac{n-i}{2}}(-1)^b\binom{(n - i) - b}{b}O_k^*(\lambda^i, n - i - 2b).$$

Proof. Let $\mathcal{Q}_k^*(\lambda^i, n - i, j)$ denote the set of $*$-tableaux of shape λ^i of length $(n - i)$ having at most $(k - 1)$ rows containing exactly j pairs $(+\square_1, -\square_1)$ and set $Q_k^*(\lambda^i, n - i, j) = |\mathcal{Q}_k^*(\lambda^i, n - i, j)|$. Let $(\lambda^s)_{s=0}^{(n-2b)-i}$ be an $*$-tableaux of shape λ^i. We select from the set $\{0, \ldots, (n - 2b) - i - 1\}$ an increasing sequence of labels $(r_1, \ldots, r_b)$. For each r_s we insert a pair $(+\square_1, -\square_1)$ after the corresponding shape λ^{r_s}; see Fig. 5.5. This insertion generates an $*$-tableaux of length $(n - i)$ of shape λ^i.

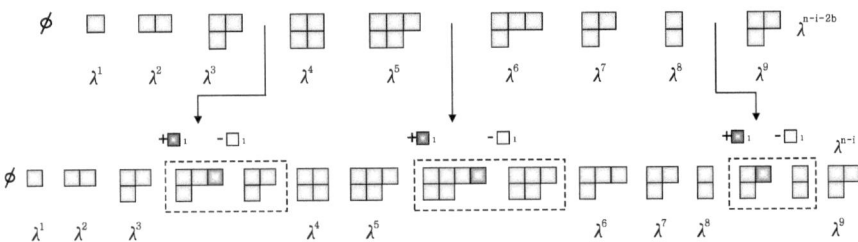

Fig. 5.5. Illustration of the proof idea: pairs $(+\square_1, -\square_1)$ are inserted at positions $3, 5,$ and 8, respectively.

Considering the above insertion for all sequences $(r_1, \ldots, r_b)$, we arrive at a family $\mathcal{F}_b$ of $*$-tableaux of length $(n - i)$ containing at least b pairs, $(+\square_1, -\square_1)$. Since we can insert at any position $0 \le h \le ((n-i) - 2b - 1)$, $\mathcal{F}_b$ has cardinality $\binom{(n-i)-b}{b} \mathsf{O}_k^*(\lambda^i, n - i - 2b)$. By construction, each $*$-tableaux $(\lambda^s)_{s=0}^{n-i} \in \mathcal{F}_b$ that exhibits exactly j pairs $(+\square_1, -\square_1)$ appears with multiplicity $\binom{j}{b}$, whence

$$\sum_{j \ge b} \binom{j}{b} \mathsf{Q}_k^*(\lambda^i, n - i, j) = \binom{(n-i)-b}{b} \mathsf{O}_k^*(\lambda^i, n - i - 2b).$$

We consider $F_k(x) = \sum_{j \ge 0} \mathsf{Q}_k^*(\lambda^i, n - i, j) x^j$. Taking the bth derivative and setting $x = 1$ we obtain $\frac{1}{b!} F_k^{(b)}(1) = \sum_{j \ge b} \binom{j}{b} \mathsf{Q}_k^*(\lambda^i, n - i, j) 1^{j-b}$ and computing the Taylor expansion of $F_k(x)$ at $x = 1$

$$F_k(x) = \sum_{b \ge 0} \frac{1}{b!} F_k^{(b)}(1) \, (x - 1)^b$$

$$= \sum_{b=0}^{\frac{n-i}{2}} \binom{(n-i)-b}{b} \mathsf{O}_k^*(\lambda^i, n - i - 2b) \, (x - 1)^b.$$

Since $\mathsf{W}_k^*(\lambda^i, n-i) = \mathsf{Q}_k^*(\lambda^i, n-i, 0)$ is the constant term of $F_k(x)$, the lemma follows.

Proof of Theorem 5.4:

Proof. The idea is to interpret $*$-tableaux without pairs of steps, $(+\square_1, -\square_1)$, (good $*$-tableaux) as paths of a stochastic process. To this end, we index the shapes λ^{i+1} according to their predecessors: let $i = 0, 1, \ldots, n - 1$ and

$j \in \{0, 1^+, 1^-, \ldots, (k-1)^+, (k-1)^-\}$. Setting $\lambda_j^0 = \varnothing$, we write λ_j^{i+1}, if λ^{i+1} is obtained via

- doing nothing (λ_0^{i+1}),
- adding a square in the jth row (λ_{j+}^{i+1}),
- deleting a square in the jth row (λ_{j-}^{i+1}).

With this notation, the number of good $*$-tableaux of shape λ_{1+}^{i+1} of length $(n - (i+1))$ is given as follows:

$$\mathsf{V}_k^*(\lambda_{1+}^{i+1}, n - (i+1)) = \mathsf{W}_k^*(\lambda_{1+}^{i+1}, n - (i+1)) - \mathsf{W}_k^*(\lambda_{1-}^{i+2}, n - (i+2)).$$

In order to derive transition probabilities, we establish two equations: first, for any λ_j^i, where $j \neq 1^+$, we have $\mathsf{W}_k^*(\lambda_j^i, n - i) =$

$$\mathsf{V}_k^*(\lambda_{1+}^{i+1}, n - (i+1)) + \mathsf{W}_k^*(\lambda_{1-}^{i+1}, n - (i+1)) +$$

$$\sum_{h=2}^{k-1} \left(\mathsf{W}_k^*(\lambda_{h+}^{i+1}, n - (i+1)) + \mathsf{W}_k^*(\lambda_{h-}^{i+1}, n - (i+1)) \right) +$$

$$\mathsf{W}_k^*(\lambda_0^{i+1}, n - (i+1))$$

and second, in case of $j = 1^+$, we have $\mathsf{V}_k^*(\lambda_{1+}^i, n - i) =$

$$\mathsf{V}_k^*(\lambda_{1+}^{i+1}, n - (i+1)) + \mathsf{W}_k^*(\lambda_0^{i+1}, n - (i+1)) +$$

$$\sum_{h=2}^{k-1} \left(\mathsf{W}_k^*(\lambda_{h+}^{i+1}, n - (i+1)) + \mathsf{W}_k^*(\lambda_{h-}^{i+1}, n - (i+1)) \right).$$

We are now in a position to specify the process $(X^i)_{i=0}^n$:

- $X^0 = X^n = \varnothing$ and X^i is a shape having at most $(k-1)$ rows.
- For $0 \leq i \leq n-1$, X^i and X^{i+1} differ by at most one square.
- There exists no subsequence $X^i, X^{i+1}, X^{i+2} = X^i$ obtained by first adding and second removing a square in the first row.
- For $j \neq 1^+$

$$\mathbb{P}_n(X^{i+1} = \lambda_l^{i+1} \mid X^i = \lambda_j^i) = \begin{cases} \dfrac{\mathsf{W}_k^*(\lambda_l^{i+1}, n-(i+1))}{\mathsf{W}_k^*(\lambda_j^i, n-i)} & \text{for } l \neq 1^+ \\[2ex] \dfrac{\mathsf{V}_k^*(\lambda_{1+}^{i+1}, n-(i+1))}{\mathsf{W}_k^*(\lambda_j^i, n-i)} & \text{for } l = 1^+. \end{cases} \tag{5.5}$$

- For $j = 1^+$

$$\mathbb{P}_n(X^{i+1} = \lambda_l^{i+1} \mid X^i = \lambda_{1+}^i) = \begin{cases} \dfrac{\mathsf{W}_k^*(\lambda_l^{i+1}, n-(i+1))}{\mathsf{V}_k^*(\lambda_{1+}^i, n-i)}, & \\[1ex] & \text{for } l \neq 1^+, 1^- \\[1ex] \dfrac{\mathsf{V}_k^*(\lambda_{1+}^{i+1}, n-(i+1))}{\mathsf{V}_k^*(\lambda_{1+}^i, n-i)}, & \text{for } l = 1^+. \end{cases} \tag{5.6}$$

We observe that eqs. (5.5) and (5.6) imply

$$\prod_{i=0}^{n-1} \mathbb{P}_n(X^{i+1} = \lambda^{i+1} \mid X^i = \lambda^i) = \frac{\mathsf{W}_k^*(\lambda^n = \varnothing, 0)}{\mathsf{W}_k^*(\lambda^0 = \varnothing, n)} = \frac{1}{\mathsf{W}_k^*(\varnothing, n)}.$$

Consequently, the process $(X^i)_{i=0}^n$ generates random k-noncrossing structures with uniform probability in $O(n)$ time and space. According to Corollary 5.1, we can for any λ^i, having at most $(k-1)$ rows, compute $\mathsf{O}_k^0(\lambda^i, n-i)$ in $O(n)$ time. Consequently, we can generate the arrays $(\mathsf{O}_k^*(\lambda^i, n-i))_{\lambda^i, n-i}$ and $(\mathsf{W}_k^*(\lambda^i, n-i))_{\lambda^i, n-i}$ in $O(n^2) + O(n^2)O(n^{k-1})$ time and $O(n^k)$ space. A random k-noncrossing structure is then generated as an $*$-tableaux with at most $(k-1)$ rows using the array $(\mathsf{W}_k^*(\lambda^i, n-i))_{\lambda^i, n-i}$ with $O(n)$ time and space complexity.

Once the polynomial coefficients, $p_h(n+h)$, are computed, eq. (5.4) allows for the efficient computation of the transition probabilities,

$$\mathbb{P}_n(X^{i+1} = \lambda^{i+1} \mid X^i = \lambda^i),$$

for *any* n. However, for all applications n is always fixed, in which case the transition probabilities can be computed directly. To this end we use the recursiveness of the $*$-tableaux itself. Plainly, a shape λ^{i+1} is obtained from λ^i, by adding or removing a square in one row, or do nothing, whence

$$\mathsf{O}_k^*(\lambda^i, i) = \mathsf{O}_k^*(\lambda_0^{i-1}, i-1) + \sum_{j=1}^{k-1}\left(\mathsf{O}_k^*(\lambda_{j+}^{i-1}, i-1) + \mathsf{O}_k^*(\lambda_{j-}^{i-1}, i-1)\right), \quad (5.7)$$

initialized at $\mathsf{O}_k^*(\varnothing, 0) = 1$. For fixed n, recursion (5.7) facilitates the calculation of $\mathsf{O}_k^*(\lambda^i, i)$ for arbitrary λ^i and i; see Fig. 5.6. Let n be the total number of steps and set $\mathsf{O}_k^*(\varnothing, 0) = 1$. In the algorithm, we consider the subroutines

- **Step**, where we calculate all $\mathsf{O}_k^*(\lambda^i, i)$ for all λ^i and i,
- **FillArray**, consisting of $(k-1)$ **For**-loops (lines 2–5).

The output is an array whose entries are the integers, $\mathsf{O}_k^*(\lambda^i, i)$, indexed by step-labeled shapes, λ^i.

Algorithm 5.7.

```
1: FillArray (n, c, k)
2: if c < k then
3:    for x_c(n) = 0 to n − Σ_{s=1}^{c-1} x_s(n) do
4:       FillArray (n, c + 1, k)
5:    end for
6: else
7:    O_k^*(λ_0^{n+1}, n + 1) ← O_k^*(λ^n, n)
8:    for j = 1 to k − 1 do
```

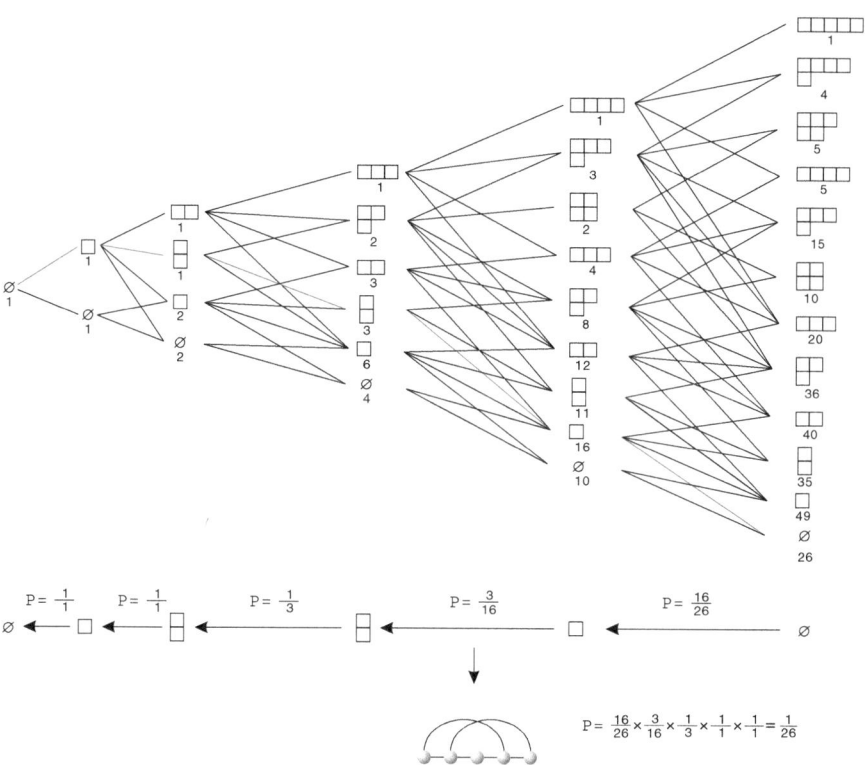

Fig. 5.6. Calculating the transition probabilities: Starting at $\lambda = \varnothing$, we inductively construct (from *left* to *right*) all possible shapes. After n steps, the quantities $O_k^*(\lambda^i, i)$ for $1 \le i \le n$ are derived and we can sample k-noncrossing structures with uniform probability.

```
 9:      if x_j(n) + 1 ≤ x_{j-1}(n) then
10:         O_k^*(λ_{j+}^{n+1}, n + 1) ← O_k^*(λ^n, n)
11:      end if
12:      if x_j(n) - 1 ≥ x_{j+1}(n) then
13:         O_k^*(λ_{j-}^{n+1}, n + 1) ← O_k^*(λ^n, n)
14:      end if
15:   end for
16: end if
```

Algorithm 5.8.

```
1: Step
2: for l = 0 to n do
3:    FillArray(l, 0, k − 1)
4: end for
```

Consequently, the subroutine `FillArray` and `Step` have $O(n^{k-1})$ and $O(n^k)$ time complexity, respectively. Since there are $O(n^{k-1})$ shapes in the ith step, `Step` has $O(n^k)$ space complexity.

5.2 Central limit theorems

In this section we study statistical properties of k-noncrossing RNA structures with minimum arc length $\lambda \geq 2$ and stack-length $\tau \geq 2$. The results presented here are due to or derived from [60, 73, 77, 79]. We shall prove here that the number of arcs and stacks in RNA pseudoknot structures are, in the limit of long sequences, Gaussian distributed. This allows us to conclude that neutral networks, i.e., the sets of sequences folding into a given structure, are exponentially smaller than sequence space. As mentioned in Chapter 1, these findings have profound implications for sequence to structure maps into RNA pseudoknot structures: they imply molecular diversity, i.e., the existence of exponentially many distinct molecular phenotypes.

In order to derive the statistics of arcs, stacks, hairpin loops, interior loops, and bulges in k-noncrossing structures, we use a specific parameterization of the bivariate generating functions. We show that it is the shift of the singularities in this parametrization that determines the limit distribution. We base our analysis on a theorem of Bender (Theorem 5.10); see also the quasi-powers theorem [42].

Let us begin by illustrating the key idea of the approach. Suppose we are given a set A_n (of size a_n). For instance, let A_n be the set of subsets of $\{1, \ldots, n\}$. Suppose further we are given $A_{n,k}$ (of size $a_{n,k}$), $k \in \mathbb{N}$ representing a disjoint set partition of A_n. For instance, let $A_{n,k}$ be the number of subsets of $\{1, \ldots, n\}$ with exactly k elements. Consider the random variable ξ_n having the probability distribution $\mathbb{P}(\xi_n = k) = a_{n,k}/a_n$, then the corresponding probability generating function is given by

$$\sum_{k \geq 0} \mathbb{P}(\xi_n = k) w^k = \sum_{k \geq 0} \frac{a_{n,k}}{a_n} w^k = \frac{\sum_{k \geq 0} a_{n,k} w^k}{\sum_{k \geq 0} a_{n,k} 1^k}.$$

Let $\varphi_n(w) = \sum_{k \geq 0} a_{n,k} w^k$, then $\frac{\varphi_n(w)}{\varphi_n(1)}$ is the probability generating function of ξ_n and

$$f(z, w) = \sum_{n \geq 0} \varphi_n(w) z^n = \sum_{n \geq 0} \sum_{k \geq 0} a_{n,k} w^k z^n$$

is called the bivariate generating function. For instance, in the above example we have $\mathbb{P}(\xi_n) = \binom{n}{k}/2^n$ and the resulting bivariate generating function is given by

$$\sum_{n \geq 0} \sum_{k \leq n} \binom{n}{k} w^k z^n = \frac{1}{1 - z(1 + w)}.$$

The key idea now is to consider $f(z, w)$ as being parametrized by w and to study the change of its singularity in an ϵ-disc centered at $w = 1$. Indeed the moment generating function is

$$E(e^{s\xi_n}) = \sum_{k \geq 0} \frac{a_{n,k}}{a_n} e^{sk} = \frac{\varphi_n(e^s)}{\varphi_n(1)} = \frac{[z^n]f(z, e^s)}{[z^n]f(z, 1)}$$

and $\frac{[z^n]f(z, e^{it})}{[z^n]f(z,1)} = E(e^{it\xi_n})$ is the characteristic function of ξ_n. We observe that the coefficients of $f(z, w)$ control the distribution, which can, for large n, be obtained via singularity analysis. The resulting computation can be surprisingly simple.

Let us make this explicit for the binomial distribution, where we have the bivariate generating function

$$\sum_{n \geq 0} \sum_{k \leq n} \binom{n}{k} w^k z^n = \frac{1}{1 - z(1 + w)}.$$

The unique singularity of $f(z, e^s)$ is the simple pole $r(s) = \frac{1}{1+e^s}$, parametrized in s. The crucial point is now

$$\frac{\varphi_n(e^s)}{\varphi_n(1)} \sim \left(\frac{r(0)}{r(s)}\right)^n \tag{5.8}$$

for s uniformly in a neighborhood of 0, which is a simple observation in this particular example. However, to prove this for RNA pseudoknot structures, this requires more work; see Theorem 2.21. Back to eq. (5.8), Taylor expansion shows

$$\frac{\varphi_n(e^{it})}{\varphi_n(1)} \sim \exp(i \cdot \frac{n}{2} \cdot t - \frac{1}{2} \cdot \frac{n}{4} \cdot t^2 + O(t^3))$$

uniformly for t taken from any arbitrary finite interval.

We can now apply the Lévy–Cramér theorem (Theorem 5.9) to the normalized characteristic function of the random variable

$$\frac{\xi_n - \frac{n}{2}}{\sqrt{\frac{n}{4}}},$$

which implies the asymptotic normality of ξ_n. Thus $\binom{n}{k}$ is asymptotically normally distributed with mean $\frac{n}{2}$ and variance $\frac{n}{4}$.

5.2.1 The central limit theorem

The main result of this section is a central limit theorem for distributions given in terms of bivariate generating functions. The central limit theorem is due to Bender [11] and based on the following classic result on limit distributions [40]:

Theorem 5.9. (Lévy–Cramér)   *Let $\{\xi_n\}$ be a sequence of random variables and let $\{\varphi_n(x)\}$ and $\{F_n(x)\}$ be the corresponding sequences of characteristic and distribution functions. If there exists a function $\varphi(t)$, such that $\lim_{n\to\infty} \varphi_n(t) = \varphi(t)$ uniformly over an arbitrary finite interval enclosing the origin, then there exists a random variable ξ with distribution function $F(x)$ such that*

$$F_n(x) \Longrightarrow F(x)$$

uniformly over any finite or infinite interval of continuity of $F(x)$.

We come now to the central limit theorem. It analyzes the characteristic function via the above Lévy–Cramér theorem.

Theorem 5.10. *Suppose we are given the bivariate generating function*

$$f(z, u) = \sum_{n,m \geq 0} f(n, m)\, z^n\, u^m,$$

where $f(n, m) \geq 0$ and $f(n) = \sum_t f(n, t)$. Let $\mathbb{X}_n$ be an r.v. such that $\mathbb{P}(\mathbb{X}_n = t) = f(n, t)/f(n)$. Suppose

$$[z^n]f(z, e^s) \sim c(s)\, n^\alpha\, \gamma(s)^{-n} \tag{5.9}$$

uniformly in s in a neighborhood of 0, where $c(s)$ is continuous and nonzero near 0, α is a constant, and $\gamma(s)$ is analytic near 0. Then there exists a pair (μ, σ) such that the normalized random variable

$$\mathbb{X}_n^* = \frac{\mathbb{X}_n - \mu\, n}{\sqrt{n\, \sigma^2}}$$

has asymptotically normal distribution with parameter $(0, 1)$, that is, we have

$$\lim_{n\to\infty} \mathbb{P}\left(\mathbb{X}_n^* < x\right) = \frac{1}{\sqrt{2\pi}} \int_{-\infty}^{x} e^{-\frac{1}{2}c^2}\, dc, \tag{5.10}$$

where μ and σ^2 are given by

$$\mu = -\frac{\gamma'(0)}{\gamma(0)} \quad and \quad \sigma^2 = \left(\frac{\gamma'(0)}{\gamma(0)}\right)^2 - \frac{\gamma''(0)}{\gamma(0)}. \tag{5.11}$$

Proof. Suppose we are given the random variable (r.v.) ξ_n with mean μ_n and variance σ_n^2. We consider the rescaled r.v. $\eta_n = (\xi_n - \mu_n)\sigma_n^{-1}$ and the characteristic function of η_n:

$$f_{\eta_n}(c) = \mathbb{E}[e^{ic\eta_n}] = \mathbb{E}[e^{ic\frac{\xi_n}{\sigma_n}}]e^{-i\frac{\mu_n}{\sigma_n}c}.$$

We derive substituting for the term $\mathbb{E}[e^{ic\eta_n}]$

$$f_{\mathbb{X}_n}(c) = \left(\sum_{t \geq 0} \frac{f(n,t)}{f(n)} e^{ic\frac{t}{\sigma_n}} \right) e^{-i\frac{\mu_n}{\sigma_n}c}.$$

Since $[z^n]f(z,e^s) = \sum_t f(n,t)e^{ts}$ we have

$$[z^n]f(z,0) = \sum_t f(n,t) \quad \text{and} \quad [z^n]f\left(z,\frac{ic}{\sigma_n}\right) = \sum_t f(n,t)e^{t\frac{ic}{\sigma_n}}.$$

We accordingly obtain

$$f_{\mathbb{X}_n}(c) = \left[\frac{[z^n]f\left(z,\frac{ic}{\sigma_n}\right)}{[z^n]f(z,0)} \right] e^{-i\frac{\mu_n}{\sigma_n}c}.$$

By assumption we have

$$[z^n]f(z,e^s) \sim c(s)\, n^\alpha\, \gamma(s)^{-n} \tag{5.12}$$

uniformly in s in a neighborhood of 0, where $c(s)$ is continuous and nonzero near 0, α is a constant, and $\gamma(s)$ is analytic near 0. Therefore we arrive at

$$f_{\mathbb{X}_n}(c) \sim \frac{c(\frac{ic}{\sigma_n})}{c(0)} \left[\frac{\gamma(\frac{ic}{\sigma_n})}{\gamma(0)} \right]^{-n} e^{-i\frac{\mu_n}{\sigma_n}c},$$

uniformly in c, where c is contained in an arbitrary bounded interval. Taking the logarithm we obtain

$$\ln f_{\mathbb{X}_n}(c) \sim \ln \frac{c(\frac{ic}{\sigma_n})}{c(0)} - n \ln \frac{\gamma(\frac{ic}{\sigma_n})}{\gamma(0)} - i\frac{\mu_n}{\sigma_n}c.$$

Expanding $g(s) = \ln(\gamma(s)/\gamma(0))$ in its Taylor series at $s = 0$ (note that $g(0) = 0$ holds) yields

$$\ln \frac{\gamma(\frac{ic}{\sigma_n})}{\gamma(0)} = \frac{\gamma'(0)}{\gamma(0)} \frac{ic}{\sigma_n} - \left[\frac{\gamma''(0)}{\gamma(0)} - \left(\frac{\gamma'(0)}{\gamma(0)} \right)^2 \right] \frac{c^2}{2\sigma_n^2} + O\left(\left(\frac{ic}{\sigma_n} \right)^3 \right) \tag{5.13}$$

and $\ln f_{\mathbb{X}_n}(c)$ becomes asymptotically

$$\ln \frac{c(\frac{ic}{\sigma_n})}{c(0)} - n \left\{ \frac{\gamma'(0)}{\gamma(0)} \frac{ic}{\sigma_n} - \frac{1}{2} \left[\frac{\gamma''(0)}{\gamma(0)} - \left(\frac{\gamma'(0)}{\gamma(0)} \right)^2 \right] \frac{c^2}{\sigma_n^2} + O\left(\left(\frac{ic}{\sigma_n} \right)^3 \right) \right\}$$

$$- \frac{i\mu_n c}{\sigma_n}.$$

$$\tag{5.14}$$

$f(z, e^s)$ is analytic in s where s is contained in a disc of radius ϵ around 0 and therefore in particular continuous in s for $|s| < \epsilon$. In view of eq. (5.14) we set

$$\mu = -\frac{\gamma'(0)}{\gamma(0)}, \qquad \sigma^2 = \left(\frac{\gamma'(0)}{\gamma(0)}\right)^2 - \frac{\gamma''(0)}{\gamma(0)}.$$

Setting $\mu_n = n\mu$ and $\sigma_n^2 = n\sigma^2$ we can conclude that for fixed $c \in\,]-\infty, \infty[$

$$\lim_{n\to\infty} \left(\ln c((ic)/(\sigma_n)) - \ln c(0)\right) = 0$$

and (5.14) becomes

$$\ln f_{\mathbb{X}_n}(c) \sim -c^2/2 + O(((ic)/\sigma_n)^3)$$

with uniform error term for c contained in any bounded interval. This is equivalent to

$$\lim_{n\to\infty} f_{\mathbb{X}_n}(c) = \exp(-c^2/2),$$

uniformly in c. Theorem 5.9 implies now eq. (5.10) and the proof of Theorem 5.10 is complete.

The crucial points for applying Theorem 5.10 are

- eq. (5.9)
$$[z^n]f(z, e^s) \sim c(s)\, n^\alpha\, \gamma(s)^{-n},$$

  uniformly in s in a neighborhood of 0, where $c(s)$ is continuous and nonzero near 0 and α is a constant,
- $\gamma(s)$ is analytic in s.

In the following, we encounter generating functions of the form $\mathbf{F}_k(\psi(z, s))$. In this situation, Theorem 2.21 guarantees under specific conditions

$$[z^n]\mathbf{F}_k(\psi(z, s)) \sim A(s)\, n^{-((k-1)^2+(k-1)/2)} \left(\frac{1}{\gamma(s)}\right)^n,$$

where $A(s)$ being continuous, whence $\alpha = \alpha_k = -((k-1)^2 + (k-1)/2)$. The analyticity of $\gamma(s)$ is guaranteed by the analytic implicit function theorem [42].

The conditions that need to be verified in order to apply Theorem 2.21 are

- $\psi(z, s)$ is analytic function in some domain $\mathcal{D} = \{(z, s)|\, |z| \leq r, |s| < \epsilon\}$ and $\psi(0, s) = 0$,
- $\gamma(s)$ is the unique dominant singularity of $\mathbf{F}_k(\psi(z, s))$ and solution of $\psi(\gamma(s), s) = \rho_k^2$,
- $|\gamma(s)| \leq r$ as well as $\frac{\partial}{\partial z}\psi(\gamma(s), s) \neq 0$ for $|s| < \epsilon$.

5.2.2 Arcs and stacks

In this section we study the distribution of the numbers of arcs and stacks in k-noncrossing, τ-canonical structures. Let $\mathbb{A}_{n,k,\tau}(S)$ denote the number of arcs in a k-noncrossing, τ-canonical structure, S, and let $\mathcal{A}_{k,\tau}(n,h)$ and $\mathsf{A}_{k,\tau}(n,h)$ denote the set and number of k-noncrossing, τ-canonical structures, having exactly h arcs. Analogously, let $\mathbb{S}_{n,k,\tau}(S)$ be the number of stacks in a k-noncrossing, τ-canonical structure, S, and $\mathcal{S}_{k,\tau}(n,h)$ ($\mathsf{S}_{k,\tau}(n,h)$) denote the set (number) of k-noncrossing, τ-canonical structures, having exactly h stacks. In this section we study the r.vs.

- $\mathbb{A}_{n,k,\tau}$, where $\mathbb{P}\left(\mathbb{A}_{n,k,\tau}=h\right)=\frac{\mathsf{A}_{k,\tau}(n,h)}{\mathsf{T}_{k,\tau}(n)}$,
- $\mathbb{S}_{n,k,\tau}$, where $\mathbb{P}\left(\mathbb{S}_{n,k,\tau}=h\right)=\frac{\mathsf{S}_{k,\tau}(n,h)}{\mathsf{T}_{k,\tau}(n)}$.

Let us first consider arcs in k-noncrossing, τ-canonical structures, i.e., the r.v. $\mathbb{A}_{n,k,\tau}$. The first step is to compute the bivariate generating function

$$\mathbf{A}_{k,\tau}(z,u)=\sum_{n\geq 0}\sum_{0\leq h\leq \frac{n}{2}}\mathsf{A}_{k,\tau}(n,h)\,u^h z^n.$$

Recall that $\mathcal{I}_k(n,m)$ ($i_k(n,m)$) denote the set (number) of shapes of length $2n$ with m 1-arcs and

$$\mathbf{I}_k(z,u)=\sum_{n\geq 0}\sum_{m=0}^{n}i_k(n,m)z^n u^m=\frac{1+z}{1+2z-zu}\mathbf{F}_k\left(\frac{z(1+z)}{(1+2z-zu)^2}\right).$$

Furthermore, $\mathcal{I}_k(m)$ denotes the set of shapes having exactly m 1-arcs.

Theorem 5.11. *Let $k,\tau\in\mathbb{N}$ $k\geq 2$ and let u,x,y,z be indeterminants. Then we have the identity of formal power series*

$$\mathbf{A}_{k,\tau}(z,u)=\frac{1}{u_\tau(z,u)z^2-z+1}\mathbf{F}_k\left(\left(\frac{\sqrt{u_\tau(z,u)}\,z}{u_\tau(z,u)z^2-z+1}\right)^2\right), \quad (5.15)$$

where $u_\tau(z,u)$ is given by

$$u_\tau(z,u)=\frac{u\left(uz^2\right)^{\tau-1}}{(uz^2)^\tau-uz^2+1}.$$

Considered as a relation between analytic functions, eq. (5.15) holds for $u=e^s$ and $|s|\leq\epsilon$ for ϵ sufficiently small and $|z|\leq 1/2$.

Proof. Let $\mathcal{T}_{k,\tau}(\gamma,h,n)$ denote the set of k-noncrossing, τ-canonical structures, having length n and h arcs, contained in the preimage of a fixed shape, $\gamma\in\mathcal{I}_k(m)$. Then $\mathcal{T}_{k,\tau}=\dot\cup\varphi^{-1}(\gamma)$ and $\varphi^{-1}(\gamma)=\dot\cup_{n,h}\mathcal{T}_{k,\tau}(\gamma,h,n)$ where $\varphi\colon\mathcal{T}_{k,\tau}\to\mathcal{I}_k$ is the surjective projection into V_k-shapes. Then

$$\mathbf{A}_{k,\tau}(z,u) = \sum_{m \geq 0} \sum_{\gamma \in \mathcal{I}_k(m)} \underbrace{\sum_{n,h} |\mathcal{T}_{k,\tau}(\gamma,h,n)| z^n u^h}_{\mathbf{A}_\gamma(z,u)},$$

where $\mathbf{A}_\gamma(z,u)$ is the bivariate generating function of k-noncrossing, τ-canonical structures having the shape γ. A structure inflated from γ has s stems and $(2s+1)$ intervals of isolated vertices, m of which contain at least one isolated vertex. We build these structures in a modular way via the combinatorial classes $\mathcal{M}$ (stems), $\mathcal{K}^\tau$ (stacks), $\mathcal{N}^\tau$ (induced stacks), $\mathcal{L}$ (isolated vertices), $\mathcal{R}$ (labeled arcs), and $\mathcal{Z}$ (vertices), where $\mathbf{Z}(z) = z$ and $\mathbf{R}(z,u) = uz^2$. We proceed in complete analogy to the proof of Theorem 4.9, in fact all we have to is to substitute $\mathbf{R}(z,u) = uz^2$, i.e., the bivariate generating function of labeled arcs for $\mathbf{R}(z) = z^2$. Accordingly we generate the following:

- Isolated segments, i.e., sequences of isolated vertices $\mathcal{L} = \mathrm{SEQ}(\mathcal{Z})$, where

$$\mathbf{L}(z) = \frac{1}{1-z}.$$

- Stacks, i.e., pairs consisting of the minimal sequence of arcs $\mathcal{R}^\tau$ and an arbitrary extension consisting of arcs of arbitrary finite length $\mathcal{K}^\tau = \mathcal{R}^\tau \times \mathrm{SEQ}(\mathcal{R})$, with generating function

$$\mathbf{K}^\tau(z,u) = \frac{(uz^2)^\tau}{1-uz^2}.$$

- Induced stacks, i.e., stacks together with at least one nonempty interval of isolated vertices on either or both its sides

$$\mathcal{N}^\tau = \mathcal{K}^\tau \times \left(\mathcal{Z} \times \mathcal{L} + \mathcal{Z} \times \mathcal{L} + (\mathcal{Z} \times \mathcal{L})^2\right),$$

having the generating function

$$\mathbf{N}^\tau(z,u) = \frac{(uz^2)^\tau}{1-uz^2}\left(2\frac{z}{1-z} + \left(\frac{z}{1-z}\right)^2\right).$$

- Stems, that is, pairs consisting of stacks $\mathcal{K}^\tau$ and an arbitrarily long sequence of induced stacks

$$\mathcal{M}^\tau = \mathcal{K}^\tau \times \mathrm{SEQ}(\mathcal{N}^\tau),$$

where

$$\mathbf{M}^\tau(z,u) = \frac{\mathbf{K}^\tau(z,u)}{1-\mathbf{N}^\tau(z,u)} = \frac{\frac{(uz^2)^\tau}{1-uz^2}}{1 - \frac{(uz^2)^\tau}{1-uz^2}\left(2\frac{z}{1-z} + \left(\frac{z}{1-z}\right)^2\right)}.$$

Plainly, the second inflation is identical to that of Theorem 4.9. Combining steps I and II we derive

$$\mathcal{A}_\gamma = (\mathcal{M}^\tau)^s \times \mathcal{L}^{2s+1-m} \times (\mathcal{Z} \times \mathcal{L})^m$$

and compute

$$\mathbf{A}_\gamma(z, u) = \left(\frac{\frac{(uz^2)^\tau}{1-uz^2}}{1 - \frac{(uz^2)^\tau}{1-uz^2}\left(2\frac{z}{1-z} + \left(\frac{z}{1-z}\right)^2\right)} \right)^s \left(\frac{1}{1-z}\right)^{2s+1-m} \left(\frac{z}{1-z}\right)^m$$

$$= (1-z)^{-1} \left(\frac{(uz^2)^\tau}{(1-z)^2(1-uz^2) - (2z-z^2)(uz^2)^\tau} \right)^s z^m.$$

Since for any $\gamma, \gamma_1 \in \mathcal{I}_k(s, m)$, $\mathbf{A}_\gamma(z, u) = \mathbf{A}_{\gamma_1}(z, u)$ holds we obtain

$$\mathbf{A}_{k,\tau}(z, u) = \sum_{m \geq 0} \sum_{\gamma \in \mathcal{I}_k(m)} \mathbf{A}_\gamma(z, u) = \sum_{s \geq 0} \sum_{m=0}^{s} i_k(s, m) \mathbf{A}_\gamma(z, u).$$

We set

$$\eta_\tau(z, u) = \frac{(uz^2)^\tau}{(1-z)^2(1-uz^2) - (2z-z^2)(uz^2)^\tau}.$$

Then we have $\mathbf{A}_{k,\tau}(z, u) = \sum_{s \geq 0} \sum_{m=0}^{s} i_k(s, m) \mathbf{A}_\gamma(z, u)$ and according to Theorem 4.6

$$\sum_{s \geq 0} \sum_{m=0}^{s} i_k(s, m) x^s y^m = \frac{1+x}{1+2x-xy} \sum_{s \geq 0} f_k(2s, 0) \left(\frac{x(1+x)}{(1+2x-xy)^2} \right)^s.$$

Therefore we arrive, setting $x = \eta_\tau(z, u)$ and $y = z$, at

$$\frac{(1-z)w_\tau(z, u)}{(1-z)^2 w_\tau(z, u) + (uz^2)^\tau(1-z)} \mathbf{F}_k \left(\frac{(uz^2)^\tau(1-z)^2 w_\tau(z, u)}{((1-z)^2 w_\tau(z, u) + (uz^2)^\tau(1-z))^2} \right),$$

where $w_\tau(z, u) = (uz^2)^\tau - uz^2 + 1$. Accordingly we have

$$\mathbf{A}_{k,\tau}(z, u) = \frac{1}{u_\tau(z, u)z^2 - z + 1} \mathbf{F}_k \left(\left(\frac{\sqrt{u_\tau(z, u)}\, z}{u_\tau(z, u)z^2 - z + 1} \right)^2 \right)$$

and the proof of the theorem is complete.

For structure classes with minimum arc length $\lambda > 2$ we observe that Theorems 4.10 and 5.11 immediately imply for $\lambda \leq \tau + 1$:

Theorem 5.12. *Let* $k, \tau \in \mathbb{N}$ $k \geq 2$, u, z *be indeterminants and suppose* $\lambda \leq \tau + 1$. *Then we have the identity of formal power series*

$$\mathbf{A}_{k,\tau}^{[\lambda]}(z, u) = \frac{1}{v_\lambda(z, u)} \mathbf{F}_k \left(\left(\frac{\sqrt{u_\tau(z, u)} \, z}{v_\lambda(z, u)} \right)^2 \right),$$

where $u_\tau(z, u)$ *and* $v_\lambda(z, u)$ *are given by*

$$u_\tau(z, u) = \frac{u \, (uz^2)^{\tau - 1}}{(uz^2)^\tau - uz^2 + 1},$$

$$v_\lambda(z, u) = 1 - z + u_\tau(z, u) \sum_{h=2}^{\lambda} z^h.$$

Theorem 5.11 puts us in position to use singularity analysis in order to compute the asymptotic distribution of the r.v. $\mathbb{A}_{n,k,\tau}$. We next study the singularities of a specific parametrization of $\mathbf{A}_{k,\tau}(z, u)$. We set $u = e^s$ and consider

$$\mathbf{A}_{k,\tau}^*(z, s) = \sum_{n \geq 0} \alpha_{n,k,\tau}(s) z^n,$$

where $\alpha_{n,k,\tau}(s) = \sum_{h \leq \frac{n}{2}} \mathbf{A}_{k,\tau}(n, h) e^{sh}$. The following analysis of $\mathbf{A}_{k,\tau}^*(z, s)$ puts us in position to use Theorem 2.21 in order to establish the central limit theorem, Theorem 5.14, for the distribution of the numbers of arcs.

Proposition 5.13. *Suppose* $\epsilon > 0$, $k \geq 2$ *and* $u = e^s$, *where* $|s| < \epsilon$.
(a) *Any dominant singularity of* $\mathbf{A}_{k,\tau}^*(z, s)$ *is a singularity of*

$$\mathbf{F}_k \left(\left(\frac{\sqrt{u_\tau(z, u)} \, z}{u_\tau(z, u) z^2 - z + 1} \right)^2 \right).$$

Let $\gamma_{k,\tau}(s)$ *be a solution of the equation*

$$\left(\frac{\sqrt{u_\tau(z, u)} \, z}{u_\tau(z, u) z^2 - z + 1} \right)^2 - \rho_k^2 = 0, \tag{5.16}$$

such that $\gamma_{k,\tau}(0)$ *is the minimal real positive solution of eq. (5.16). Then* $\gamma_{k,\tau}(s)$ *is analytic in* s *and a dominant singularity of* $\mathbf{A}_{k,\tau}^*(z, s)$.
(b) *Suppose* $2 \leq k \leq 9$ *and* $2 \leq \tau \leq 7$. *Then* $\gamma_{k,\tau}(s)$ *is the unique dominant singularity of* $\mathbf{A}_{k,\tau}^*(z, s)$ *and*

$$[z^n] \mathbf{A}_{k,\tau}^*(z, s) \sim a_{k,\tau}(s) \, n^{-((k-1)^2 + \frac{k-1}{2})} \left(\frac{1}{\gamma_{k,\tau}(s)} \right)^n, \tag{5.17}$$

for some $a_{k,\tau}(s) \in \mathbb{C}$, *uniformly in* s *contained in a neighborhood of* 0. *In particular, the subexponential factors of the coefficients of* $\mathbf{A}_{k,\tau}^*(z, s)$ *coincide with those of* $\mathbf{F}_k(z)$ *and are independent of* s.

Proof. In order to prove assertion (a) we establish the existence of $\gamma_{k,\tau}(s)$. For this purpose we consider the equations

$$\forall\, 2 \le i \le k; \qquad F_{i,\tau}(z,s) = \left(\frac{\sqrt{u_\tau(z,e^s)}\, z}{u_\tau(z,e^s)z^2 - z + 1}\right)^2 - \rho_i^2,$$

where $\rho_i = 1/(2i-2)$. Theorem 5.11 and Proposition 2.22 imply that the singularities of $\mathbf{A}_{k,\tau}^*(z,e^s)$ are contained in the set of roots of

$$F_{i,\tau}(z,s) = 0 \quad \text{and} \quad (u_\tau(z,e^s)z^2 - z + 1) = 0,$$

where $i \le k$. For $s = 0$ there exists a unique minimal real solution $r_{i,\tau}$, satisfying $F_{i,\tau}(z,0) = 0$. For $|s| < \epsilon$, ϵ being sufficiently small we observe

- $\frac{\partial}{\partial z} F_{i,\tau}(r_{i,\tau}, 0) \neq 0$,
- $\frac{\partial}{\partial z} F_{i,\tau}(z,s)$ and $\frac{\partial}{\partial s} F_{i,\tau}(z,s)$ are continuous.

According to the analytic implicit function theorem [42], there exist for $2 \le i \le k$ unique analytic functions $\gamma_{i,\tau}(s)$ for s in a neighborhood of 0 that satisfy

$$F_{i,\tau}(\gamma_{i,\tau}(s), s) = 0 \quad \text{and} \quad \gamma_{i,\tau}(0) = r_{i,\tau},$$

which proves that $\gamma_{k,\tau}(s)$ exists satisfying $\gamma_{k,\tau}(0) = r_{k,\tau}$. Let

$$\mathbf{W}_{k,\tau}(z,s) = \mathbf{F}_k\left(\left(\frac{\sqrt{u_\tau(z,e^s)}\, z}{u_\tau(z,e^s)z^2 - z + 1}\right)^2\right).$$

Claim 1. For $|s| < \epsilon$, all dominant singularities of $\mathbf{A}_k^*(z,s)$ are singularities of $\mathbf{W}_{k,\tau}(z,s)$ and $\gamma_{k,\tau}(s)$ is the unique dominant singularity.
Let $\zeta(s)$ be a dominant singularity of $\mathbf{A}_{k,\tau}^*(z,s)$. Clearly $\zeta(s)$ is a dominant singularity of either $\mathbf{W}_{k,\tau}(z,s)$ or $(u_\tau(z,e^s)z^2 - z + 1)^{-1}$. If $\zeta(s)$ is a singularity of the latter, then, by construction, $\zeta(s)$ is also a singularity of

$$\psi_\tau(z,s) = \left(\frac{\sqrt{u_\tau(z,e^s)}\, z}{u_\tau(z,e^s)z^2 - z + 1}\right)^2,$$

implying that $\mathbf{W}_{k,\tau}(z,s)$ is non-finite at $\zeta(s)$, which is impossible. We now set $s = 0$ and compute for $2 \le \tau \le 7$ and $2 \le k \le 9$ the minimum positive real solutions of $\psi_\tau(z,s) = \rho_i^2$ for $2 \le i \le k$. We observe that $\gamma_{k,\tau}(0)$, the minimum positive real solution of $\psi_\tau(z,0) = \rho_k^2$ satisfies $\gamma_{k,\tau}(0) < \gamma_{i,\tau}(0)$ for $2 \le i < k$. Therefore, $\gamma_{k,\tau}(0)$ is the unique dominant singularity of $\mathbf{A}_{k,\tau}^*(z,0)$. By construction, for ϵ sufficiently small and $|s| < \epsilon$ the singularities of $(u_\tau(z,e^s)z^2 - z + 1)^{-1}$ and $\gamma_{k,\tau}(s)$ are continuous in s. Therefore, for sufficiently small ϵ,

$$|\zeta(s)| > |\gamma_{k,\tau}(s)|$$

holds and we have proved that for $|s| < \epsilon$ and ϵ sufficiently small, all dominant singularities of $\mathbf{A}^*_{k,\tau}(z,s)$ are singularities of $\mathbf{W}_{k,\tau}(z,s)$. By construction, $\gamma_{k,\tau}(s)$ is a singularity of $\mathbf{A}^*_{k,\tau}(z,s)$ and $\gamma_{k,\tau}(0)$ is the unique dominant singularity of $\mathbf{A}^*_{k,\tau}(z,0)$. Since $\gamma_{k,\tau}(s)$ is continuous in s and $\gamma_{k,\tau}(0) < \gamma_{i,\tau}(0)$ for $2 \leq i < k$, we can conclude that, for ϵ sufficiently small, $\gamma_{k,\tau}(s)$ is the unique dominant singularity of $\mathbf{A}^*_{k,\tau}(z,s)$. This proves Claim 1 and assertion (a) follows.

It remains to prove (b). We observe that $\psi_\tau(z,s)$ is algebraic and analytic in some domain $\mathcal{D} = \{(z,s) | |z| \leq r, |s| < \epsilon\}$ such that $\psi_\tau(0,s) = 0$. According to (a), $\gamma_{k,\tau}(s)$ is the unique dominant singularity satisfying

$$F_{k,\tau}(\gamma_{k,\tau}(s), s) = 0 \ , \quad \frac{\partial}{\partial z} F_{k,\tau}(\gamma_{k,\tau}(s), s) \neq 0, \quad \text{and} \quad |\gamma_{k,\tau}(s)| \leq r$$

in s in a neighborhood of 0. Assertion (a) guarantees that, uniformly in s, in a neighborhood of 0,

$$\mathbf{A}^*_{k,\tau}(z,s) \sim b_{k,\tau}(s)\, \mathbf{F}_k(\psi_\tau(z,s)), \quad \text{for } z \to \gamma_{k,\tau}(s),$$

where $b_k(s) \in \mathbb{C}$. Verifying $\frac{\partial}{\partial z}\psi_\tau(z,s) \neq 0$ for sufficiently small s allows us to employ Theorem 2.21 which guarantees

$$[z^n]\, \mathbf{A}^*_{k,\tau}(z,s) \sim a_{k,\tau}(s)\, n^{-((k-1)^2+(k-1)/2)} \left(\frac{1}{\gamma_{k,\tau}(s)}\right)^n, \tag{5.18}$$

for some $a_{k,\tau}(s) \in \mathbb{C}$, uniformly in s contained in a neighborhood of 0. Therefore, the asymptotic expansion is uniform in s and eq. (5.17) follows. In addition, the subexponential factors of the coefficients of $\mathbf{A}^*_{k,\tau}(z,s)$ coincide with those of $\mathbf{F}_k(z)$ and are consequently independent of s and τ, whence the proposition.

As a consequence of the results presented in Section 5.2.1, in particular, Theorem 5.10 and Proposition 5.13 we derive

Theorem 5.14. *Let $k, \tau \in \mathbb{N}$, $k \geq 2$ and let $\mathbb{A}_{n,k,\tau}(S)$ be the number of arcs in a k-noncrossing, τ-canonical structure, S. Then there exists a pair $(\mu_{k,\tau}, \sigma_{k,\tau})$ such that the normalized random variable $\mathbb{A}^*_{n,k,\tau}$ has asymptotically normal distribution with parameter $(0,1)$, where $\mu_{k,\tau}$ and $\sigma^2_{k,\tau}$ are given by*

$$\mu_{k,\tau} = -\frac{\gamma'_{k,\tau}(0)}{\gamma_{k,\tau}(0)}, \qquad \sigma^2_{k,\tau} = \left(\frac{\gamma'_{k,\tau}(0)}{\gamma_{k,\tau}(0)}\right)^2 - \frac{\gamma''_{k,\tau}(0)}{\gamma_{k,\tau}(0)}, \tag{5.19}$$

where $\gamma_{k,\tau}(s)$ is the unique dominant singularity of $\mathbf{A}_{k,\tau}(z, e^s)$; see Table 5.1.

Let us next analyze stacks in k-noncrossing, τ-canonical structures. To this end we compute in Theorem 5.15 the generating function

$$\mathbf{S}_{k,\tau}(z,u) = \sum_{0 \leq n} \sum_{0 \leq h \leq \frac{n}{2}} \mathbf{S}_{k,\tau}(n,h)\, u^h z^n.$$

	$k = 2$		$k = 3$		$k = 4$	
	$\mu_{k,\tau}$	$\sigma^2_{k,\tau}$	$\mu_{k,\tau}$	$\sigma^2_{k,\tau}$	$\mu_{k,\tau}$	$\sigma^2_{k,\tau}$
$\tau = 1$	0.276393	0.0447214	0.390891	0.0415653	0.425464	0.0314706
$\tau = 2$	0.317240	0.0643144	0.381701	0.0559928	0.403574	0.0470546
$\tau = 3$	0.336417	0.0791378	0.383555	0.0670987	0.400288	0.0559818
$\tau = 4$	0.348222	0.0916871	0.386408	0.0767872	0.400412	0.0667094
$\tau = 5$	0.356484	0.1028563	0.389134	0.0855937	0.401402	0.0748305
$\tau = 6$	0.362717	0.1130777	0.391573	0.0937749	0.402640	0.0823440
$\tau = 7$	0.367658	0.1225974	0.393733	0.1014803	0.403908	0.0894075
	$k = 5$		$k = 6$		$k = 7$	
	$\mu_{k,\tau}$	$\sigma^2_{k,\tau}$	$\mu_{k,\tau}$	$\sigma^2_{k,\tau}$	$\mu_{k,\tau}$	$\sigma^2_{k,\tau}$
$\tau = 1$	0.443020	0.0251601	0.453775	0.0209395	0.461750	0.0179291
$\tau = 2$	0.416068	0.0413361	0.424531	0.0373179	0.430788	0.0342976
$\tau = 3$	0.410087	0.0517052	0.416860	0.0474929	0.421957	0.0443150
$\tau = 4$	0.408701	0.0603242	0.414487	0.0558238	0.418872	0.0524231
$\tau = 5$	0.408741	0.0680229	0.413886	0.0632201	0.417800	0.0595864
$\tau = 6$	0.409306	0.0751211	0.413996	0.0700206	0.417575	0.0661575
$\tau = 7$	0.410071	0.0817830	0.414421	0.0763943	0.417747	0.0723092

Table 5.1. Arcs: central limit theorem for the numbers of arcs in k-noncrossing, τ-canonical structures. We list $\mu_{k,\tau}$ and $\sigma^2_{k,\tau}$ as derived from eq. (5.19). Note that $\mu_{k,\tau}$ drops from $\tau = 1$ to $\tau = 2$ for $k > 2$ (blue entries), indicating that canonical pseudoknot structures have less arcs, while for $k = 2$ we have $\mu_{2,1} < \mu_{2,2}$ (red entries). In other words, canonical secondary structures contain on average more arcs than arbitrary secondary structures.

Theorem 5.15. *Let $k, \tau \in \mathbb{N}$, $k \geq 2$, and suppose u, z are indeterminants. Then we have the identity of formal power series*

$$\mathbf{S}_{k,\tau}(z, u) = \frac{1}{g_\tau(z, u) z^2 - z + 1} \mathbf{F}_k \left(\left(\frac{\sqrt{g_\tau(z, u)} z}{g_\tau(z, u) z^2 - z + 1} \right)^2 \right),$$

where

$$g_\tau(z, u) = \frac{u z^{2(\tau-1)}}{u z^{2\tau} - z^2 + 1}.$$

Proof. Let $\mathcal{T}_{k,\tau}(\gamma, h, n)$ denote the set of k-noncrossing, τ-canonical structures, having length n and h stacks, contained in the preimage of a fixed shape, $\gamma \in \mathcal{I}_k(m)$. Then $\mathcal{T}_{k,\tau} = \dot{\cup} \varphi^{-1}(\gamma)$ and $\varphi^{-1}(\gamma) = \dot{\cup}_{n,h} \mathcal{T}_{k,\tau}(\gamma, h, n)$, where $\varphi \colon \mathcal{T}_{k,\tau} \to \mathcal{I}_k$ is the surjective projection into V_k-shapes. We derive

$$\mathbf{S}_{k,\tau}(z, u) = \sum_{m \geq 0} \sum_{\gamma \in \mathcal{I}_k(m)} \underbrace{\sum_{n,h} |\mathcal{T}_{k,\tau}(\gamma, h, n)| z^n u^h}_{\mathbf{S}_\gamma(z,u)},$$

where $\mathbf{S}_\gamma(z, u)$ is the bivariate generating function of k-noncrossing, τ-canonical structures having the shape γ. A structure inflated from γ has s stems and $(2s+1)$ intervals of isolated vertices, m of which contain at least one isolated vertex. We consider the classes $\mathcal{M}, \mathcal{K}^\tau, \mathcal{N}^\tau, \mathcal{L}, \mathcal{R}$, and $\mathcal{Z}$ and proceed in analogy to Theorem 4.9. Notice that the only difference occurs when considering the class of stack which we intend to account for specifically: we therefore generate the following:

- Labeled stacks $\mathcal{K}_\mu^\tau = \mu \times (\mathcal{R}^\tau \times \mathrm{SEQ}\,(\mathcal{R}))$, with generating function

$$\mathbf{K}_\mu^\tau(z, u) = u \cdot \frac{z^{2\tau}}{1 - z^2}.$$

- Labeled induced stacks, that is, stacks together with some nonempty intervals of isolated vertices

$$\mathcal{N}_\mu^\tau = \mathcal{K}_\mu^\tau \times \left(\mathcal{Z} \times \mathcal{L} + \mathcal{Z} \times \mathcal{L} + (\mathcal{Z} \times \mathcal{L})^2 \right),$$

where

$$\mathbf{N}_\mu^\tau(z, u) = u \cdot \frac{z^{2\tau}}{1 - z^2} \left(2\frac{z}{1-z} + \left(\frac{z}{1-z}\right)^2 \right).$$

- Stems, that is, pairs consisting of labeled stacks $\mathcal{K}^\tau$ and an arbitrarily long sequence of labeled induced stacks

$$\mathcal{M}_\mu^\tau = \mathcal{K}_\mu^\tau \times \mathrm{SEQ}\,\left(\mathcal{N}_\mu^\tau\right),$$

having the generating function

$$\mathbf{M}_\mu^\tau(z, u) = \mathbf{K}_\mu^\tau(z, u) \cdot \frac{1}{1 - \mathbf{N}_\mu^\tau(z, u)} = \frac{u \cdot \frac{z^{2\tau}}{1-z^2}}{1 - u \cdot \frac{z^{2\tau}}{1-z^2}\left(2\frac{z}{1-z} + \left(\frac{z}{1-z}\right)^2\right)}.$$

Considering the second inflation step as in Theorem 4.9 we arrive at

$$S_\gamma = \left(\mathcal{M}_\mu^\tau\right)^s \times \mathcal{L}^{2s+1-m} \times (\mathcal{Z} \times \mathcal{L})^m, \qquad (5.20)$$

where μ is the combinatorial marker for stacks. We compute

$$\mathbf{S}_\gamma(z, u) = \left(\frac{u \cdot \frac{z^{2\tau}}{1-z^2}}{1 - u \cdot \frac{z^{2\tau}}{1-z^2}\left(2\frac{z}{1-z} + \left(\frac{z}{1-z}\right)^2\right)} \right)^s \left(\frac{1}{1-z}\right)^{2s+1-m} \left(\frac{z}{1-z}\right)^m$$

$$= (1-z)^{-1} \left(\frac{uz^{2\tau}}{(1-z)^2(1-z^2) - (2z-z^2)uz^{2\tau}} \right)^s z^m.$$

Since for any $\gamma, \gamma_1 \in \mathcal{I}_k(s, m)$, $\mathbf{S}_\gamma(z, u) = \mathbf{S}_{\gamma_1}(z, u)$ holds we derive

$$\mathbf{S}_{k,\tau}(z, u) = \sum_{m \geq 0} \sum_{\gamma \in \mathcal{I}_k(m)} \mathbf{S}_\gamma(z, u)$$

$$= \sum_{s \geq 0} \sum_{m=0}^{s} i_k(s, m) \mathbf{S}_\gamma(z, u).$$

We set

$$\eta_\tau(z, u) = \frac{u z^{2\tau}}{(1 - z)^2 (1 - z^2) - (2z - z^2) u z^{2\tau}}.$$

We then have

$$\mathbf{S}_{k,\tau}(z, u) = \sum_{s \geq 0} \sum_{m=0}^{s} i_k(s, m) \mathbf{S}_\gamma(z, u),$$

$$\sum_{s \geq 0} \sum_{m=0}^{s} i_k(s, m) \, x^s \, y^m = \frac{1 + x}{1 + 2x - xy} \sum_{s \geq 0} f_k(2s, 0) \left(\frac{x(1 + x)}{(1 + 2x - xy)^2} \right)^s.$$

Therefore, substituting $x = \eta_\tau(z, u)$ and $y = z$, we derive

$$\frac{1 - z^2 + u z^{2\tau}}{1 - z - z^2 + z^3 + 2u z^{2\tau} - u z^{1+2\tau}} \mathbf{F}_k \left(\frac{u z^{2\tau} (1 - z^2 + u z^{2\tau})}{(1 - z - z^2 + z^3 + 2u z^{2\tau} - u z^{1+2\tau})^2} \right).$$

Setting

$$g_\tau(z, u) = \frac{u z^{2(\tau-1)}}{u z^{2\tau} - z^2 + 1}$$

we arrive at

$$\mathbf{S}_{k,\tau}(z, u) = \frac{1}{g_\tau(z, u) z^2 - z + 1} \mathbf{F}_k \left(\left(\frac{\sqrt{g_\tau(z, u)} z}{g_\tau(z, u) z^2 - z + 1} \right)^2 \right),$$

whence the theorem.

Of course we have for minimum arc length $\lambda > 2$.

Theorem 5.16. *Let* $k, \tau \in \mathbb{N}$ $k \geq 2$, u, z *be indeterminants and suppose* $\lambda \leq \tau + 1$. *Then we have the identity of formal power series*

$$\mathbf{S}_{k,\tau}^{[\lambda]}(z, u) = \frac{1}{d_\lambda(z, u)} \mathbf{F}_k \left(\left(\frac{\sqrt{g_\tau(z, u)} \, z}{d_\lambda(z, u)} \right)^2 \right),$$

where $g_\tau(z, u)$ *and* $d_\lambda(z, u)$ *are given by*

$$g_\tau(z, u) = \frac{u z^{2(\tau-1)}}{u z^{2\tau} - z^2 + 1},$$

$$d_\lambda(z, u) = 1 - z + g_\tau(z, u) \sum_{n=2}^{\lambda} z^n.$$

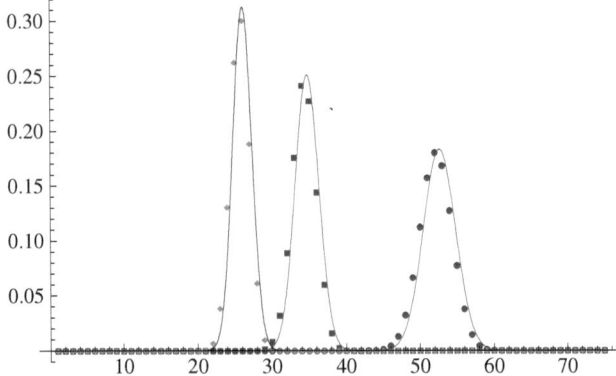

Fig. 5.7. Central limit theorems of Theorem 5.17 versus exact enumeration data for 3-noncrossing, 2-, 3-, and 4-canonical RNA structures with arc length ≥ 2 of length $n = 300$. We display the asymptotic stack distributions (*solid curves: red/blue/black*) and actual frequencies (*circle, box, diamond*) as computed for $n = 300$.

Proving the analogue of Proposition 5.13 for stacks we eventually derive the central limit theorem for stacks in k-noncrossing structures; see Fig. 5.7.

Theorem 5.17. *Let $k, \tau \in \mathbb{N}$, $k \geq 2$ and let $\mathbb{S}_{n,k,\tau}(S)$ be the number of stacks in a k-noncrossing, τ-canonical structure, S. Then there exists a pair $(\mu_{k,\tau}, \sigma_{k,\tau})$ such that the normalized random variable $\mathbb{S}^*_{n,k,\tau}$ has asymptotically normal distribution with parameter $(0, 1)$, where $\mu_{k,\tau}$ and $\sigma^2_{k,\tau}$ are given by*

$$\mu_{k,\tau} = -\frac{\gamma'_{k,\tau}(0)}{\gamma_{k,\tau}(0)}, \qquad \sigma^2_{k,\tau} = \left(\frac{\gamma'_{k,\tau}(0)}{\gamma_{k,\tau}(0)}\right)^2 - \frac{\gamma''_{k,\tau}(0)}{\gamma_{k,\tau}(0)}, \qquad (5.21)$$

where $\gamma_{k,\tau}(s)$ is the unique dominant singularity of $\mathbf{S}_{k,\tau}(z, e^s)$; see Table 5.2.

5.2.3 Hairpin loops, interior loops, and bulges

In this section we study three specific types of basic building blocks, called loops, of k-noncrossing, τ-canonical structures. We consider in the following hairpin, interior, and bulge loops of k-noncrossing, τ-canonical structures; see Fig. 5.8. Here a *bulge loop* is a either a triple of the form $((i_1, j_1), [i_1 + 1, i_2 - 1], (i_2, j_1 - 1))$ or $((i_1, j_1), (i_1 + 1, j_2), [j_2 + 1, j_1 - 1])$. We will eventually complete the above picture by discussing in Chapter 6 the two remaining loop types in k-noncrossing structures: multi- and pseudoknot-loops. In Fig. 5.9 we compare the distribution of hairpins and bulges in 3-noncrossing structures of length $n = 200$ obtained by Theorem 5.20 with uniformly generated structures; see Theorem 5.4.

For fixed k-noncrossing, τ-canonical structure, S, let $\mathbb{H}_{n,k,\tau}(S)$, $\mathbb{I}_{n,k,\tau}(S)$, and $\mathbb{B}_{n,k,\tau}(S)$ denote the number of hairpin loops, interior loops, and bulges in S. Then we have the r.vs.

	$k=2$		$k=3$		$k=4$	
	$\mu_{k,\tau}$	$\sigma^2_{k,\tau}$	$\mu_{k,\tau}$	$\sigma^2_{k,\tau}$	$\mu_{k,\tau}$	$\sigma^2_{k,\tau}$
$\tau=1$	0.236068	0.036260	0.373864	0.047201	0.416408	0.036366
$\tau=2$	0.135106	0.014758	0.175455	0.015860	0.190231	0.013993
$\tau=3$	0.095730	0.008494	0.115767	0.008430	0.123519	0.007604
$\tau=4$	0.074552	0.005708	0.086881	0.005435	0.091807	0.004935
$\tau=5$	0.061253	0.004186	0.069769	0.003889	0.073251	0.003541
$\tau=6$	0.052094	0.003248	0.058416	0.002969	0.061047	0.002707
$\tau=7$	0.045386	0.002621	0.050316	0.002369	0.052397	0.002162
	$k=5$		$k=6$		$k=7$	
	$\mu_{k,\tau}$	$\sigma^2_{k,\tau}$	$\mu_{k,\tau}$	$\sigma^2_{k,\tau}$	$\mu_{k,\tau}$	$\sigma^2_{k,\tau}$
$\tau=1$	0.437411	0.028803	0.449961	0.023671	0.458314	0.020032
$\tau=2$	0.198709	0.012520	0.204413	0.011389	0.208594	0.010498
$\tau=3$	0.128130	0.006976	0.131323	0.006494	0.133721	0.006110
$\tau=4$	0.094786	0.004569	0.096875	0.004290	0.098459	0.004069
$\tau=5$	0.075377	0.003292	0.076878	0.003104	0.078023	0.002956
$\tau=6$	0.062665	0.002523	0.063813	0.002385	0.064691	0.002277
$\tau=7$	0.053684	0.002018	0.054600	0.001911	0.055303	0.001827

Table 5.2. Stacks: central limit theorem for the numbers of stacks in k-noncrossing, τ-canonical structures. We list $\mu_{k,\tau}$ and $\sigma^2_{k,\tau}$ derived from eq. (5.21).

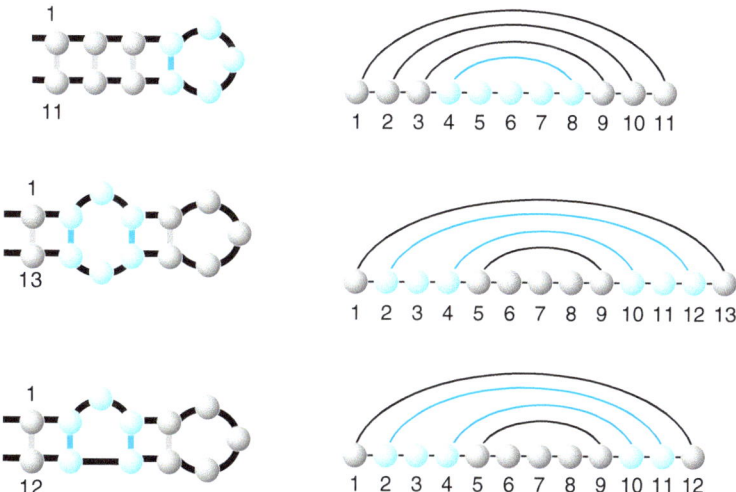

Fig. 5.8. Hairpin loop (*top*), interior loop (*middle*), and bulge (*bottom*).

- $\mathbb{H}_{n,k,\tau}$, where $\mathbb{P}\left(\mathbb{H}_{n,k,\tau}=t\right)=\frac{h_{k,\tau}(n,t)}{\mathsf{T}_{k,\tau}(n)}$,
- $\mathbb{I}_{n,k,\tau}$, where $\mathbb{P}\left(\mathbb{I}_{n,k,\tau}=t\right)=\frac{i_{k,\tau}(n,t)}{\mathsf{T}_{k,\tau}(n)}$,
- $\mathbb{B}_{n,k,\tau}$, where $\mathbb{P}\left(\mathbb{B}_{n,k,\tau}=t\right)=\frac{b_{k,\tau}(n,t)}{\mathsf{T}_{k,\tau}(n)}$.

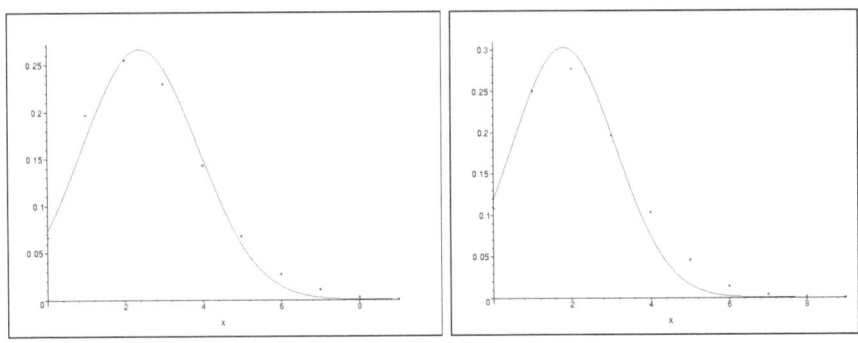

Fig. 5.9. The distribution of hairpins (*left*) and bulges (*right*) in 3-noncrossing structures of length $n = 200$. The *solid curves* are derived from the central limit theorem, Theorem 5.20. The *dots* are obtained via uniformly generating 3-noncrossing structures; see Theorem 5.4.

Here $h_{k,\tau}(n,t)$, $i_{k,\tau}(n,t)$, and $b_{k,\tau}(n,t)$ are the numbers of k-noncrossing, τ-canonical structures of length n with t hairpin loops, interior loops, and bulges. The key for computing the distributions of the above r.vs. is the bivariate generating functions

$$\mathbf{H}_{k,\tau}(z,u_1) = \sum_{n \geq 0} \sum_{t \geq 0} h_{k,\tau}(n,t) z^n u_1^t,$$

$$\mathbf{I}_{k,\tau}(z,u_2) = \sum_{n \geq 0} \sum_{t \geq 0} i_{k,\tau}(n,t) z^n u_2^t,$$

$$\mathbf{B}_{k,\tau}(z,u_3) = \sum_{n \geq 0} \sum_{t \geq 0} b_{k,\tau}(n,t) z^n u_3^t.$$

By construction, V_k-shapes, as introduced in Section 4.1.2, do not preserve stack-length, interior loops, bulges, and unpaired regions. When projecting into a V_k-shape, each stem, i.e., each sequence of nested stacks, is mapped into a single arc and all hairpin loops project into 1-arcs. Recall that $\mathcal{I}_k(m)$ denotes the set of shapes γ having m 1-arcs.

Plainly, for any shape we can construct its unique k-noncrossing, τ-canonical structure of minimal length by adding arcs to each shape-arc such that every stack consists of exactly τ arcs and inserting exactly one isolated vertex into each 1-arc.

Theorem 5.18. *Suppose $k, \tau \in \mathbb{N}, k \geq 2, \tau \geq 1$. Then*

$$\mathbf{H}_{k,\tau}(z,u_1) = \frac{(1-z)(1-z^2+z^{2\tau})}{(1-z)^2(1-z^2+z^{2\tau})+z^{2\tau}-z^{2\tau+1}u_1}$$

$$\mathbf{F}_k\left(\frac{z^{2\tau}(1-z)^2(1-z^2+z^{2\tau})}{\left((1-z)^2(1-z^2+z^{2\tau})+z^{2\tau}-z^{2\tau+1}u_1\right)^2}\right),$$

$$\mathbf{I}_{k,\tau}(z, u_2) = \frac{(1-z^2)(1-z)^2 - u_2 z^{2\tau+2} + (2z^2 - 2z + 1)z^{2\tau}}{(1-z)\left((1-z^2)(1-z)^2 - u_2 z^{2\tau+2} + (2z^2 - 3z + 2)z^{2\tau}\right)}$$

$$\mathbf{F}_k \left(\frac{z^{2\tau}\left((1-z^2)(1-z)^2 - u_2 z^{2\tau+2} + (2z^2 - 2z + 1)z^{2\tau}\right)}{\left((1-z^2)(1-z)^2 - u_2 z^{2\tau+2} + (2z^2 - 3z + 2)z^{2\tau}\right)^2} \right),$$

$$\mathbf{B}_{k,\tau}(z, u_3) = \frac{(1-z^2)(1-z) - 2u_3 z^{2\tau+1} + (z+1)z^{2\tau}}{(1-z)\left((1-z^2)(1-z) - 2u_3 z^{2\tau+1} + (z+2)z^{2\tau}\right)}$$

$$\mathbf{F}_k \left(\frac{z^{2\tau}\left((1-z^2)(1-z) - 2u_3 z^{2\tau+1} + (z+1)z^{2\tau}\right)}{(1-z)\left((1-z^2)(1-z) - 2u_3 z^{2\tau+1} + (z+2)z^{2\tau}\right)^2} \right).$$

Proof. We prove the theorem via symbolic enumeration representing a k-noncrossing, τ-canonical structure as the inflation of a shape, γ. A structure inflated from $\gamma \in \mathcal{I}_k(s, m)$ has exactly s stems, $(2s+1)$ (possibly empty) intervals of isolated vertices, and m nonempty such intervals. We use the notation and overall strategy of Theorem 4.9 and proceed by considering the combinatorial classes $\mathcal{M}$ (stems), $\mathcal{K}^\tau$ (stacks), $\mathcal{N}^\tau$ (induced stacks), $\mathcal{L}$ (isolated vertices), $\mathcal{R}$ (arcs), and $\mathcal{Z}$ (vertices), where $\mathbf{Z}(z) = z$ and $\mathbf{R}(z) = z^2$. Let μ_1, μ_2, and μ_3 be the combinatorial markers for hairpin loops, interior loops, and bulge loops, respectively. Then

$$\mathcal{T}_\gamma = (\mathcal{M})^s \times \mathcal{L}^{2s+1-m} \times (\mu_1 \times [\mathcal{Z} \times \mathcal{L}])^m,$$
$$\mathcal{M} = \mathcal{K}^\tau \times \text{SEQ}(\mathcal{N}^\tau),$$
$$\mathcal{N}^\tau = \mathcal{K}^\tau \times \left(\mu_3 \times [\mathcal{Z} \times \mathcal{L}] + \mu_3 \times [\mathcal{Z} \times \mathcal{L}] + \mu_2 \times [(\mathcal{Z} \times \mathcal{L})^2] \right),$$
$$\mathcal{K}^\tau = \mathcal{R}^\tau \times \text{SEQ}(\mathcal{R}),$$
$$\mathcal{L} = \text{SEQ}(\mathcal{Z}).$$

Consequently, translating the above relations into generating functions, we derive the following expression for $\mathbf{T}_\gamma(z, u_1, u_2, u_3)$:

$$\left(\frac{\frac{z^{2\tau}}{1-z^2}}{1 - \frac{z^{2\tau}}{1-z^2}\left(2\frac{u_3 z}{1-z} + u_2\left(\frac{z}{1-z}\right)^2\right)} \right)^s \left(\frac{1}{1-z} \right)^{2s+1-m} \left(\frac{u_1 z}{1-z} \right)^m$$

$$= (1-z)^{-1} \left(\frac{z^{2\tau}}{(1-z^2)(1-z)^2 - (2u_3 z(1-z) + u_2 z^2)z^{2\tau}} \right)^s (u_1 z)^m,$$

where the indeterminants u_i ($i = 1, 2, 3$) correspond to the combinatorial markers μ_i, $i = 1, 2, 3$, i.e., the occurrences of hairpin loops, interior loops, and bulges. Since for any two $\gamma, \gamma' \in \mathcal{I}_k(s, m)$

$$\mathbf{T}_\gamma(z, u_1, u_2, u_3) = \mathbf{T}_{\gamma'}(z, u_1, u_2, u_3)$$

holds we derive

$$\mathbf{H}_{k,\tau}(z, u_1) = \sum_{m \geq 0} \sum_{\gamma \in \mathcal{I}_k(m)} \mathbf{T}_\gamma(z, u_1, 1, 1) = \sum_{s \geq 0} \sum_{m=0}^{s} i_k(s, m) \mathbf{T}_\gamma(z, u_1, 1, 1),$$

$$\mathbf{I}_{k,\tau}(z, u_2) = \sum_{m \geq 0} \sum_{\gamma \in \mathcal{I}_k(m)} \mathbf{T}_\gamma(z, 1, u_2, 1) = \sum_{s \geq 0} \sum_{m=0}^{s} i_k(s, m) \mathbf{T}_\gamma(z, 1, u_2, 1),$$

$$\mathbf{B}_{k,\tau}(z, u_3) = \sum_{m \geq 0} \sum_{\gamma \in \mathcal{I}_k(m)} \mathbf{T}_\gamma(z, 1, 1, u_3) = \sum_{s \geq 0} \sum_{m=0}^{s} i_k(s, m) \mathbf{T}_\gamma(z, 1, 1, u_3).$$

We furthermore set

$$\eta(u_2, u_3) = \frac{z^{2\tau}}{(1 - z^2)(1 - z)^2 - (2u_3 \, z \, (1 - z) + u_2 \, z^2) z^{2\tau}}.$$

As in the proof of Theorem 4.9 it now remains to observe

$$\sum_{s \geq 0} \sum_{m=0}^{s} i_k(s, m) \, x^s \, y^m = \frac{1 + x}{1 + 2x - xy} \mathbf{F}_k \left(\frac{x(1 + x)}{(1 + 2x - xy)^2} \right)$$

and to subsequently substitute $x = \eta(1, 1)$ and $y = u_1 z$ for deriving $\mathbf{H}_{k,\tau}(z, u_1)$. Substituting $x = \eta(u_2, 1)$ and $y = z$, we obtain $\mathbf{I}_{k,\tau}(z, u_2)$ and finally $x = \eta(1, u_3)$ and $y = z$ produces the expression for $\mathbf{B}_{k,\tau}(z, u_3)$.

The next proposition is the analogue of Proposition 5.13. It is based on Theorem 2.21 and facilitates the application of Theorem 5.10.

Proposition 5.19. *Suppose* $2 \leq k \leq 7$, $1 \leq \tau \leq 10$. *There exists a unique dominant* $\mathbf{H}_{k,\tau}(z, e^s)$-*singularity,* $\gamma_{k,\tau}(s)$, *such that for* $|s| < \epsilon$, *where* $\epsilon > 0$:
(1) $\gamma_{k,\tau}(s)$ *is analytic,*
(2) $\gamma_{k,\tau}(s)$ *is the solution of minimal modulus of*

$$\frac{z^{2\tau}(1 - z)^2(1 - z^2 + z^{2\tau})}{((1 - z)^2(1 - z^2 + z^{2\tau}) + z^{2\tau} - z^{2\tau+1}e^s)^2} - \rho_k^2 = 0$$

and

$$[z^n]\mathbf{H}_{k,\tau}(z, e^s) \sim C(s) \, n^{-((k-1)^2 + \frac{k-1}{2})} \left(\frac{1}{\gamma_{k,\tau}(s)} \right)^n,$$

uniformly in s *in a neighborhood of* 0 *and continuous* $C(s)$.

Proof. The first step is to establish the existence and uniqueness of the dominant singularity $\gamma_{k,\tau}(s)$. We denote

$$\vartheta(z, s) = (1 - z)^2(1 - z^2 + z^{2\tau}) + z^{2\tau} - z^{2\tau+1}e^s,$$
$$\psi_\tau(z, s) = z^{2\tau}(1 - z)^2(1 - z^2 + z^{2\tau})\vartheta(z, s)^{-2},$$
$$\omega_\tau(z, s) = (1 - z)(1 - z^2 + z^{2\tau})\vartheta(z, s)^{-1},$$

and consider, following the strategy of Proposition 5.13, the equations

$$\forall\, 2 \leq i \leq k; \qquad F_{i,\tau}(z,s) = \psi_\tau(z,s) - \rho_i^2,$$

where $\rho_i = 1/(2i - 2)$. Theorem 5.18 and Proposition 2.22 imply that the singularities of $\mathbf{H}_{k,\tau}(z, e^s)$ are contained in the set of roots of

$$F_{i,\tau}(z,s) = 0 \quad \text{and} \quad \vartheta(z,s) = 0,$$

where $i \leq k$. Let $r_{i,\tau}$ denote the solution of minimal modulus of

$$F_{i,\tau}(z,0) = \psi_\tau(z,0) - \rho_i^2 = 0.$$

We next verify that, for sufficiently small $\epsilon_i > 0$, $|z - r_{i,\tau}| < \epsilon_i$, $|s| < \epsilon_i$, the following assertions hold:

- $\frac{\partial}{\partial z} F_{i,\tau}(r_{i,\tau}, 0) \neq 0$,
- $\frac{\partial}{\partial z} F_{i,\tau}(z,s)$ and $\frac{\partial}{\partial s} F_{i,\tau}(z,s)$ are continuous.

The analytic implicit function theorem guarantees the existence of a unique analytic function $\gamma_{i,\tau}(s)$ such that, for $|s| < \epsilon_i$,

$$F_{i,\tau}(\gamma_{i,\tau}(s), s) = 0 \quad \text{and} \quad \gamma_{i,\tau}(0) = r_{i,\tau}.$$

Analogously, we obtain the unique analytic function $\delta(s)$ satisfying $\vartheta(z,s) = 0$ and where $\delta(0)$ is the minimal solution of $\vartheta(z,0) = 0$ for $|s| < \epsilon_\delta$, for some $\epsilon_\delta > 0$. We next verify that the unique dominant singularity of $\mathbf{H}_{k,\tau}(z,1) = \mathbf{T}_{k,\tau}(z)$ is the minimal positive solution $r_{k,\tau}$ of $F_{k,\tau}(z,0) = 0$ and subsequently use the continuity argument as in the proof of Proposition 5.13. Therefore, for sufficiently small ϵ where $\epsilon < \epsilon_i$ and $\epsilon < \epsilon_\delta$, $|s| < \epsilon$, the moduli of $\gamma_{i,\tau}(s)$, $i < k$, and $\delta(s)$ are all strictly larger than the modulus of $\gamma_{k,\tau}(s)$. Consequently, $\gamma_{k,\tau}(s)$ is the unique dominant singularity of $\mathbf{H}_{k,\tau}(z, e^s)$.

Claim. There exists some continuous $C(s)$ such that, uniformly in s, for s in a neighborhood of 0

$$[z^n]\mathbf{H}_{k,\tau}(z, e^s) \sim C(s)\, n^{-((k-1)^2 + \frac{k-1}{2})} \left(\frac{1}{\gamma_{k,\tau}(s)} \right)^n.$$

To prove the claim, let r be some positive real number such that $r_{k,\tau} < r < \delta(0)$. For sufficiently small $\epsilon > 0$ and $|s| < \epsilon$,

$$|\gamma_{k,\tau}(s)| \leq r \quad \text{and} \quad |\delta(s)| > r.$$

Then $\psi_\tau(z,s)$ and $\omega_\tau(z,s)$ are all analytic in $\mathcal{D} = \{(z,s)\,|\,|z| \leq r, |s| < \epsilon\}$ and $\psi_\tau(0,s) = 0$. Since $\gamma_{k,\tau}(s)$ is the unique dominant singularity of

$$\mathbf{H}_{k,\tau}(z, e^s) = \omega_\tau(z,s)\, \mathbf{F}_k(\psi_\tau(z,s)),$$

satisfying

$$\psi_\tau(\gamma_{k,\tau}(s), s) = \rho_k^2 \quad \text{and} \quad |\gamma_{k,\tau}(s)| \leq r,$$

for $|s| < \epsilon$. For sufficiently small $\epsilon > 0$, $\frac{\partial}{\partial z} F_{k,\tau}(z,s)$ is continuous and $\frac{\partial}{\partial z} F_{k,\tau}(r_{k,\tau}, 0) \neq 0$. Thus there exists some $\epsilon > 0$, such that for $|s| < \epsilon$, $\frac{\partial}{\partial z} F_{k,\tau}(\gamma_{k,\tau}(s), s) \neq 0$. According to Theorem 2.21, we therefore derive

$$[z^n]\mathbf{H}_{k,\tau}(z, e^s) \sim C(s)\, n^{-((k-1)^2 + \frac{k-1}{2})} \left(\frac{1}{\gamma_{k,\tau}(s)} \right)^n,$$

uniformly in s in a neighborhood of 0 and continuous $C(s)$.

After establishing the analogues of Proposition 5.19 for $\mathbf{I}_{k,\tau}(z, u)$ and $\mathbf{B}_{k,\tau}(z, u)$, Theorem 5.10 eventually implies the following central limit theorems for the distributions of hairpin loops, interior loops, and bulges in k-noncrossing structures.

Theorem 5.20. *Let $k, \tau \in \mathbb{N}$, $k \geq 2$ and suppose the random variable $\mathbb{X}$ denotes either $\mathbb{H}_{n,k,\tau}$, $\mathbb{I}_{n,k,\tau}$ or $\mathbb{B}_{n,k,\tau}$. Then there exists a pair*

$$(\mu_{k,\tau,\mathbb{X}}, \sigma^2_{k,\tau,\mathbb{X}})$$

such that the normalized random variable $\mathbb{X}^*$ has asymptotically normal distribution with parameter $(0, 1)$, where $\mu_{k,\tau,\mathbb{X}}$ and $\sigma^2_{k,\tau,\mathbb{X}}$ are given by

$$\mu_{k,\tau,\mathbb{X}} = -\frac{\gamma'_{k,\tau,\mathbb{X}}(0)}{\gamma_{k,\tau,\mathbb{X}}(0)}, \qquad \sigma^2_{k,\tau,\mathbb{X}} = \left(\frac{\gamma'_{k,\tau,\mathbb{X}}(0)}{\gamma_{k,\tau,\mathbb{X}}(0)} \right)^2 - \frac{\gamma''_{k,\tau,\mathbb{X}}(0)}{\gamma_{k,\tau,\mathbb{X}}(0)}, \qquad (5.22)$$

where $\gamma_{k,\tau,\mathbb{X}}(s)$ represents the unique dominant singularity of $\mathbf{H}_{k,\tau}(z, e^s)$, $\mathbf{I}_{k,\tau}(z, e^s)$, and $\mathbf{B}_{k,\tau}(z, e^s)$, respectively.

In Tables 5.3, 5.4, and 5.5 we present the values of the pairs $(\mu_{k,\tau,\mathbb{X}}, \sigma^2_{k,\tau,\mathbb{X}})$.

	$k = 2$		$k = 3$		$k = 4$	
	$\mu_{k,\tau}$	$\sigma^2_{k,\tau}$	$\mu_{k,\tau}$	$\sigma^2_{k,\tau}$	$\mu_{k,\tau}$	$\sigma^2_{k,\tau}$
$\tau = 1$	0.105573	0.032260	0.012013	0.011202	0.003715	0.003641
$\tau = 2$	0.061281	0.018116	0.009845	0.008879	0.003734	0.003602
$\tau = 3$	0.043900	0.012752	0.007966	0.007060	0.003200	0.003060
$\tau = 4$	0.034477	0.009896	0.006680	0.005854	0.002757	0.002622
	$k = 5$		$k = 6$		$k = 7$	
	$\mu_{k,\tau}$	$\sigma^2_{k,\tau}$	$\mu_{k,\tau}$	$\sigma^2_{k,\tau}$	$\mu_{k,\tau}$	$\sigma^2_{k,\tau}$
$\tau = 1$	0.001626	0.001612	0.000855	0.000852	0.000505	0.000504
$\tau = 2$	0.001897	0.001864	0.001123	0.001111	0.000731	0.000726
$\tau = 3$	0.001693	0.001655	0.001035	0.001021	0.000692	0.000686
$\tau = 4$	0.001486	0.001448	0.000922	0.000907	0.000624	0.000618

Table 5.3. Hairpin loops: The central limit theorem for the numbers of hairpin loops in k-noncrossing, τ-canonical structures. We list $\mu_{k,\tau}$ and $\sigma^2_{k,\tau}$ derived from eq. (5.22).

	$k = 2$		$k = 3$		$k = 4$	
	$\mu_{k,\tau}$	$\sigma^2_{k,\tau}$	$\mu_{k,\tau}$	$\sigma^2_{k,\tau}$	$\mu_{k,\tau}$	$\sigma^2_{k,\tau}$
$\tau = 1$	0.015403	0.013916	0.001185	0.011759	0.000264	0.000264
$\tau = 2$	0.012959	0.011395	0.001823	0.001793	0.000603	0.000599
$\tau = 3$	0.011075	0.009570	0.001878	0.001837	0.000693	0.000688
$\tau = 4$	0.009682	0.008261	0.001803	0.001755	0.000700	0.000693
	$k = 5$		$k = 6$		$k = 7$	
	$\mu_{k,\tau}$	$\sigma^2_{k,\tau}$	$\mu_{k,\tau}$	$\sigma^2_{k,\tau}$	$\mu_{k,\tau}$	$\sigma^2_{k,\tau}$
$\tau = 1$	0.000090	0.000090	0.000039	0.000039	0.000019	0.000019
$\tau = 2$	0.000275	0.000274	0.000149	0.000149	0.000090	0.000090
$\tau = 3$	0.000343	0.000341	0.000198	0.000198	0.000126	0.000126
$\tau = 4$	0.000359	0.000357	0.000214	0.000213	0.000140	0.000140

Table 5.4. Interior loops: The central limit theorem for the numbers of interior loops in k-noncrossing, τ-canonical structures. We list $\mu_{k,\tau}$ and $\sigma^2_{k,\tau}$ derived from eq. (5.22).

	$k = 2$		$k = 3$		$k = 4$	
	$\mu_{k,\tau}$	$\sigma^2_{k,\tau}$	$\mu_{k,\tau}$	$\sigma^2_{k,\tau}$	$\mu_{k,\tau}$	$\sigma^2_{k,\tau}$
$\tau = 1$	0.049845	0.042310	0.008982	0.008684	0.003094	0.003058
$\tau = 2$	0.025088	0.021785	0.005789	0.005597	0.002457	0.002422
$\tau = 3$	0.015859	0.013979	0.003936	0.003814	0.001762	0.001737
$\tau = 4$	0.011197	0.009980	0.002878	0.002795	0.001318	0.001301
	$k = 5$		$k = 6$		$k = 7$	
	$\mu_{k,\tau}$	$\sigma^2_{k,\tau}$	$\mu_{k,\tau}$	$\sigma^2_{k,\tau}$	$\mu_{k,\tau}$	$\sigma^2_{k,\tau}$
$\tau = 1$	0.001422	0.001414	0.000770	0.000767	0.000463	0.000462
$\tau = 2$	0.001326	0.001316	0.000817	0.000813	0.000547	0.000546
$\tau = 3$	0.000991	0.000984	0.000632	0.000629	0.000436	0.000435
$\tau = 4$	0.000755	0.000750	0.000489	0.000486	0.000342	0.000341

Table 5.5. Bulges: Central limit theorems for the numbers of bulge-loops in k-noncrossing, τ-canonical structures. We list $\mu_{k,\tau}$ and $\sigma^2_{k,\tau}$ derived from eq. (5.22).

5.3 Discrete limit laws

The correspondence between secondary structures and Motzkin paths shows that any notion of "irreducibility" is related to the number of nontrivial returns, i.e., the number of non-endpoints, for which the Motzkin path, meets the x-axis; see Fig. 1.4.

For Dyck paths, this question has been studied by Shapiro [30], who showed that the expected number of nontrivial returns of Dyck paths of length $2n$ equals $\frac{2n-2}{n+2}$. Shapiro and Cameron [20] derived expectation and variance of the number of nontrivial returns for generalized Dyck paths from $(0,0)$ to $((t+1)n, 0)$

$$\mathbb{E}[\xi_t] = \frac{2n-2}{tn+2} \quad \text{and} \quad \mathbb{V}[\xi_t] = \frac{2tn(n-1)((t+1)n+1)}{(tn+2)^2(tn+3)}. \tag{5.23}$$

One approach to obtain eq. (5.23) is to use the Riordan matrix [121], an infinite, lower triangular matrix $L = (l_{n,k})_{n,k\geq 0} = (g, f)$, where $g(z) = \sum_{n\geq 0} g_n z^n$, $f(z) = \sum_{n\geq 0} f_n z^n$ with $f_0 = 0, f_1 \neq 0$, such that $\sum_{n\geq k} l_{n,k} z^n = g(z)f^k(z)$. Clearly,

$$\mathbf{C}(z) = \sum_{n\geq 0} C_n z^n = \frac{1-\sqrt{1-4z}}{2z} \quad \text{where} \quad C_n = \frac{1}{n+1}\binom{2n}{n}$$

is the generating function of Dyck paths. Let $\zeta_{n,j}$ denote the number of Dyck paths of length $2n$ with j nontrivial returns. We consider the Riordan matrix $L = (\zeta_{n,j})_{n,j\geq 0} = (z\mathbf{C}(z), z\mathbf{C}(z))$ and extract the coefficients $\zeta_{n,j}$ from its generating function $(z\mathbf{C}(z))^{j+1}$ by Lagrange inversion. Setting $f(z) = zG(f(z))$ with $f(z) = \mathbf{C}(z) - 1$ and $G(z) = (1+z)^2$, we obtain

$$\zeta_{n,j} = [z^{n-j-1}](f(z)+1)^{j+1} = \frac{j+1}{2n-j-1}\binom{2n-j-1}{n},$$

where $\sum_{j\geq 0} \zeta_{n,j} = C_n$. From this we immediately compute

$$\mathbb{E}[\xi_1] = \sum_{j\geq 1} j \cdot \frac{\zeta_{n,j}}{C_n},$$

$$\mathbb{V}[\xi_1] = \sum_{j\geq 1} j^2 \cdot \frac{\zeta_{n,j}}{C_n} - \left(\sum_{j\geq 1} j \cdot \frac{\zeta_{n,j}}{C_n}\right)^2,$$

from which the expression of eq. (5.23) for $t = 1$ follows.

In Section 5.3.1, we consider the bivariate generating function directly, which relates to the Riordan matrix in case of generalized Dyck paths as follows:

$$\sum_{n\geq 0}\sum_{j\geq 0} \zeta_{n,j} w^j z^n = \sum_{j\geq 0} z^{j+1}\mathbf{C}(z)^{j+1}w^j = \frac{z\mathbf{C}(z)}{1-wz\mathbf{C}(z)}.$$

The main idea is to derive the bivariate generating functions from the Riordan matrix employing irreducible paths and to establish via singularity analysis discrete limit laws. The continuity theorem of discrete limit laws stated below will be used in the proofs of Theorems 5.23 and 5.25. It ensures that under certain conditions the pointwise convergence of probability generating functions implies the convergence of its coefficients.

Theorem 5.21. (Flajolet and Sedgewick [42]) *Let u be an indeterminant and Ω be a set contained in the unit disc, having at least one accumulation point in the interior of the disc.*
Assume $P_n(u) = \sum_{k \geq 0} p_{n,k} u^k$ and $q(u) = \sum_{k \geq 0} q_k u^k$ such that

$$\forall\, u \in \Omega; \quad \lim_{n \to \infty} P_n(u) = q(u).$$

Then we have for any finite k,

$$\lim_{n \to \infty} p_{n,k} = q_k \quad and \quad \lim_{n \to \infty} \sum_{j \leq k} p_{n,j} = \sum_{j \leq k} q_j. \tag{5.24}$$

In Section 2.1.3 we showed in Theorem 2.2 that there exists a bijection between k-noncrossing partial matchings and walks of length n in $\mathbb{Z}^{k-1}$ which start and end at $a = (k-1, k-2, \ldots, 1)$, having steps $0, \pm e_i$, $1 \leq i \leq k-1$ such that $0 < x_{k-1} < \cdots < x_1$ at any step. These walks correspond to $*$-tableaux where we identify the rth coordinate of a lattice point with the number of squares in the corresponding shape of the $*$-tableaux. In Chapter 4 we computed various generating functions of k-noncrossing structures. Let $\mathbf{T}_k(z)$ denote the generating function of such a fixed class $\mathcal{S}_k$. Clearly, since each structure is in particular a k-noncrossing partial matching, $\mathcal{S}_k$ corresponds via the bijection of Theorem 2.2 to a subset of $*$-tableaux. We refer to this class of tableaux as $\sharp$-tableaux. We remark that our results hold for various classes of Motzkin paths. For instance, we have for nontrivial returns of Motzkin paths with height ≥ 3 and plateau length ≥ 3:

$$\lim_{n \to \infty} \mathbb{E}[\eta_n] \approx 5.4526 \quad \text{and} \quad \lim_{n \to \infty} \mathbb{V}[\eta_n] \approx 20.3179.$$

We next come to irreducible subdiagrams. A subdiagram of a k-noncrossing diagram is a subgraph over a subset $M \subset [n]$ of consecutive vertices that corresponds to some $\sharp$-tableaux. We are now in position to introduce gaps and irreducible subdiagrams; see Fig. 5.10:

- A gap of length r (r-gap) is a maximal sequence of consecutive $\varnothing$-shapes $(\lambda^j, \ldots, \lambda^{j+r})$. In particular, a gap consisting of a single $\varnothing$-shape is a 0-gap.
- An irreducible subdiagram is a subdiagram whose corresponding $\sharp$-tableaux has exactly two 0-gaps, λ^0 and λ^n, respectively.

Plainly, any k-noncrossing diagram corresponds to a unique (alternating) sequence of gaps and irreducible subdiagrams. Note that a k-noncrossing diagram without any unpaired vertices is not necessarily irreducible. A sequence of r isolated vertices in a k-noncrossing diagram corresponds via the bijection of Theorem 2.2 to a sequence of consecutive $\varnothing$-shapes of length $(r+1)$, $(\varnothing, \ldots, \varnothing)$. Obviously, any $\sharp$-tableaux can uniquely be decomposed into a sequence of gaps and irreducible $\sharp$-tableaux. Note that $\varnothing$-shapes directly preceding and following nonempty shapes are considered for gaps as well as for irreducible $\sharp$-tableaux; see Fig. 5.11.

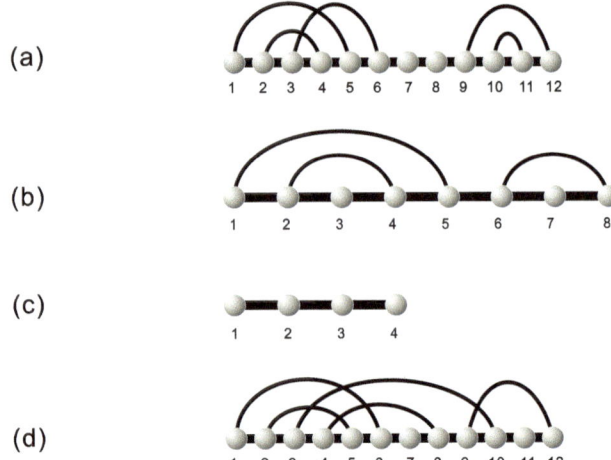

Fig. 5.10. Subdiagrams, gaps, and irreducibility: A subdiagram (**a**) is decomposed into the irreducible subdiagram over $(1, 6)$, the gap $(7, 8)$, and the irreducible subdiagram over $(9, 12)$. A subdiagram (**b**) decomposes into the irreducible subdiagram over $(1, 5)$, the 0-gap, and the irreducible subdiagram over $(6, 8)$. Finally we display a gap (**c**) and an irreducible diagram over $(1, 12)$ (**d**).

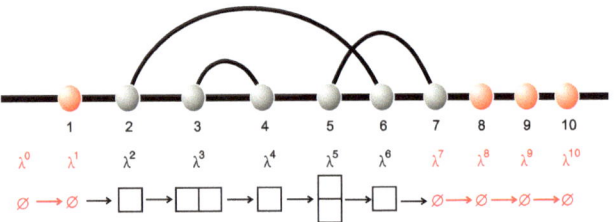

Fig. 5.11. Gaps in diagrams and their associated $\sharp$-tableaux: here we have the 1-gap (λ^0, λ^1), the irreducible $\sharp$-tableaux over $(\lambda^1, \ldots, \lambda^7)$, and the 3-gap $(\lambda^7, \ldots, \lambda^{10})$.

5.3.1 Irreducible substructures

Let $\delta_{n,j}^{(k)}$ denote the number of $\sharp$-tableaux of length n with less than k rows, containing exactly j irreducible $\sharp$-tableaux. Furthermore, let

$$\mathbf{U}_k(z, u) = \sum_{n \geq 0} \sum_{j \geq 0} \delta_{n,j}^{(k)} u^j z^n$$

and $\delta_n^{(k)} = \sum_{j \geq 0} \delta_{n,j}^{(k)}$. Plainly, $\mathbf{T}_k(z) = \sum_{n \geq 0} \delta_n^{(k)} z^n$ and we denote the generating function of irreducible $\sharp$-tableaux by $\mathbf{Irr}_k(z)$.

Lemma 5.22. *The bivariate generating function of $\sharp$-tableaux of length n with less than k rows, which contain exactly i irreducible $\sharp$-tableaux, is given by*

$$\mathbf{Irr}_k(z) = 1 - z - \frac{1}{\mathbf{T}_k(z)}, \tag{5.25}$$

$$\mathbf{U}_k(z, u) = \frac{\frac{1}{1-z}}{1 - u\left(1 - \frac{1}{(1-z)\mathbf{T}_k(z)}\right)}. \tag{5.26}$$

Proof. Since each ♯-tableaux can uniquely be decomposed into a sequence of gaps and irreducible ♯-tableaux, we obtain for fixed j

$$\sum_{n \geq j} \delta_{n,j}^{(k)} z^n = \mathbf{Irr}_k(z)^j \left(\frac{1}{1-z}\right)^{j+1}.$$

As a result, the bivariate generating function of $\delta_{n,j}$ is given by

$$\begin{aligned}
\mathbf{U}_k(z, u) &= \sum_{j \geq 0} \sum_{n \geq j} \delta_{n,j}^{(k)} z^n u^j \\
&= \sum_{j \geq 0} \mathbf{Irr}_k(z)^j \left(\frac{1}{1-z}\right)^{j+1} u^j \\
&= \frac{1}{1 - z - u\mathbf{Irr}_k(z)}.
\end{aligned}$$

Setting $u = 1$, we derive

$$\mathbf{T}_k(z) = \mathbf{U}_k(z, 1) = \frac{1}{1 - z - \mathbf{Irr}_k(z)}, \tag{5.27}$$

whence

$$\mathbf{Irr}_k(z) = 1 - z - \frac{1}{\mathbf{T}_k(z)}.$$

Consequently, $\mathbf{U}_k(z, u)$ is given by

$$\mathbf{U}_k(z, u) = \frac{1}{1 - z - u\mathbf{Irr}_k(z)} = \frac{\frac{1}{1-z}}{1 - u\left(1 - \frac{1}{(1-z)\mathbf{T}_k(z)}\right)}$$

and the lemma follows.

Setting $g(z) = \frac{1}{1-z}$ and $h(z) = 1 - \frac{1}{(1-z)\mathbf{T}_k(z)}$, Lemma 5.22 implies

$$\mathbf{U}_k(z, u) = g(z) \cdot \frac{1}{1 - uh(z)} = g(z) \cdot g(uh(z)). \tag{5.28}$$

Let $\xi_n^{(k)}$ be a r.v. such that

$$\mathbb{P}(\xi_n^{(k)} = i) = \frac{\delta_{n,i}^{(k)}}{\delta_n^{(k)}}$$

and let ρ_v, ρ_w denote the radius of convergence of the power series $v(z)$ and $w(z)$, respectively. We denote $\tau_w = \lim_{z \to \rho_w^-} w(z)$ and call a D-finite function $F(z) = v(w(z))$ subcritical if and only if $\tau_w < \rho_v$ [42].

Theorem 5.23. *Let α_k be the real positive dominant singularity of $\mathbf{T}_k(z)$ and set $\tau_k = (1 - \alpha_k)\mathbf{T}_k(\alpha_k)$. Then the r.v. $\xi_n^{(k)}$ satisfies the discrete limit law*

$$\lim_{n \to \infty} \mathbb{P}(\xi_n^{(k)} = i) = q_i, \quad where \quad q_i = \frac{i}{\tau_k^2}\left(\frac{\tau_k - 1}{\tau_k}\right)^{i-1},$$

that is, $\xi_n^{(k)}$ is determined by the density function of a $\Gamma(\ln \frac{\tau_k}{\tau_k - 1}, 2)$-distribution. Furthermore, the probability generating function of the limit distribution $q(u) = \sum_{i \geq 1} q_i u^i$ is given by

$$q(u) = \frac{u}{((1 - u)\tau_k + u)^2}. \tag{5.29}$$

Proof. Since $g(z) = \frac{1}{1-z}$ and $h(z) = 1 - \frac{1}{(1-z)\mathbf{T}_k(z)}$ have non-negative coefficients and $h(0) = 0$, the composition $g(h(z))$ is well defined as a formal power series. According to eq. (5.28) we may express $\mathbf{U}_k(z, u)$ as

$$\mathbf{U}_k(z, u) = g(z) \cdot g(uh(z)).$$

Furthermore, for $z = \alpha_k$ we have $1 - \frac{1}{(1-\alpha_k)\mathbf{T}_k(\alpha_k)} < 1 = \rho_g$.

Claim 1. $h(z)$ has a singular expansion at its dominant singularity $z = \alpha_k$ and there exists some constant $\tilde{c}_k > 0$ such that

$$h(z) = \begin{cases} \tilde{P}\left(1 - \frac{z}{\alpha_k}\right) + \tilde{c}_k\left(1 - \frac{z}{\alpha_k}\right)^\mu \ln\left(\frac{1}{1 - \frac{z}{\alpha_k}}\right)(1 + o(1)) & \text{for } k \text{ odd,} \\ \tilde{P}\left(1 - \frac{z}{\alpha_k}\right) + \tilde{c}_k\left(1 - \frac{z}{\alpha_k}\right)^\mu (1 + o(1)) & \text{for } k \text{ even,} \end{cases}$$

for $z \to \alpha_k$ and $\mu = (k - 1)^2 + \frac{k-1}{2} - 1$, where $\tilde{P}$ is a polynomial of degree not larger than μ.

To prove Claim 1 we notice that the D-finiteness of $\mathbf{T}_k(z)$ guarantees the existence of an analytic continuation $\mathbf{T}_k^*(z)$ for which $\mathbf{T}_k(z) = \mathbf{T}_k^*(z)$ holds for some simply connected Δ_{α_k}-domain [125]. Equation (5.27) implies $\mathbf{T}_k^*(z) > 0$ for $z \in \Delta_{\alpha_k}$, from which we conclude that

$$h^*(z) = 1 - \frac{1}{(1 - z)\mathbf{T}_k^*(z)}$$

is an analytic continuation of $h(z)$ to Δ_{α_k}. To obtain the singular expansion of $h(z)$, we consider the singular expansion of $(1 - z)h(z) + z = 1 - \frac{1}{\mathbf{T}_k(z)}$ and rewrite it as

$$(1 - z)h(z) + z = f(v(z)),$$

where $f(v) = 1 - \frac{1}{1-v}$ and $v(z) = 1 - \mathbf{T}_k(z)$. Then the composition $f(v(z))$ belongs to the subcritical case. The singular expansion of $f(v(z))$ is then given by combining the regular expansion of $f(v)$ with the singular expansion of $v(z)$ at α_k. Setting $v = v(z)$ and $\beta_k = v(\alpha_k) < 1$ we compute

$$f(v(z)) = \frac{-\beta_k}{1 - \beta_k} + \frac{1}{(1 - \beta_k)^2} \cdot (v - \beta_k)(1 + o(1))$$

$$= \frac{-\beta_k}{1 - \beta_k} + \frac{1}{(1 - \beta_k)^2} \cdot (\mathbf{T}_k(z) - \mathbf{T}_k(\alpha_k))(1 + o(1)).$$

Recall that $\mathbf{T}_k(z)$ is a composition of the form $\mathbf{F}_k(\vartheta(z))$, where $\vartheta(z)$ is algebraic and $\vartheta(0) = 0$. Furthermore, we are given the supercritical case of singularity analysis, see Theorem 2.21, i.e., the subexponential factors of the asymptotic expressions of $[z^n]\mathbf{T}_k(z)$ coincide with those of $[z^n]\mathbf{F}_k(z)$. Consequently

$$\mathbf{T}_k(z) = \begin{cases} P\left(1 - \frac{z}{\alpha_k}\right) + c_k(1 - \frac{z}{\alpha_k})^\mu \ln\left(\frac{1}{1 - \frac{z}{\alpha_k}}\right)(1 + o(1)) & k \text{ odd}, z \to \alpha_k \\ P\left(1 - \frac{z}{\alpha_k}\right) + c_k(1 - \frac{z}{\alpha_k})^\mu(1 + o(1)) & k \text{ even}, z \to \alpha_k, \end{cases}$$

where P is a polynomial of degree not larger than μ. Consequently, $h(z)$ has a singular expansion at $z = \alpha_k$, given by

$$h(z) = \frac{1}{1 - z} \cdot f(v(z)) - \frac{z}{1 - z}$$

$$= \begin{cases} \tilde{P}\left(1 - \frac{z}{\alpha_k}\right) + \tilde{c}_k\left(1 - \frac{z}{\alpha_k}\right)^\mu \ln\left(\frac{1}{1 - \frac{z}{\alpha_k}}\right)(1 + o(1)) & k \text{ odd}, z \to \alpha_k \\ \tilde{P}\left(1 - \frac{z}{\alpha_k}\right) + \tilde{c}_k\left(1 - \frac{z}{\alpha_k}\right)^\mu(1 + o(1)) & k \text{ even}, z \to \alpha_k, \end{cases}$$

where $\tilde{P}$ is a polynomial of degree not larger than μ, Claim 1 is proved. Note that Claim 1 and Theorem 2.19 imply

$$[z^n]h(z) \sim \tilde{c}_k n^{-\mu-1}\alpha_k^{-n}(1 + o(1)).$$

According to Claim 1, $\mathbf{U}_k(z, u) = g(z)g(uh(z))$, for $u \in (0, 1)$ has the unique dominant singularity α_k and a singular expansion. Without loss of generality, we restrict our analysis in the following to the case $k \equiv 1 \mod 2$. We consider first $\mathbf{U}_k(z, 1) = \mathbf{T}_k(z)$. For $k \equiv 1 \mod 2$, Theorem 2.19 implies

$$[z^n]\mathbf{U}_k(z, 1) = \tilde{c}_k \alpha_k^{-n} n^{-\mu-1}(1 + o(1)). \tag{5.30}$$

Second, we consider the bivariate generating function $\mathbf{U}_k(z, u)$. For any fixed $u \in (0, 1)$, we write

$$\mathbf{U}_k(z, u) = g(z) \cdot v_u(w(z)),$$

where $v_u(z) = \frac{z}{z - u(z-1)}$ and $w(z) = (1-z)\mathbf{T}_k(z)$. We focus on the composition $v_u(w(z))$ which belongs to the subcritical case of singularity analysis [42] VI.9., p. 411. See also Prop. IX.1, p. 629, therein. In the subcritical case, the inner function, $w(z)$, has a singular expansion at its unique dominant singularity having strictly smaller modulus than that of the singularity of the outer function, v_u. The singular expansion of $v_u(w(z))$ is then given by combining the regular expansion of v_u with the singular expansion of $w(z)$ at α_k. Setting $w = w(z)$ and $\tau_k = w(\alpha_k) > 1$ we compute

$$
\begin{aligned}
\mathbf{U}_k(z, u) &= g(z) \cdot \frac{w(z)}{w(z) - u(w(z) - 1)} \\
&= \frac{g(\alpha_k) \cdot \tau_k}{\tau_k - u(\tau_k - 1)} + g(\alpha_k) \frac{d}{dw} \left(\frac{w}{w - u(w - 1)} \right) \Bigg|_{w=\tau_k} (w - \tau_k) + \cdots \\
&= \frac{g(\alpha_k) \cdot \tau_k}{\tau_k - u(\tau_k - 1)} + g(\alpha_k) \frac{u}{((1 - u)\tau_k + u)^2} (w - \tau_k)(1 + o(1)).
\end{aligned}
$$

The transfer theorem, Theorem 2.19, guarantees

$$
\begin{aligned}
[z^n]\mathbf{U}_k(z, u) &= g(\alpha_k) \frac{u}{((1 - u)\tau_k + u)^2} (1 - \alpha_k)[z^n]\mathbf{T}_k(z)(1 + o(1)) \\
&= \frac{u}{((1 - u)\tau_k + u)^2} \tilde{c}_k \alpha_k^{-n} n^{-\mu - 1}(1 + o(1)).
\end{aligned}
$$

We consequently arrive at

$$
\lim_{n \to \infty} \frac{[z^n]\mathbf{U}_k(z, u)}{[z^n]\mathbf{U}_k(z, 1)} = \frac{u}{((1 - u)\tau_k + u)^2} = q(u). \tag{5.31}
$$

In view of eq. (5.31) and

$$
[u^i]q(u) = \frac{i}{\tau_k^2} \left(\frac{\tau_k - 1}{\tau_k} \right)^{i-1} = q_i,
$$

Theorem 5.21 implies the discrete limit law

$$
\lim_{n \to \infty} \mathbb{P}(\xi_n^{(k)} = i) = \lim_{n \to \infty} \frac{\delta_{n,i}^{(k)}}{\delta_n^{(k)}} = q_i, \qquad \text{where} \quad q_i = \frac{i}{\tau_k^2} \left(\frac{\tau_k - 1}{\tau_k} \right)^{i-1}.
$$

Since the density function of a $\Gamma(\lambda, r)$-distribution is given by

$$
f_{\lambda, r}(x) = \begin{cases} \frac{\lambda^r}{\Gamma(r)} x^{r-1} e^{-\lambda x}, & x > 0 \\ 0 & , x \leq 0, \end{cases} \tag{5.32}
$$

where $\lambda > 0$ and $r > 0$, we obtain, setting $r = 2$ and $\lambda = \ln \frac{\tau_k}{\tau_k - 1} > 0$

$$\lim_{n\to\infty} \mathbb{P}(\xi_n^{(k)} = i) = \frac{i}{\tau_k^2}\left(\frac{\tau_k - 1}{\tau_k}\right)^{i-1}$$

$$= \frac{1}{\tau_k(\tau_k - 1)}\left(\ln\frac{\tau_k}{\tau_k - 1}\right)^{-2}\left(\ln\frac{\tau_k}{\tau_k - 1}\right)^2 \cdot i\left(\frac{\tau_k - 1}{\tau_k}\right)^i$$

$$= \frac{1}{\tau_k(\tau_k - 1)}\left(\ln\frac{\tau_k}{\tau_k - 1}\right)^{-2} f_{\ln\frac{\tau_k}{\tau_k - 1},2}(i)$$

and the proof of the theorem is complete.

5.3.2 The limit distribution of nontrivial returns

Let $\beta_n^{(k)}$ denote the number of $\sharp$-tableaux of length n. Let $\beta_{n,i}^{(k)}$ denote the number of $\sharp$-tableaux of length n, having exactly i $\varnothing$-shapes contained in the sequence $(\lambda^1, \dots, \lambda^n)$. Let $\mathbf{W}_k(z, u)$ denote the bivariate generating function of $\beta_{n,i}^{(k)}$. Then $\beta_{n,i}^{(k)} = [z^n u^i]\mathbf{W}_k(z, u)$ and

$$\mathbf{W}_k(z, u) = \sum_{i \geq 0}\sum_{n \geq i} \beta_{n,i}^{(k)} z^n u^i.$$

Furthermore, we set $\beta_n^{(k)} = [z^n]\mathbf{W}_k(z, 1)$.

Lemma 5.24. *The bivariate generating function of the number of $\sharp$-tableaux of length n, with less than k rows, containing exactly i $\varnothing$-shapes, is given by*

$$\mathbf{W}_k(z, u) = \frac{1}{1 - u\left(1 - \frac{1}{\mathbf{T}_k(z)}\right)}.$$

Proof. Suppose the $\sharp$-tableaux $(\lambda^1, \dots, \lambda^n)$ contains exactly i $\varnothing$-shapes. These $\varnothing$-shapes split $(\lambda^1, \dots, \lambda^n)$ uniquely into exactly i $\sharp$-tableaux, each of which either being a gap of length 2 or an irreducible $\sharp$-tableaux. We conclude from this that for fixed i

$$\sum_{n \geq i} \beta_{n,i}^{(k)} z^n = (z + \mathbf{Irr}_k(z))^i$$

holds. Therefore, the bivariate generating function $\mathbf{W}_k(z, u)$ satisfies

$$\mathbf{W}_k(z, u) = \sum_{i \geq 0}\sum_{n \geq i} \beta_{n,i}^{(k)} z^n u^i = \sum_{i \geq 0}(z + \mathbf{Irr}_k(z))^i u^i$$

$$= \frac{1}{1 - u(z + \mathbf{Irr}_k(z))}$$

$$= \frac{1}{1 - u\left(1 - \frac{1}{\mathbf{T}_k(z)}\right)},$$

where the last equality follows from eq. (5.25), proving the lemma.

We set $g(z) = \frac{1}{1-z}$, $h(z) = 1 - \frac{1}{\mathbf{T}_k(z)}$ and let $\eta_n^{(k)}$ denote the random variable having probability distribution

$$\mathbb{P}(\eta_n^{(k)} = i) = \frac{\beta_{n,i}^{(k)}}{\beta_n^{(k)}}.$$

In our next theorem, we prove that the limit distribution of $\eta_n^{(k)}$ is determined by the density function of a $\Gamma(\lambda, r)$-distribution.

Theorem 5.25. *Let α_k denote the real, positive, dominant singularity of $\mathbf{T}_k(z)$ and let $\tau_k = \mathbf{T}_k(\alpha_k)$. Then the r.v. $\eta_n^{(k)}$ satisfies the discrete limit law*

$$\lim_{n \to \infty} \mathbb{P}(\eta_n^{(k)} = i) = q_i, \quad where \quad q_i = \frac{i}{\tau_k^2} \left(\frac{\tau_k - 1}{\tau_k} \right)^{i-1},$$

that is, $\eta_n^{(k)}$ is determined by the density function of a $\Gamma(\ln \frac{\tau_k}{\tau_k - 1}, 2)$-distribution and the limit distribution has the probability generating function (Fig. 5.12)

$$q(u) = \sum_{i \geq 1} q_i u^i = \frac{u}{(\tau_k(1 - u) + u)^2}.$$

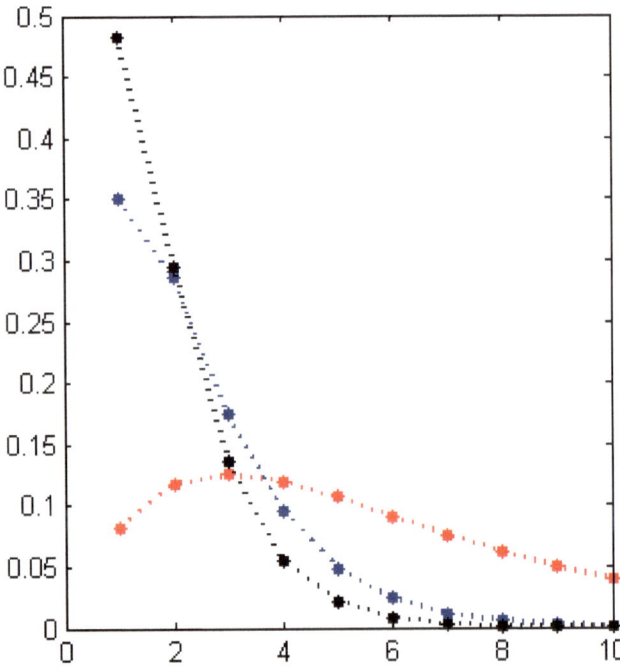

Fig. 5.12. Illustration of Theorem 5.25: the discrete distributions of $\eta_n^{(k)}$ for k-noncrossing, 2-canonical structures having minimal arc length 2. We display the distributions for $k = 2$ (*red*), $k = 3$ (*blue*), and $k = 4$ (*black*).

Proof. Since $g(z) = \frac{1}{1-z}$ and $h(z) = 1 - \frac{1}{\mathbf{T}_k(z)}$ have non-negative coefficients and $h(0) = 0$, the composition $g(h(z))$ is again a power series. $\mathbf{W}_k(z, u) = g(uh(z))$ has its unique dominant singularity at $z = \alpha_k$. Furthermore, we observe that irrespective of potential singularities arising from $\mathbf{T}_k(z) = 0$, the dominant singularity of $h(z) = 1 - \frac{1}{\mathbf{T}_k(z)}$ equals the dominant singularity of $\mathbf{T}_k(z)$, i.e. $z = \alpha_k$.

Claim 1. $h(z)$ has a singular expansion at $z = \alpha_k$ and there exists some constant $c_k > 0$ such that

$$h(z) = \begin{cases} P\left(1 - \frac{z}{\alpha_k}\right) + c_k\left(1 - \frac{z}{\alpha_k}\right)^{\mu} \ln\left(\frac{1}{1-\frac{z}{\alpha_k}}\right)(1 + o(1)) & \text{for } k \text{ odd} \\ P\left(1 - \frac{z}{\alpha_k}\right) + c_k\left(1 - \frac{z}{\alpha_k}\right)^{\mu}(1 + o(1)) & \text{for } k \text{ even,} \end{cases}$$

where P is the polynomial of degree not larger than μ, for $z \to \alpha_k$ and $\mu = (k-1)^2 + \frac{k-1}{2} - 1$.

The proof of Claim 1 is analogous to that of Theorem 5.23. First, we observe that h has an analytic continuation and second we compute its order via the subcritical case of singularity analysis. In the following, we restrict our analysis to the case $k \equiv 1 \mod 2$. The coefficients of $\mathbf{W}_k(z, 1) = \mathbf{T}_k(z)$ are, according to Theorem 2.19, asymptotically given by

$$[z^n]\mathbf{W}_k(z, 1) = \tilde{c}_k\alpha_k^{-n}n^{-\mu-1}(1 + o(1)).$$

Claim 1 implies that, for any fixed $u \in (0, 1)$, $\mathbf{W}_k(z, u) = g(uh(z))$ has a singular expansion at its unique dominant singularity $z = \alpha_k$. We proceed by expressing $\mathbf{W}_k(z, u) = v_u(w(z))$, where $v_u(z) = \frac{z}{z(1-u)+u}$ and $w(z) = \mathbf{T}_k(z)$. Setting $\tau_k = \mathbf{T}_k(\alpha_k)$, the singular expansion of $\mathbf{W}_k(z, u) = v_u(w(z))$ is according to the subcritical paradigm [42] derived by combining the regular expansion of v_u and the singular expansion of w:

$$\mathbf{W}_k(z, u) = \frac{w}{w(1-u) + u}$$

$$= \frac{\tau_k}{\tau_k(1-u) + u} + \frac{u}{(\tau_k(1-u) + u)^2} \cdot (w - \tau_k)(1 + o(1)).$$

Accordingly, Theorem 2.19 implies

$$[z^n]\mathbf{W}_k(z, u) = \frac{u}{((1-u)\tau_k + u)^2}\tilde{c}_k\alpha_k^{-n}n^{-\mu-1}(1 + o(1)).$$

Consequently we arrive at

$$\lim_{n\to\infty} \frac{[z^n]\mathbf{W}_k(z, u)}{[z^n]\mathbf{W}_k(z, 1)} = \frac{u}{(\tau_k(1-u) + u)^2},$$

where $\tau_k = \mathbf{T}_k(\alpha_k)$. In view of $[u^i]q(u) = \frac{i}{\tau_k^2}\left(\frac{\tau_k-1}{\tau_k}\right)^{i-1} = q_i$, Theorem 5.21 implies the discrete limit law

$$\lim_{n\to\infty} \mathbb{P}(\eta_n^{(k)} = i) = \lim_{n\to\infty} \frac{\beta_{n,i}^{(k)}}{\beta_n^{(k)}} = q_i.$$

Using eq. (5.32), setting $r = 2$ and $\lambda = \ln \frac{\tau_k}{\tau_k - 1} > 0$, we analogously obtain

$$\lim_{n\to\infty} \mathbb{P}(\eta_n^{(k)} = i) = \frac{i}{\tau_k^2} \left(\frac{\tau_k - 1}{\tau_k} \right)^{i-1}$$

$$= \frac{1}{\tau_k(\tau_k - 1)} \left(\ln \frac{\tau_k}{\tau_k - 1} \right)^{-2} \left(\ln \frac{\tau_k}{\tau_k - 1} \right)^2 \cdot i \left(\frac{\tau_k - 1}{\tau_k} \right)^i$$

$$= \frac{1}{\tau_k(\tau_k - 1)} \left(\ln \frac{\tau_k}{\tau_k - 1} \right)^{-2} f_{\ln \frac{\tau_k}{\tau_k - 1}, 2}(i)$$

and Theorem 5.25 is proved.

5.4 Exercises

5.1. Let $F(n, h)$ denote the number of paths of length n starting at $(0,0)$ and ending at (n, h), having up- and down-steps and that stay within the first quadrant. Prove [110]

$$F(n, h) = \binom{n}{\frac{n-h}{2}} - \binom{n}{\frac{n-h-2}{2}}.$$

5.2. Explicit formulas for $O_3^*(\lambda^i, n - i)$ and $O_3^0(\lambda^i, n - i)$ [48, 52]:
Let λ_{h_1,h_2}^i denote the shape with at most two rows, where $x_1^{\lambda^i}(n) + x_2^{\lambda^i}(n) = h_1$ and $x_1^{\lambda^i}(n) - x_2^{\lambda^i}(n) = h_2$. Then we have

$$O_3^0(\lambda_{h_1,h_2}^i, n - i) = t(n - i, h_1, h_2)$$
$$= F((n - i) + 2, h_1 + 2)F((n - i), h_2) -$$
$$F((n - i) + 2, h_2)F(n - i, h_1 + 2),$$

$$O_3^*(\lambda_{h_1,h_2}^i, n - i) = \begin{cases} \sum_{l=0}^{\frac{n}{2}} \binom{n-i}{2l} t(n - i - 2l, h_1, h_2), \\ \qquad \text{for } (n - i) \text{ even} \\ \sum_{l=0}^{\frac{n}{2}} \binom{n-i}{2l+1} t(n - i - 2l - 1, h_1, h_2), \\ \qquad \text{for } (n - i) \text{ odd.} \end{cases} \qquad (5.33)$$

5.3. Prove the analogue of Proposition 5.13 for stacks.

5.4. Prove a central limit theorem on the distribution of stems in k-noncrossing, τ canonical structures.

6

Folding

The theory presented in Chapter 4 provides key information on folding maps into RNA pseudoknot structures. These maps generate exponentially many structures and each of these has a neutral network of exponential size. One question in this context is that of generating a particular sequence to structure map, via the ab initio folding of RNA pseudoknot structures. This chapter is based on Waterman's original papers [70, 144], Rivas and Eddy's algorithm and subsequent analysis [111, 112], and finally the combinatorial fold `cross` [72].

Let us review mfe folding starting with RNA secondary structures. The first mfe-folding algorithms for RNA secondary structure are due to [29, 46, 81]. We discussed in Chapter 1 the key idea of the dynamic programming (DP) folding routine for secondary structures, where the underlying energy was obtained by independent base pair contributions. Waterman et al. [96, 142, 144, 150] subsequently presented the prediction of the loop-based mfe secondary structure [132] via a DP routine having $O(n^3)$-time and $O(n^2)$-space complexity. We discuss this algorithm in detail in Section 6.1.

As discussed in Chapter 1, the DP routine serves also a paradigm for pseudoknot folding algorithms [3, 21, 112]: Rivas and Eddy's [111] gap-matrix variant of Waterman's DP folding routine for secondary structures [70, 96, 142–144]. But there is a fundamental difference between applying the DP routine to secondary versus pseudoknot structures: in the context of cross-serial dependencies, the DP routine does no longer match the combinatorics of the output class. It is exactly for this reason, why it is so difficult specifying exactly which pseudoknot types these folding algorithms can generate [112]. In Section 6.1 we have a closer look at the DP paradigm used for pseudoknot folding by means of analyzing the algorithm of Rivas and Eddy [34, 111, 112].

A different approach towards folding of RNA pseudoknot structures is that of combinatorial folds, i.e., folding algorithms having an a priori output class, considering only RNA pseudoknot structures of a specific type. In Section 6.2 we present the ab initio folding algorithm `cross`, which generates 3-noncrossing, 3-canonical RNA pseudoknot structures having arc length

C. Reidys, *Combinatorial Computational Biology of RNA*,
DOI 10.1007/978-0-387-76731-4_6,
© Springer Science+Business Media, LLC 2011

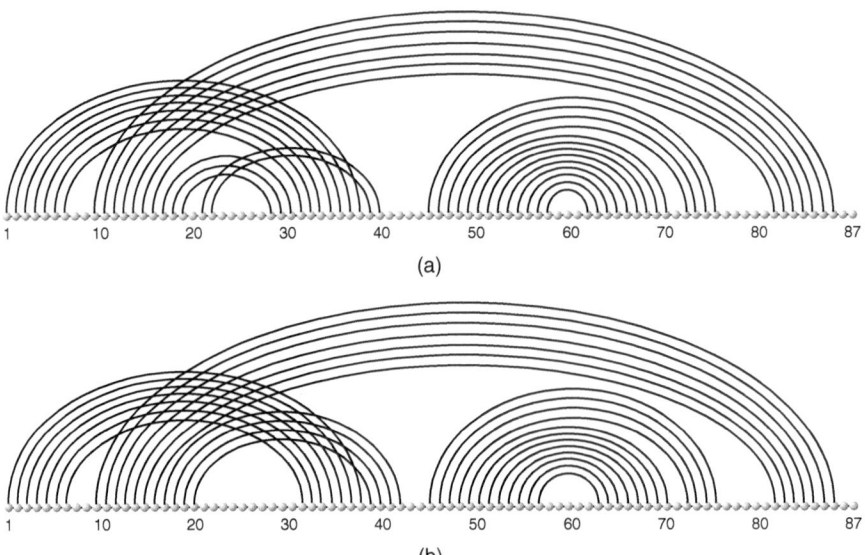

Fig. 6.1. The HDV virus pseudoknot structure: the natural structure (**a**)
`http://www.ekevanbatenburg.nl/PKBASE/PKB00075.HTML` versus the structure
folded by `cross` (**b**). The structure generated by `cross` differs from the natural
structure displayed in (**a**) by seven base pairs.

$\lambda \geq 4$; see Fig. 6.1. We remark that `cross` is by no means conceptually restricted to the case $k = 3$ [72].

In order to generate mfe pseudoknot structures we need a concept of
pseudoknot loops and their associated energies. Suppose we are given a k-
noncrossing, σ-canonical structure, S. Let α, β be S-arcs. We denote the set
of S-arcs that cross β by $\mathcal{A}_S(\beta)$. Clearly we have

$$\beta \in \mathcal{A}_S(\alpha) \quad \Longleftrightarrow \quad \alpha \in \mathcal{A}_S(\beta).$$

$\alpha \in \mathcal{A}_S(\beta)$ is called minimal, β-crossing if there exists no $\alpha' \in \mathcal{A}_S(\beta)$ such
that $\alpha' \prec \alpha$. Here $\prec$ denotes the partial order over the set of arcs (written as
(i, j), $i < j$) of a k-noncrossing diagram

$$(i_1, j_1) \prec (i_2, j_2) \quad \Longleftrightarrow \quad i_2 < i_1 \wedge j_1 < j_2.$$

Note that $\alpha \in \mathcal{A}_S(\beta)$ can be minimal β-crossing, while β is *not* minimal
α-crossing. We call a pair of mutually crossing arcs (α, β) balanced, if α is
minimal, β-crossing and β is minimal α-crossing, respectively.

Let d be a diagram. Then $L(d)$ [63] is the graph obtained by considering
d-arcs as $L(d)$-vertices and in which two vertices are adjacent if their corresponding arcs are crossing in d; see Fig. 6.2.

Now we are in position to discuss multi- and pseudoknot loops in k-
noncrossing structures:

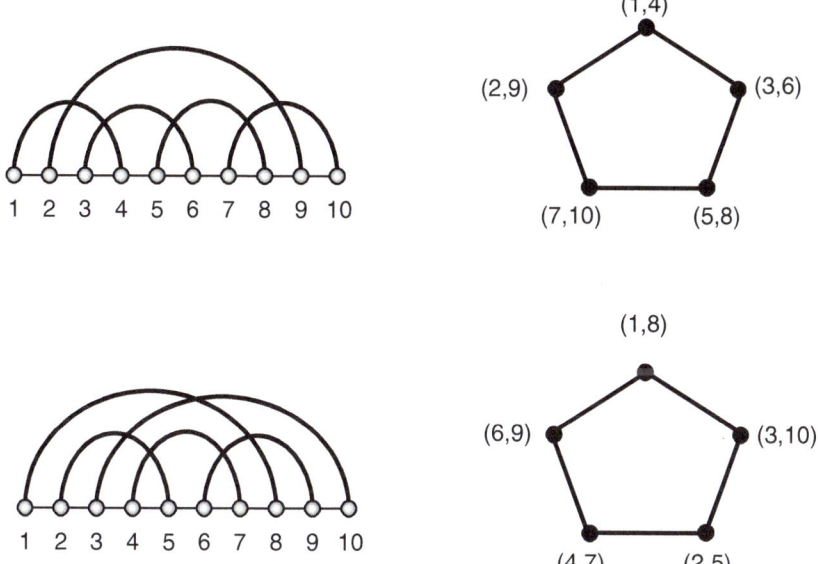

Fig. 6.2. The L-graphs of two nonplanar, 3-noncrossing structures.

- A *multi*-loop, see Fig. 6.3, is a sequence

$$((i_1, j_1), [i_1 + 1, \omega_1 - 1], S_{\omega_1}^{\tau_1}, [\tau_1 + 1, \omega_2 - 1], S_{\omega_2}^{\tau_2}, \dots),$$

where $S_{\omega_h}^{\tau_h}$ denotes a k-noncrossing structure over $[\omega_h, \tau_h]$ (i.e., nested in (i_1, j_1)) and subject to the following condition: if all $S_{\omega_h}^{\tau_h} = (\omega_h, \tau_h)$, i.e., all substructures are simply arcs, for all h, then $h \geq 2$.

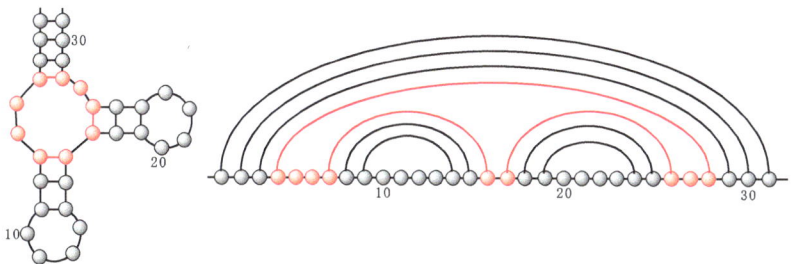

Fig. 6.3. A multi-loop in a secondary structure. We display the loop in the planar graph (*left*) and the diagram (*right*) representation of the structure, respectively.

- A *pseudoknot*, see Fig. 6.4, consists of the following data:
  (P1) a set of arcs

$$P = \{(i_1, j_1), (i_2, j_2), \dots, (i_t, j_t)\},$$

where $i_1 = \min\{i_s\}$ and $j_t = \max\{j_s\}$, subject to the following conditions:
(i) the diagram induced by the arc set P, d_P, is irreducible, i.e., $L(d_P)$, is connected and
(ii) for each $(i_s, j_s) \in P$ there exists some arc β (not necessarily contained in P) such that (i_s, j_s) is minimal β-crossing
(P2) all vertices $i_1 < r < j_t$, not contained in hairpinloops, interiorloops, or multi-loops.
We call a pseudoknot balanced if its arc set can be decomposed into pairs of balanced arcs.

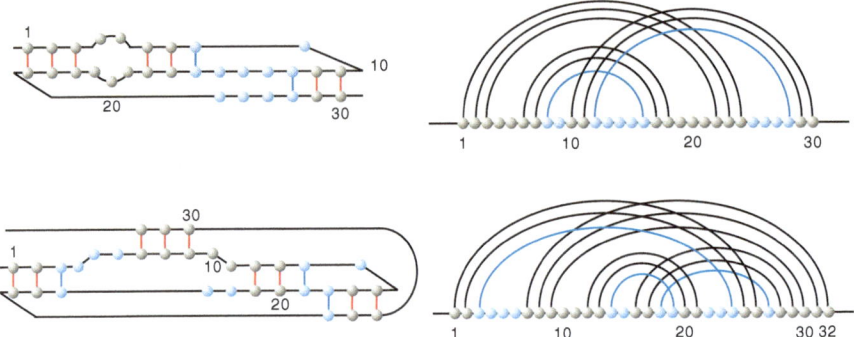

Fig. 6.4. Pseudoknots: we display a balanced (*top*) and an unbalanced pseudoknot (*bottom*). The latter contains the stack over $(3, 24)$, which is minimal for the arc $(9, 30)$, which is *not* contained in the pseudoknot.

Our energy model is a generalization of [31, 111], see Figs. 6.5 and 6.6. We remark that we do not consider dangles. It is indeed a generalization, since we consider 3-noncrossing nonplanar structures. As for the pseudoknot energy parameters we have

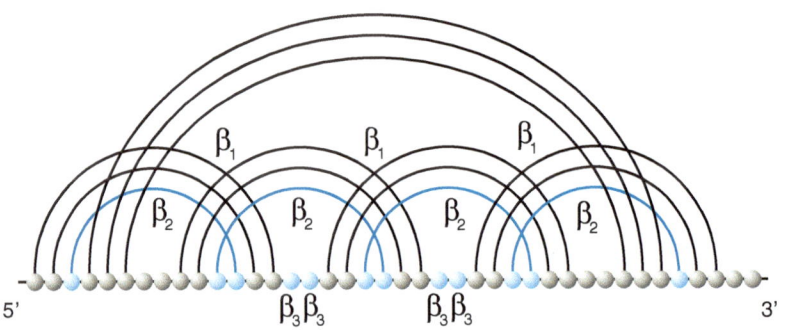

Fig. 6.5. A nonplanar 3-noncrossing pseudoknot and its energy $3\beta_1 + 4\beta_2 + 4\beta_3$. This configuration cannot be inductively generated by pairs of gap matrices.

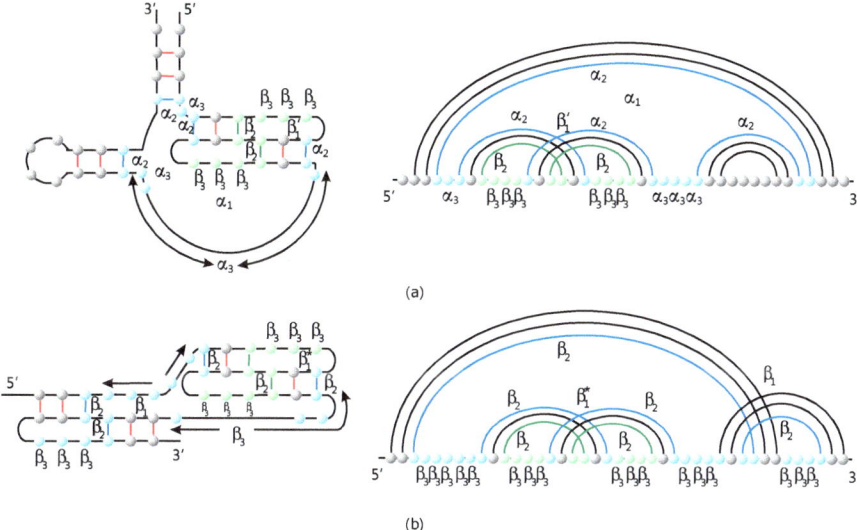

(a)

(b)

Fig. 6.6. (**a**) A multi-loop containing a pseudoknot: as in the case of standard loops, pseudoknot base pairs contained in the multi-loop are assigned the energy contribution α_2. The penalty for the formation of a pseudoknot within a multi-loop is given by β'_1. (**b**) A pseudoknot within pseudoknot: the formation of a pseudoknot in a pseudoknot contributes $\beta^\star_1$.

$$G^{\text{pseudo}} = \beta_1 + B \cdot \beta_2 + U \cdot \beta_3,$$

where β_1, β_2, and β_3 parameterize specific penalties; B is the number of base pairs; and U is the number of unpaired bases therein.

6.1 DP folding based on loop energies

6.1.1 Secondary structures

In Chapter 1, we discussed the folding of secondary structures with respect to an energy model, in which individual base pairs contributed additively. It is well known, however, that the mfe energy of a secondary structure, derived on the basis of loops, is more accurate [89].

The DP routine in the loop-based model requires *two* matrices, $vx(i,j)$ and $wx(i,j)$. In $vx(i,j)$, an entry represents the optimal score of a structure (over $[i,j]$) in which positions i and j form a base pair. In $wx(i,j)$ an entry represents the optimal score of a structure, regardless of whether i is paired with j or not.

The matrix $wx(i,j)$ is not new, in fact it coincides with $S(i,j)$, the matrix of the DP routine based on individual base pair contributions of Chapter 1.

As a result, we have

$$
wx(i,j) = \text{opt} \begin{cases} P + vx(i,j) \\ Q + wx(i+1,j) \\ Q + wx(i,j-1) \\ wx(i,k) + wx(k+1,j), \end{cases}
$$

where $i < k < j$, P denotes the score for external base pair, and Q is the score of an unpaired base; see Fig. 6.7.

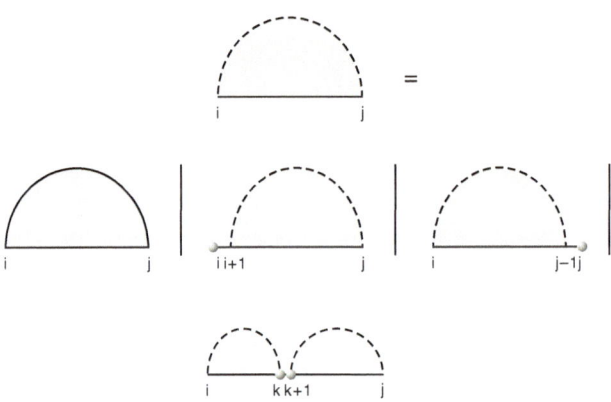

Fig. 6.7. The recursion for wx. The *dashed line* represents that it is indetermined whether or not i and j are paired, while the *solid line* means that i and j form a base pair.

In the loop-based energy model, the addition of a new base pair induces a new loop. Thus, the energy score for adding the base pair (i,j) depends on the sub-structure nested in (i,j). In order to formulate the recursion for $vx(i,j)$ we introduce a partial order over the arcs of k-noncrossing diagrams. We call the number of $\prec$-maximal S-arcs, see Fig. 6.8, the *order* of a structure S, denoted by $\omega(S)$. Suppose we are given a base pair (i,j) with nested sub-structure, S. In case of $\omega(S) = 0$, we have $S = \varnothing$ and (i,j) forms a hairpin

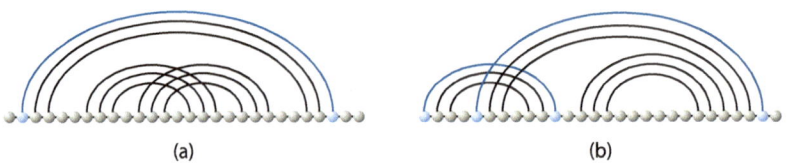

Fig. 6.8. The order of a structure: In (**a**) we display a structure of order 1, i.e., having one $\prec$-maximal arc (*blue*) and (**b**) showcases a structure of order 2, having two $\prec$-maximal arcs.

loop. For $\omega(S) = 1$, we have an interior loop and a multi-loop in the case of $\omega(S) \geq 2$. Let ϵ^r denote the energy contribution for adding the base pair (i, j) enclosing a sub-structure of order r. Then we have, see Fig. 6.9,

$$
vx(i, j) = \text{opt} \begin{cases}
\epsilon^0(i, j) \\
\epsilon^1(i, j, r, s) + vx(r, s) \\
\epsilon^2(i, j, r, s, m, n) + vx(r, s) + vx(m, n) \\
\epsilon^3(i, j, r, s, m, n, p, q) + vx(r, s) + vx(m, n) + vx(p, q) \\
\vdots
\end{cases}
\tag{6.1}
$$

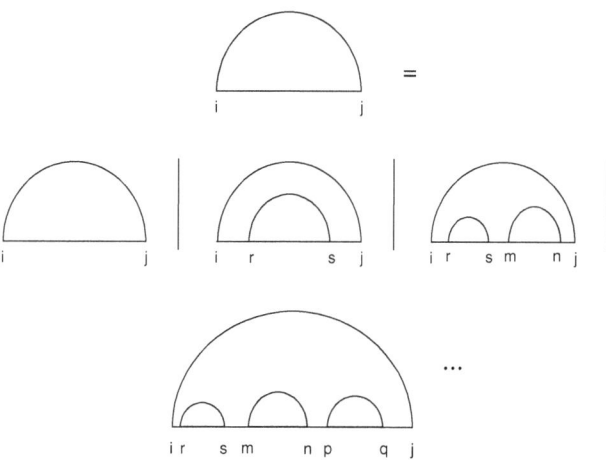

Fig. 6.9. Illustration of eq. (6.1), the vx-recursion.

In lack of detailed energy parameters of multi-loops, one truncates the recursion of eq. (6.1) and computes the score of multi-loops via

$$
M + P_I \cdot B + Q_I \cdot U,
$$

where M is the penalty of forming a multi-loop, P_I is the energy score of a closing base pair, and Q_I is the energy score of an unpaired base. B and U denote the number of closing base pairs and unpaired bases, respectively. We derive

$$
vx(i, j) = \text{opt} \begin{cases}
\epsilon^0(i, j) \\
\epsilon^1(i, j, r, s) + vx(r, s) \\
M + P_I + wx_I(i + 1, r) + wx_I(r + 1, j - 1),
\end{cases}
\tag{6.2}
$$

where $i \leq r < s \leq j$; see Fig. 6.10. Here wx_I presents the optimal score of structure which is nested in a multi-loop. Therefore, in wx_I, Q_I will be used

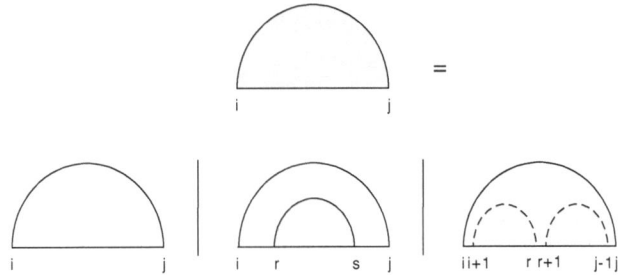

Fig. 6.10. The truncation: the vx-recursion truncated at order 2.

as the penalty for an unpaired base in a multi-loop and P_I denotes the penalty for a closing base pair in a multi-loop. In case of wx, Q represents the score for an unpaired base and P presents the score for an external base pair. We remark that in Turner's energy model, Q and P are always set to be zero.

In particular, setting $vx(i,j) = P_{i,j} + wx_I(i + 1, j - 1)$, we recover the case of independent contributions of base pairs. Via the above recursions we can inductively compute the matrices $wx(i,j)$ and $vx(i,j)$, starting with the diagonals as exercised in Chapter 1. Once the matrices are computed, we construct a structure having optimal score, by tracing back.

6.1.2 Pseudoknot structures

In this section we discuss Rivas and Eddy's beautiful idea for folding RNA pseudoknot structures [111]. The key observation here is the use of gap matrices in addition to the wx and vx, discussed above; see Fig. 6.11. There are four gap matrices, $whx(i,j,r,s)$, $vhx(i,j,r,s)$, $yhx(i,j,r,s)$ and $zhx(i,j,r,s)$, as specified in Table 6.1.

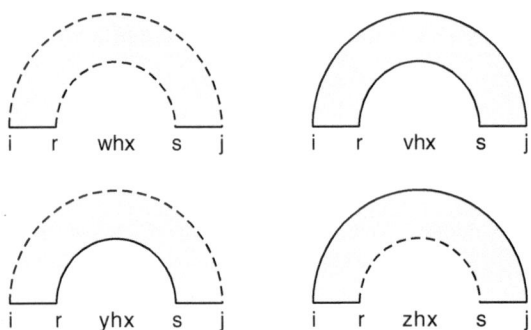

Fig. 6.11. The four gap matrices whx, vhx, yhx, and zhx. The *dashed line* is used if the relation of two vertices is unknown, while the *solid line* denotes that the two vertices form a base pair.

Matrices	(i,j)	(r,s)	Matrices	(i,j)	(r,s)
$whx(i,j;r,s)$	Unknown	Unknown	$vhx(i,j;r,s)$	Paired	Paired
$yhx(i,j;r,s)$	Unknown	Paired	$zhx(i,j;r,s)$	Paired	Unknown

Table 6.1. The gap matrices whx, vhx, yhx, and zhx.

In Fig. 6.12 we exemplify how two gap matrices generate pseudoknots. The algorithm coincides with the DP routine for secondary structures in case of gaps of size zero, that is, $r = s - 1$. Then

$$whx(i,j;r,r+1) = wx(i,j),$$
$$zhx(i,j;r,r+1) = vx(i,j),$$

for $i \leq r \leq j$. In principle, any number of gap matrices can be employed. However, the algorithm, in its current implementation, is truncated at

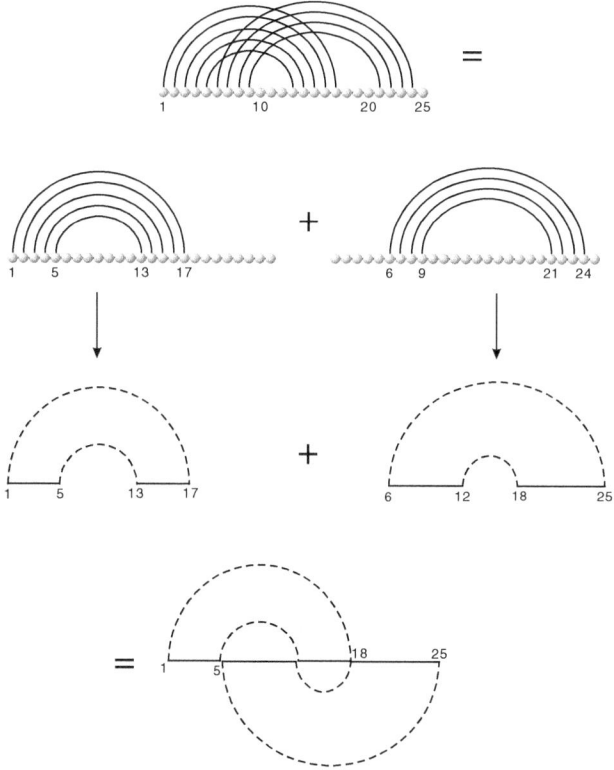

Fig. 6.12. Constructing a pseudoknot via two gap matrices.

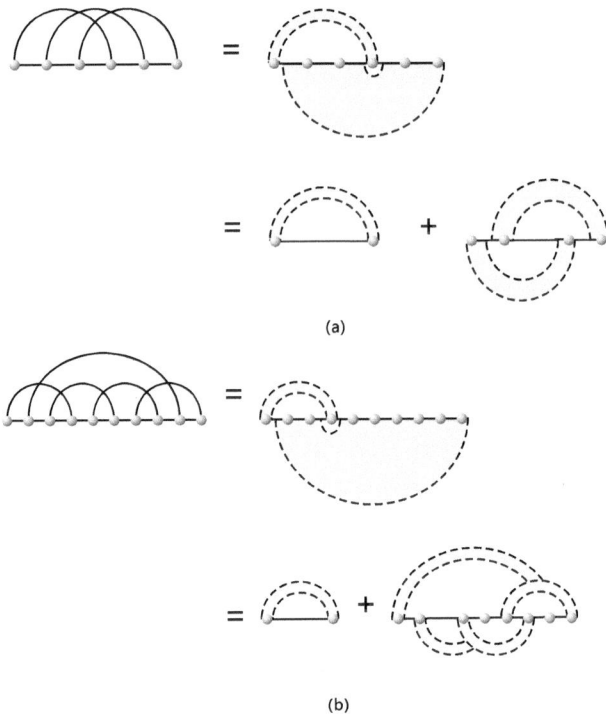

Fig. 6.13. A 4-noncrossing structure which can be generated by two gap matrices (**a**) and a 3-noncrossing structure, which cannot be generated using two gap matrices (**b**).

$O(whx + whx + whx)$, that is, at each step at most two gap matrices are used. It is not obvious at all, which structures the algorithm can generate, see Fig. 6.13, where we showcase a nonplanar 3-noncrossing structure, which cannot be generated by two gap matrices. The recursions for $vx(i,j)$ and $wx(i,j)$ in case of the two gap-matrix truncation are displayed in Fig. 6.14 and are given by

$$vx(i,j) = \text{opt} \begin{cases} \epsilon^0(i,j) \quad \text{[hairpin-loop]} \\ \epsilon^1(i,j,r,s) + vx(r,s) \quad \text{[interior-loop]} \\ M + P_I + wx_I(i+1,r) + wx_I(r+1,j-1) \\ \quad \text{[nested multi-loop]} \\ G_{wI} + M_{pk} + P_{pk} + whx(i+1,l;r,s) \\ \quad + whx(r+1,j-1;s-1,l+1) \\ \quad \text{[non-nested multi-loop]} \end{cases},$$

$$wx(i,j) = \mathrm{opt} \begin{cases} P + vx(i,j) \quad [\text{base pair } (i,j)] \\ Q + wx(i+1,j) \quad [\text{single-stranded}] \\ Q + wx(i,j-1) \quad [\text{single-stranded}] \\ wx_I(i,k) + wx_I(k+1,j) \quad [\text{nested bifurcation}] \\ G_w + whx(i,l,r,s) + whx(r+1,j,s-1,l+1) \\ \quad [\text{non-nested bifurcation}] \end{cases} .$$

Here, G_{wI} is the penalty for forming an internal pseudoknot, which is nested in a multi-loop, and G_w is the penalty for forming an external pseudoknot. M_{pk} is the score for a multi-loop containing a pseudoknot and P_{pk} is the score for closing base pair in such a multi-loop; see Fig. 6.14.

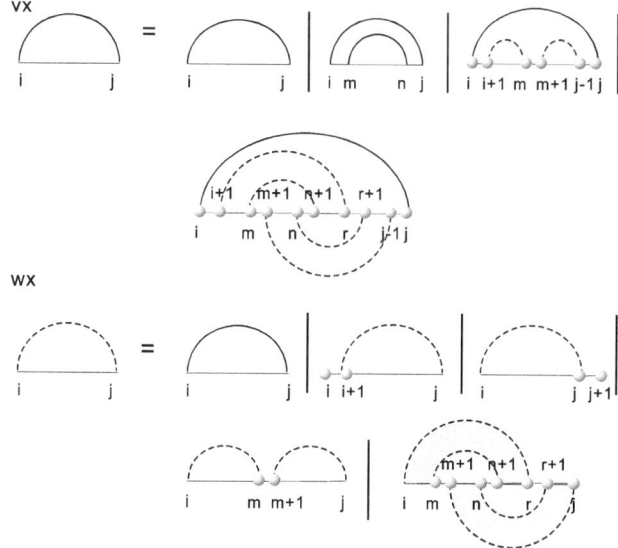

Fig. 6.14. The basic recursions: recursion for vx and wx truncated at $O(whx + whx + whx)$ in Rivas and Eddy's algorithm.

While the inductive formation of two (or more) gap matrices generates arbitrarily high numbers of mutually crossing arcs, see Fig. 6.13, this method fails to generate nonplanar, 3-noncrossing pseudoknots. In Fig. 6.5, we give an example of a 3-noncrossing structure that cannot be constructed using two gap matrices. It is clear that gap matrices can and will generate nonplanar arc configurations. However, they can only facilitate this via increasing the crossing number.

By displaying two nonplanar, 3-noncrossing structures, Fig. 6.2 makes the point that the situation is more complex: nonplanarity is not tied to crossings – there are planar as well as nonplanar 3-noncrossing structures. The situation becomes much more involved for higher crossing numbers.

6.2 Combinatorial folding

In this section we present the pseudoknot folding algorithm `cross`. The algorithm decomposes into three distinct phases, detailed in Sections 6.2.2, 6.2.3, and 6.2.4. In Fig. 6.15 we present an overview of `cross`. The input of `cross` is an RNA primary sequence and its output is a 3-noncrossing σ-canonical RNA pseudoknot structure; see Fig. 6.1. There are three key ideas. The first consists in generating all irreducible shadows. The key point is here the *recursive* generation of the motifs via Proposition 6.3, which then in turn induce the shadows. The second is to build skeleta trees: irreducible shadows serve as roots for the latter, constructed in Propositions 6.5 and 6.6. Similar ideas can be found in [91]. The important property of skeleta is that they encapsulate exactly all non-inductive arc configurations in k-noncrossing structures. Third, the skeleta are saturated via the context-sensitive DP routines detailed in Section 6.2.4. In Fig. 6.16 we present data on the mean folding times of `cross`.

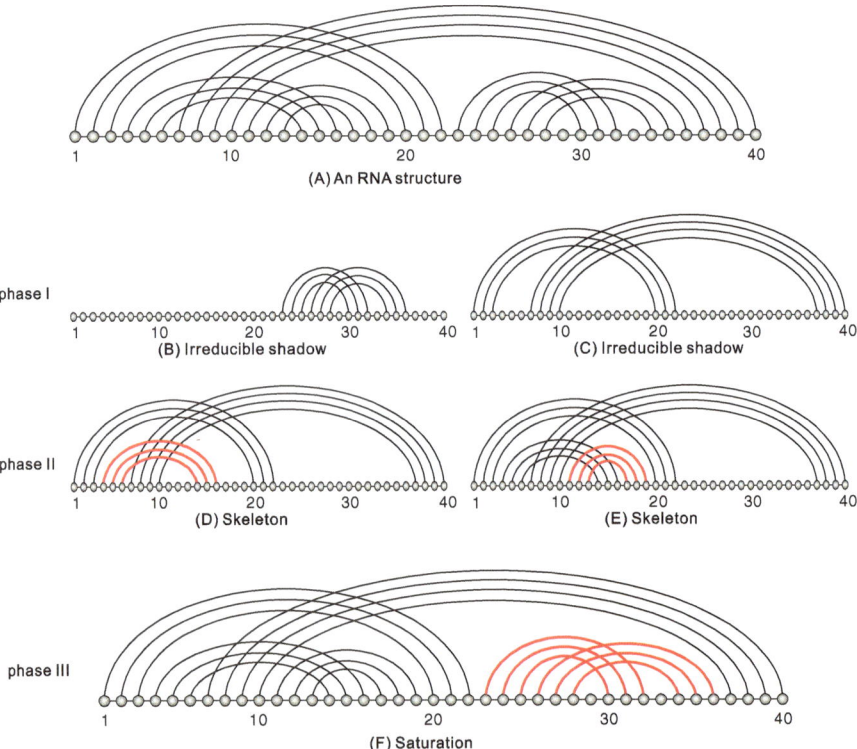

Fig. 6.15. A closer look: the generation of motifs (I), the construction of skeleta trees, rooted in irreducible shadows (II), and the saturation (III). We show in which routines the substructures are derived and how and when they are combined.

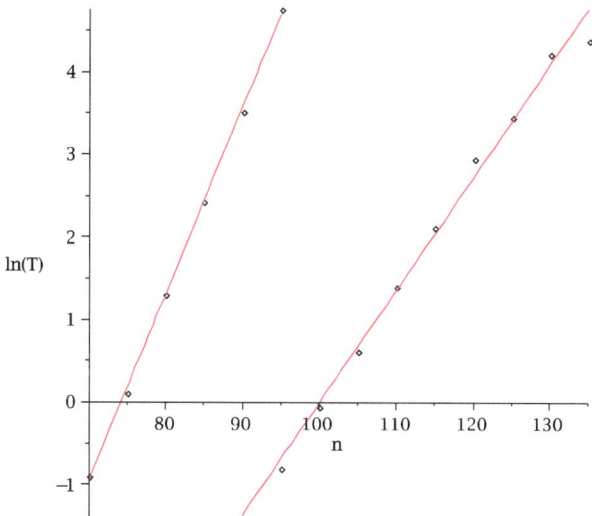

Fig. 6.16. Mean folding times: we display the logarithm of the folding times of 1000 random sequences as a function of the sequence length. For 3-canonical and 4-canonical structures the linear fits are given by $0.2263n - 19.796$ (*left*) and $0.1364n - 13.659$ (*right*), respectively, i.e., we have exponential growth rates of ≈ 1.254 and ≈ 1.146 for 3-canonical and 4-canonical structures. A random sequence of length 100 folded via a single core, 2.2 GHz CPU exhibits a mean folding time of 279 s.

6.2.1 Some basic facts

Our first objective is to introduce motifs. For this purpose recall that a k-noncrossing core is a k-noncrossing diagram without any two arcs of the form (i, j), $(i + 1, j - 1)$. We have shown in Section 4.1.1, that any k-noncrossing RNA structure, S, has a unique k-noncrossing core, $c(S)$.

Definition 6.1. (Motif) *A k-noncrossing, σ-canonical motif, M_k^σ, is a k-noncrossing, σ-canonical structure over $[n]$, having the following properties:*

- (M1) M_k^σ *has a nonnesting core.*
- (M2) *All M_k^σ-arcs are contained in stacks of length exactly $\sigma \geq 3$ and arc length $\lambda \geq 4$.*

The set of all motifs is denoted by $\mathbb{M}_k^\sigma(n)$ and we set $\mu_{k,\sigma}^*(n) = |\mathbb{M}_k^\sigma(n)|$. Property (M1) is obviously equivalent to the following: all arcs of the core, $c(M_k^\sigma)$, are $\prec$-maximal; see Fig. 6.17.

Let S be a k-noncrossing, σ-canonical structure. Suppose two k-noncrossing diagrams δ_1, δ_2 are such that δ_2 contains all δ_1-arcs and exactly one additional arc, (i, j), where $(i - 1, j + 1)$ is a δ_1-arc. We then consider δ_1 and δ_2 connected by a directed edge. With respect to this notion of edges the set of k-noncrossing diagrams over $[n]$ becomes a directed graph, which we denote by $\mathcal{G}_k(n)$.

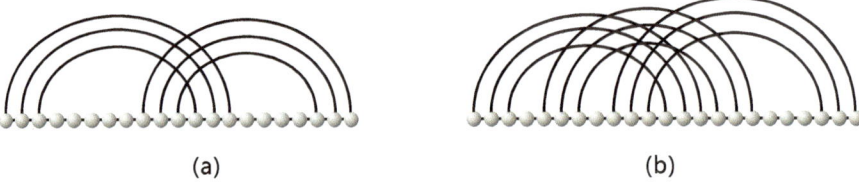

Fig. 6.17. Motifs: a 3-noncrossing, 3-canonical motif (**a**) and a 4-noncrossing, 3-canonical motif (**b**).

A shadow of S is a $\mathcal{G}_k(n)$-vertex connected to S by a $\mathcal{G}_k(n)$-path. A shadow is called irreducible if its line graph is connected.

Intuitively speaking, a shadow is derived by extending one or more stacks of a structure from top to bottom; see Fig. 6.18.

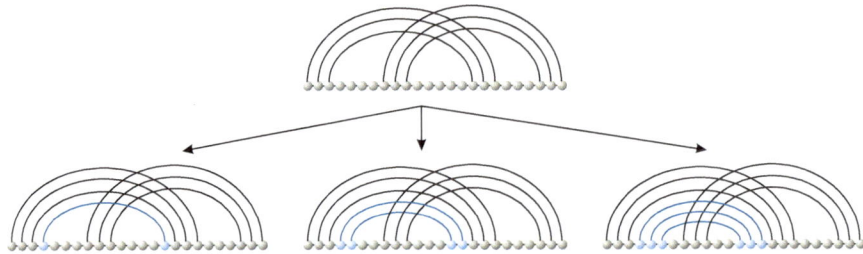

Fig. 6.18. Three shadows obtained from a given 3-noncrossing, 3-canonical motif.

We proceed by showing that k-noncrossing structures have a unique loop decomposition; see Fig. 6.19.

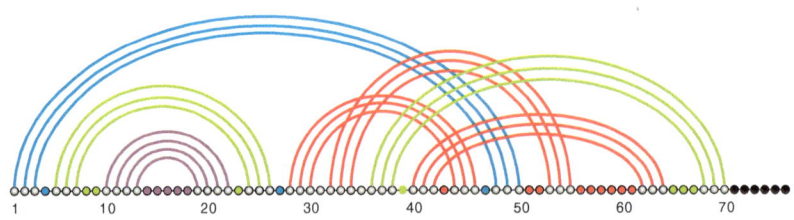

Fig. 6.19. Loop decomposition of k-noncrossing structures: We display a hairpin loop (*purple*), the noncrossing/crossing version of interior loops (*green*), a multi-loop (*blue*) and pseudoknot loops (*red*).

Proposition 6.2. *Suppose $k \geq 2, \sigma \geq 2$. Then any k-noncrossing, σ-canonical structure has a unique loop decomposition.*

Proof. Let S be a k-noncrossing, σ-canonical structure and let $c(S)$ be its core (see Section 4.1.1). We distinguish the following two scenarios:

Case (1): $\mathcal{A}_{c(S)}(\alpha) \neq \varnothing$, that is, α is a crossing arc in $c(S)$.
For any $\beta \in \mathcal{A}_{c(S)}(\alpha)$ there exists an $\prec$-minimal arc $\alpha^*(\beta) \in \mathcal{A}_{c(S)}(\beta)$ that is either nested in α or equal to α:

- If there exists some β for which $\alpha = \alpha^*(\beta)$ holds, i.e., α itself is minimal in $\mathcal{A}_{c(S)}(\beta)$, then we color α red. In other words, red arcs are minimal with respect to some crossing arc β.
- Otherwise, for any $\beta \in \mathcal{A}_{c(S)}(\alpha)$ there exists some $\alpha^*(\beta) \prec \alpha$. If $\alpha^*(\beta)$ is the unique $\prec$-maximal arc which is in the substructure nested in α, then we color α green (interior loop) and otherwise we color α blue (multi-loop).

Case (2): $\mathcal{A}_{c(S)}(\alpha) = \varnothing$, i.e., α is a noncrossing arc in $c(S)$:

- If α is $\prec$-minimal in $c(S)$, then we color α purple
- If in the substructure nested in α there exists exactly one $\prec$-maximal arc α' such that $\alpha' \prec \alpha$, we color α green (interior loop)
- Otherwise we color α blue (multi-loop)

It follows by induction on the number of $c(S)$-arcs that this procedure generates a well-defined arc coloring.

Let $i \in [n]$ be a vertex. We assign to i either the color of the $\prec$-minimal non-red $c(S)$-arc (r, s) for which $r < i < s$ holds or red if all $\prec$-minimal arcs that nest i are red and black, otherwise.

By construction, this induces a vertex arc coloring with the property of correctly identifying hairpin loops (purple), interior loops (green), multi-loops (blue), and pseudoknot loops (red) and the proposition follows.

6.2.2 Motifs

One key idea in `cross` is the identification of motifs as building blocks. Despite the fact that motifs exhibit complicated crossings, they can be *inductively* generated. This is a result of considering the "dual" of a motif which turns out to be a restricted Motzkin path. Passing from motifs to Motzkin paths can be interpreted as to exchange "first in–first out" by "first in–last out"; see the proof of Proposition 6.3.

We recall that a Motzkin path is composed by up-, down-, and horizontal steps. It starts at the origin, stays in the upper half plane, and ends on the x-axis. Let $\mathrm{Mo}_k^\sigma(n)$ denote the following set of Motzkin paths:

- The paths have height $\leq \sigma(k - 1)$
- All up- and down-steps come only in sequences of length σ
- All plateaux at height σ have length ≥ 3

Let $\mu_{k-1,\sigma}(n)$ denote the number of Motzkin paths of length n that

- have height $\leq \sigma(k - 2)$ and
- up- and down-steps come only in sequences of length σ.

We set for arbitrary $k, \sigma \geq 2$

$$G_{k,\sigma}^*(z) = \sum_{n \geq 0} \mu_{k,\sigma}^*(n) z^n,$$

$$G_{k-1,\sigma}(z) = \sum_{n \geq 0} \mu_{k-1,\sigma}(n) z^n,$$

$$G_{1,\sigma}(z) = \frac{1}{1-z}.$$

Now we are in position to make the duality between motifs and Motzkin paths precise.

Proposition 6.3. *Suppose $k, \sigma \geq 2$, then the following assertions hold:*
(a) *There exists a bijection*

$$\beta : \mathbb{M}_k^\sigma(n) \longrightarrow \mathrm{Mo}_k^\sigma(n).$$

(b) *We have the following recurrence equations:*

$$\mu_{k,\sigma}^*(n) = \mu_{k,\sigma}^*(n-1) + \sum_{s=0}^{n-(2\sigma+3)} \mu_{k-1}(n-2\sigma-s)\mu_{k,\sigma}^*(s), \quad n > 2\sigma, \quad (6.3)$$

$$\mu_{k,\sigma}(n) = \mu_{k,\sigma}(n-1) + \sum_{s=0}^{n-2\sigma} \mu_{k-1}(n-2\sigma-s)\mu_{k,\sigma}(s), \quad n > 2\sigma - 1, \quad (6.4)$$

where $\mu_{k,\sigma}^*(n) = 1$ for $0 \leq n \leq 2\sigma$ and $\mu_{k-1,\sigma}(n) = 1$ for $0 \leq n \leq 2\sigma - 1$.
(c) *We have the following formula for the generating functions:*

$$G_{k,\sigma}^*(z) = \frac{1}{1 - z - z^{2\sigma}(G_{k-1,\sigma}(z) - (z^2 + z + 1))}, \quad (6.5)$$

$$G_{k-1,\sigma}(z) = \frac{1}{1 - z - z^{2\sigma} G_{k-2,\sigma}(z)}, \quad (6.6)$$

and, in particular, for $k = 3$

$$\mu_{3,\sigma}^*(n) \sim c_\sigma \left(\frac{1}{\zeta_\sigma}\right)^n, \quad (6.7)$$

where c_σ and ζ_σ^{-1} are given in Table 6.2.

σ	2	3	4	5	6	7
ζ_σ^{-1}	1.7424	1.5457	1.4397	1.3721	1.3247	1.2894
c_σ	0.1077	0.0948	0.0879	0.0840	0.0804	0.0780

Table 6.2. The exponential growth rates of $\mu_{3,\sigma}^*(n)$.

Proof. Let M_k^σ a k-noncrossing, σ-canonical motif. We construct the bijection β as follows: reading the vertex labels of M_k^σ in increasing order we map each σ-tuple of origins and termini into a σ-tuple of up-steps and down-steps, respectively. Furthermore, isolated points are mapped into horizontal steps. The resulting paths are by construction Motzkin paths of height $\leq \sigma(k-1)$. Since motifs have arcs of length ≥ 4 the paths have at height σ plateaux of length ≥ 3. In addition we have σ-tuples of up- and down-steps. Therefore β is well defined. To see that β is bijective we construct its inverse explicitly. Consider an element $\zeta \in \mathrm{Mo}_k^\sigma(n)$. We shall pair σ-tuples of up-steps and down-steps as follows: starting from left to right we pair the first up-step with the first down-step tuple and proceed inductively; see Fig. 6.20. It is clear from the definition of Motzkin paths that this pairing procedure is well defined. Each such pair

$$((u_i, u_{i+1}, \ldots, u_{i+\sigma}), (d_j, d_{j+1}, \ldots, d_{j+\sigma}))$$

corresponds uniquely to the sequence of arcs $((i+\sigma, j), \ldots, (i, j+\sigma))$ from which we can conclude that ζ induces a unique σ-canonical diagram, δ_ζ over $[n]$. Furthermore δ_ζ has by construction a nonnesting core. A diagram contains a k-crossing if and only if it contains a sequence of arcs $(i_1, j_1), \ldots, (i_k, j_k)$ such that $i_1 < i_2 < \cdots < i_k < j_1 < j_2 < \cdots < j_k$. Therefore, δ_ζ is k-noncrossing if and only if its underlying path ζ has height $< \sigma k$. We immediately derive $\beta(\delta_\zeta) = \zeta$, whence β is a bijection. Using the Motzkin path interpretation we immediately observe that $\mathrm{Mo}_k^\sigma(n)$-paths can be constructed recursively from paths that start with a horizontal step or an up-step, respectively. The recursions of eqs. (6.3) and (6.4) and the generating functions of eqs. (6.5) and (6.6) are straightforwardly derived. As for the particular case $G_{3,\sigma}^*(z)$, we

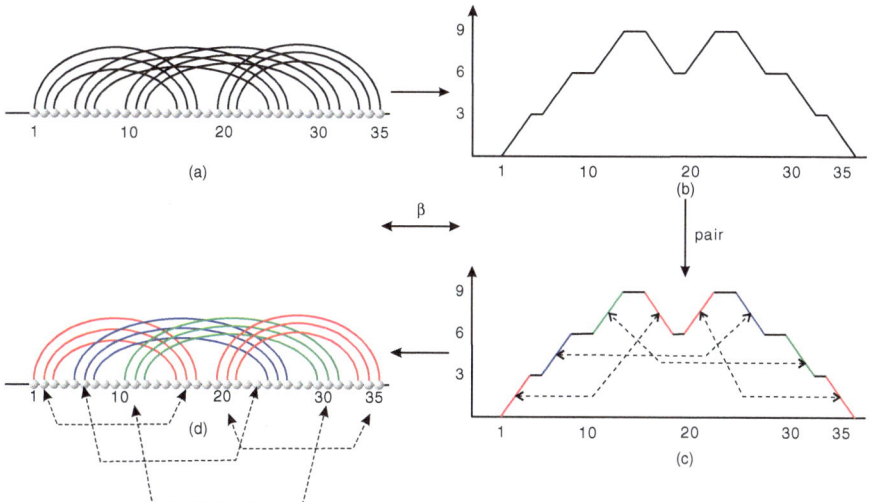

Fig. 6.20. The correspondence between motifs and Motzkin paths.

have

$$G^*_{3,\sigma}(z) = \cfrac{1}{1 - z - z^{2\sigma}\left[\cfrac{1}{1-z-z^{2\sigma}\left[\frac{1}{1-z}\right]} - (z^2 + z + 1)\right]}.$$

The unique dominant, real singularities of $G^*_{3,\sigma}(z)$ are simple poles, denoted by ζ_σ. Being a rational function, $G^*_{k,\sigma}(z)$ admits a partial fraction expansion

$$G^*_{k,\sigma}(z) = H(z) + \sum_{(\zeta,r)} \frac{c_{(\zeta,r)}}{(\zeta - z)^r}$$

and eq. (6.7) follows in view of

$$[z^n]\frac{1}{\zeta - z} = \frac{1}{\zeta}[z^n]\frac{1}{1 - z/\zeta} = \frac{1}{\zeta}\binom{n}{0}\left(\frac{1}{\zeta}\right)^n = \left(\frac{1}{\zeta}\right)^{n+1}$$

and the proof of the proposition is complete.

6.2.3 Skeleta

Definition 6.4. (Skeleton) *A skeleton, S, is a k-noncrossing structure such that*

- *its core, $c(S)$, has no noncrossing arcs and*
- *its L-graph, $L(S)$, is connected.*

We recall that $L(S)$ is obtained by considering S-arcs as vertices and two vertices are adjacent if the corresponding S-arcs are crossing; see Fig. 6.21. By construction, $L(S)$ is connected if and only if $L(c(S))$ is.

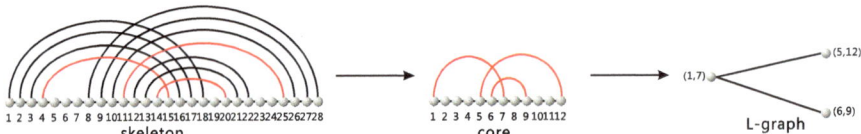

Fig. 6.21. A skeleton, its core, and its L-graph.

In addition, in a skeleton over the segment $\{i, i+1, \ldots, j-1, j\}$, $S_{i,j}$, the positions i and j are paired. Recall that an interval is a sequence of consecutive, unpaired bases $(i, i+1, \ldots, j)$, where $i - 1$ and $j + 1$ are paired and a stack of length σ is a sequence of parallel arcs

$$((i,j), (i+1, j-1), \ldots, (i + (\sigma - 1), j - (\sigma - 1))),$$

which we write as (i, j, σ). Note that $\sigma \geq \sigma_0$, where σ_0 is the minimum stack-length of the structure; see Fig. 6.22. We denote the leftmost vertex and rightmost vertex of a stack α by $\mathbf{l}(\alpha)$ and $\mathbf{r}(\alpha)$, respectively. There is an

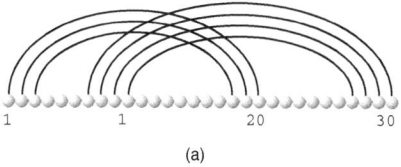
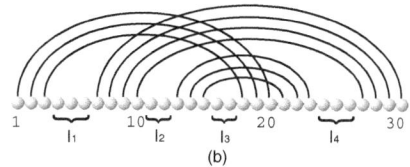

Fig. 6.22. Irreducible shadows and skeleta: an irreducible shadow (**a**), containing the stacks $(1, 20, 3)$ and $(7, 30, 4)$. (**b**) A skeleton drawn with its four induced intervals I_1, I_2, I_3, I_4.

obvious notation for a stack or interval being to the left of another stack or interval, respectively.

An irreducible shadow over $\{i, i+1, \ldots, j-1, j\}$, $IS_{i,j}$, is a skeleton whose core has no nested arcs.

We are now in position to construct the skeleta tree. Suppose we are given a k-noncrossing irreducible shadow, S_0. Let r_0 be the label number of the first paired base of S_0. We consider the pair (S_0, r_0). Suppose we obtained the pair (S_h, r_h). We next show how to derive the pair (S_{h+1}, r_{h+1}). To this end we first label the S_k intervals $\{I_1, \ldots, I_m\}$ from left to right. We construct a pair (S_{h+1}, r_{h+1}) from (S_h, r_h) where $r_{h+1} \geq r_h$ as follows: we insert into a pair of intervals (I_p, I_q), $i \in I_p, j \in I_q$, $i \geq r$ the stack $\alpha = (i, j, \sigma)$, subject to the following conditions:

- (R1) S_{h+1} is a k-noncrossing skeleton,
- (R2) $(i + \sigma - 1, j - \sigma + 1)$ is a minimal element in S_{h+1},
- (R3) r_{h+1} is the label of the first paired base preceding the interval I_p,
- (R4) $i - 1$ and $j + 1$ are not paired to each other,
- (R5) if there are some inserted stacks to the right of I_p, suppose β is the leftmost one, then α cannot cross any stack in $\mathcal{A}_{S(\beta)}(\beta)$,

where we denote the structure derived by inserting the stack β by $S(\beta)$ and the stacks that cross β in $S(\beta)$ by $\mathcal{A}_{S(\beta)}(\beta)$. In Fig. 6.23 we illustrate the process of stack insertion. We refer to the stack insertion formally by

$$(S_h, r_h) \Rightarrow_{(i,j,\sigma)} (S_{h+1}, r_{h+1})$$

and write $S_0 \sqsubseteq S$ if S is obtained from S_0 by a sequence of insertions.

Given an irreducible skeleton S_0, we consider the graph $\mathbb{G}(S_0)$. The vertices of $\mathbb{G}(S_0)$ are the set of skeleta

$$\mathbb{V}(S_0) = \{S \mid S_0 \subset S \wedge \text{the maximal } S\text{-stacks induce } S_0\}.$$

The (directed) edges of $\mathbb{G}(S_0)$ are given by

$$\mathbb{E}(S_0) = \{(S_1, S_2) \mid \exists r_1, r_2; \quad (S_1, r_1) \Rightarrow_{(i,j,\sigma)} (S_2, r_2)\}. \qquad (6.8)$$

We show that $\mathbb{G}(S_0)$ is well defined. Suppose $(S_1, r_1) \Rightarrow_{(i,j,\sigma)} (S_2, r_2)$, where $S_1 \in \mathbb{V}(S_0)$, that is, (i) S_1 contains S_0 and (ii) its maximal stacks induce

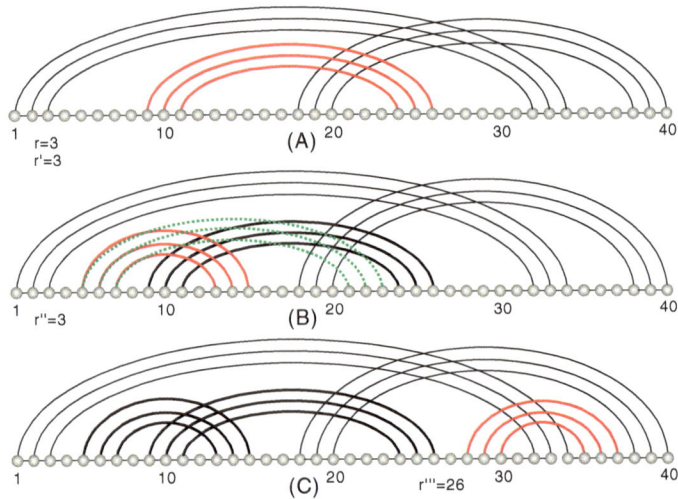

Fig. 6.23. Stack insertion: the insertion of the stacks $(9, 26, 3)$, $(5, 15, 3)$, and $(28, 37, 3)$. Currently inserted stacks are drawn in *red*. Note that in (B), where we insert $(5, 15, 3)$, it is impossible to insert $(5, 23, 3)$ (*green*). This is because in (A), $(9, 26, 3)$ crosses $(18, 40, 3)$, and (R5) implies that the newly inserted stack to the left of $(9, 26, 3)$ cannot cross $(18, 40, 3)$.

exactly S_0. We notice that the stack insertion does not affect the maximal stacks of S_1. Thus the maximal stacks of S_2 and S_1 coincide and we have by construction $S_0 \subset S_2$, whence eq. (6.8) is well defined.

We proceed by showing that the $\mathbb{G}(S_0)$ component containing S_0 is acyclic. In other words, the insertion procedure is an unambiguous grammar.

Proposition 6.5. *For any $k \geq 2$ and arbitrary k-noncrossing irreducible shadow, S_0, the $\mathbb{G}(S_0)$-component containing S_0 is a tree.*

Proof. Suppose *a contrario* that this component is not acyclic. Since all its vertices are connected to S_0 by a directed path, we may, without loss of generality, assume that we have a cycle of minimal length and of the following form:

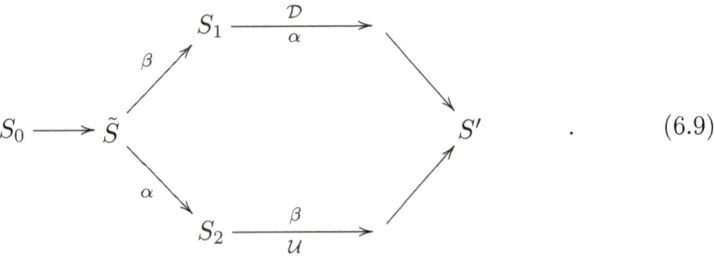

$$(6.9)$$

Therefore, the two insertion paths from $\tilde{S}$ to S', $\mathcal{U}$ and $\mathcal{D}$ in eq. (6.9) have identical length and are composed by identical sets of stacks. Let α denote the first stack inserted via $\mathcal{U}$ and β be the first inserted via $\mathcal{D}$. α and β are contained in S' and $\alpha \neq \beta$ since the cycle is minimal. Furthermore, we can, without loss of generality, assume $\mathbf{l}(\alpha) < \mathbf{l}(\beta)$.

Claim 1: α and β cross some stack γ in $\tilde{S}$.

We first claim that both $\mathbf{l}(\alpha)$ and $\mathbf{l}(\beta)$ are in the same interval of $\tilde{S}$. Otherwise, since $\mathbf{l}(\alpha) < \mathbf{l}(\beta)$, the interval containing $\mathbf{l}(\beta)$ is to the right of the interval containing $\mathbf{l}(\alpha)$; see Fig. 6.24(a). For $\mathcal{D}$, this would imply that α cannot be inserted because in view of (R3), the parameter r is weakly increasing. Consequently, $\mathbf{l}(\alpha)$ and $\mathbf{l}(\beta)$ are in the same interval of $\tilde{S}$. According to (R2), α must cross some stack in the structure $\tilde{S}$. Let γ be the first stack to the right of $\mathbf{l}(\alpha)$ in $\tilde{S}$. Then α must cross γ since otherwise α is not minimal $\tilde{S}(\alpha)$. For the same reason β must also cross γ, whence Claim 1.

Claim 2: The stacks α and β are crossing.

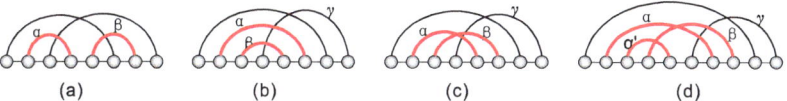

Fig. 6.24. Illustration of the different scenarios considered in Proposition 6.5: **(a)** $\mathbf{l}(\alpha)$ and $\mathbf{l}(\beta)$ are in different intervals, **(b)** β is nested in α, **(c)** α crosses β, and **(d)** α' is nested in α.

Suppose β is inserted via $\mathcal{U}$ and β does not cross α. Then β is necessarily nested in α; see Fig. 6.24(b). But then, via $\mathcal{D}$, β is inserted *before* α. Hence α cannot be minimal, contradicting (R2) and Claim 2 is proved.

Claim 3: In $\mathcal{D}$, the leftmost inserted stack to the right of α in $S(\alpha)$ is necessarily equal to β.

Suppose the leftmost inserted stack to the right of α in $S(\alpha)$ is $\alpha' \neq \beta$. Then via $\mathcal{D}$ α' is inserted before α. (R5) implies that α' cannot cross γ. Thus α' is nested in α in $S(\alpha)$; see Fig. 6.24(d). Therefore, when inserting α via $\mathcal{D}$, α is not minimal, a contradiction to (R2), and Claim 3 follows. According to the above three claims, we have reduced all scenarios to the arc configuration displayed in Fig. 6.24(c). We now consider the insertion of α via $\mathcal{D}$. Here, we first insert stack β and when inserting stack α, the leftmost inserted stack on the right of $\mathbf{l}(\alpha)$ is, according to Claim 3, equal to β. Therefore, according to (R5) α cannot cross the stack γ, since γ is contained in $\mathcal{A}_{S(\beta)}(\beta)$, a contradiction to Claim 1. This completes the proof of the proposition.

In view of Proposition 6.5 we denote the $\mathbb{G}(S_0)$-component rooted in S_0 by $\mathbb{T}(S_0)$. We next prove that our unambiguous grammar generates $\mathbb{G}(S_0)$.

Proposition 6.6. *Given an irreducible shadow S_0 and let $\mathbb{T}(S_0)$ be the skeleta tree rooted in S_0, then*

$$\mathbb{T}(S_0) = \mathbb{G}(S_0),$$

that is, every skeleton $S \in \mathbb{S}(S_0)$ can be constructed via a unique insertion path.

Proof. Clearly we have $\mathbb{T}(S_0) \subset \mathbb{G}(S_0)$. For an arbitrary k-noncrossing skeleton $S \in \mathbb{V}(S_0)$, let $\mathcal{A}_S^{\text{ne}}$ denote the set of all nested stacks in S. Since each stack is either maximal or nested we have the partition

$$\mathcal{A}_S = \mathcal{A}_{S_0} \dot{\cup} \mathcal{A}_S^{\text{ne}}.$$

Sorting the stacks in $\mathcal{A}_S^{\text{ne}}$ according to the linear ordering of their leftmost paired bases, we obtain a unique sequence $\Sigma = (\alpha_1, \alpha_2, \ldots, \alpha_n)$. We choose the first element $\alpha_s \in \Sigma$ which crosses some stack in S_0 (not necessarily α_1). Then we have

$$(S_0, i) \hookrightarrow_{\alpha_s} (S_1, r_1),$$

where, by construction, $S_1 \in \mathbb{T}(S_0)$. We then set $\mathcal{A}_S^{\text{ne}} = \mathcal{A}_S^{\text{ne}} \setminus \alpha_s$ and repeat this process until $\mathcal{A}_S^{\text{ne}} = \varnothing$. By construction, each S_h for $1 \leq h \leq n$ is contained in $\mathbb{T}(S_0)$ and $S_n = S$. This algorithm generates an insertion path in $\mathbb{T}(S_0)$ from S_0 to S and $\mathbb{V}(S_0) \subseteq \mathbb{T}(S_0)$ follows. $\qquad \square$

6.2.4 Saturation

Suppose we are given a skeleta tree $\mathbb{T}(S_0)$ with root S_0. Recall that the order of S, $\omega(S)$, denotes the number of $\prec$-maximal S-arcs. Furthermore, let $\Sigma_{i,j}$ and $\Sigma_{i,j}^{[r]}$ be some subset of structures over $\{i, i+1, \ldots, j-1, j\}$ and those of order r, respectively.

Let $\mathbb{M}_{i,j}$ denote the set of saturated skeleta over $\{i, i+1, \ldots, j-1, j\}$ and $OSM(i, j) \in \mathbb{M}_{i,j}$ be an mfe-saturated skeleton. Furthermore, let $OS(i, j)$ be an mfe structure, which is a union of disjoint $OSM(i_1, j_1), \ldots, OSM(i_r, j_r)$ and unpaired nucleotides. By $OSM^{[x]}(i, j)$ and $OS^{[x]}(i, j)$ we denote the respective OSM and OS structures of order x. We denote by $OS_{\text{mul}}(i, j)$, $OS_{\text{pk}}(i, j)$, and $OS_0(i, j)$, the mfe structures nested in a multi-loop, pseudoknot, and otherwise, respectively.

For a given skeleton $S_{i,j}$, we specify the mapping $S_{i,j} \mapsto OSM(S_{i,j})$ as follows: suppose $S_{i,j}$ has n_1 intervals, $I_1, \ldots, I_{n_1}$ labeled from left to right (Fig. 6.25). For given interval $I_r = [i_r, j_r]$ and $s_r \in \Sigma_{i_r, j_r}$ we consider the insertion of s_r into I_r, distinguishing the following four cases:

Case(1). I_r is contained in a hairpin loop.
$\omega(s_r) = 0$. That is we have $s_r = \varnothing$. The loop generated by the s_r-insertion remains obviously a hairpin loop, i.e., $((i_r - 1, j_r + 1), [i_r, j_r])$, with energy $H(i_r - 1, j_r + 1)$.
$\omega(s_r) = 1$. Let (p, q) be the unique, maximal s_r-arc. Then s_r-insertion produces the interior loop

$$((i_r - 1, j_r + 1), [i_r, p-1], (p, q), [q+1, j_r]),$$

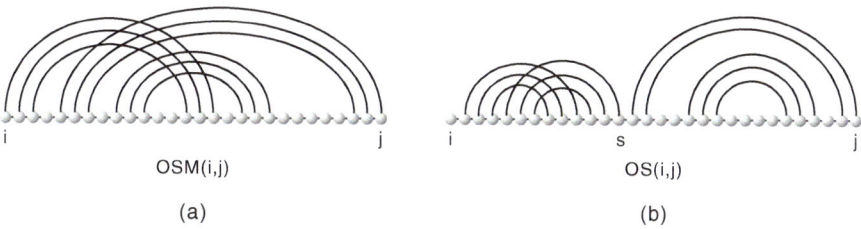

Fig. 6.25. OS versus OSM: we display an $OSM(i,j)$ (a) and an $OS(i,j)$ structure (b). The $OS(i,j)$ structure shown in (b) is evidently an union of the structures $OSM(i+1,s)$ and $OSM(s+1,j)$ and the unpaired nucleotide at position i.

with energy $I(i_r - 1, j_r + 1, p, q)$. Note that $p = i_r$ implies $q \neq j_r$ and $s_r \in OSM_0^{[1]}(p,q)$.

$\omega(s_r) \geq 2$. In this case inserting s_r into I_r creates a multi-loop in which s_r is nested. Then $s_r \in OS_{\text{mul}}^{[\geq 2]}$; see Fig. 6.26. Let $\epsilon(s)$ denote the energy of structure s. We select the set of all structures s_r such that

$$\epsilon(s_r) = \min \begin{cases} H(i_r - 1, j_r + 1) \\ I(i_r - 1, j_r + 1, p, q) + \epsilon(OSM_0^{[1]}(p,q)) \\ \qquad \text{for } i_r \leq p < q \leq j_r \text{ and } p = i_r \Rightarrow q \neq j_r \\ M + P_1 + \epsilon(OS_{\text{mul}}^{[\geq 2]}(i_r, j_r)). \end{cases}$$

Here, M is the energy penalty for forming a multi-loop and P_1 is the energy score of a closing pair in multi-loop.

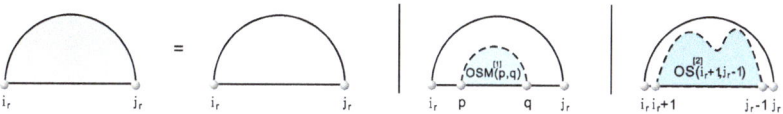

Fig. 6.26. Saturation in hairpin loops: the interval on the lhs is filled with substructures s_r such that $\omega(s_r) = 0$ (left), $\omega(s_r) = 1$ (middle), or $\omega(s_r) \geq 2$ (right).

Case(2). I_r is contained in a pseudoknot loop.
$\omega(s_r) = 0$. That is we have $s_r = \{\varnothing\}$ and the unpaired bases in I_r are considered to be contained in a pseudoknot.
$\omega(s_r) \geq 1$. In this case, s_r is a substructure which is nested in a pseudoknot; see Fig. 6.27. As a result our selection criterion is given by

$$\epsilon(s_r) = \min \begin{cases} (j_r - i_r + 1) \cdot Q_{\text{pk}} \\ \epsilon(OS_{\text{pk}}(i_r, j_r)), \end{cases}$$

where $(j_r - i_r + 1) \in \mathbb{N}$ is the number of unpaired bases in I_r and Q_{pk} is the energy score of the unpaired bases in a pseudoknot.

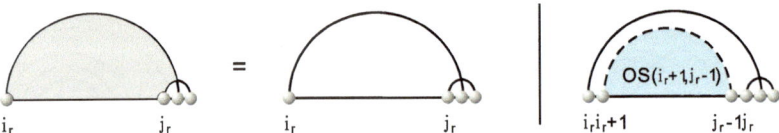

Fig. 6.27. Saturation of interval nested in a pseudoknot.

Case(3). I_r is contained in a multi-loop. In analogy to Case (2), we distinguish
$\omega(s_r) = 0$. That is we have $s_r = \{\varnothing\}$. The unpaired bases in I_r are considered to be contained in a multi-loop.
$\omega(s_r) \geq 1$. In this case, s_r is a substructure nested in a multi-loop; see Fig. 6.28. Accordingly, we select all structures satisfying

$$\epsilon(s_r) = \min \begin{cases} (j_r - i_r + 1) \cdot Q_{\mathrm{mul}} \\ \epsilon(OS_{\mathrm{mul}}(i_r, j_r)), \end{cases}$$

where Q_{mul} denotes the energy score of the unpaired bases in a multi-loop.
Case(4). I_r is contained in an interior loop. By construction, the latter is

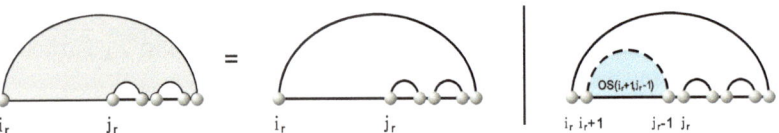

Fig. 6.28. Saturation of an interval contained in a multi-loop.

formed by the pair (I_r, I_l), where $r < l$. We then select pairs s_r in Σ_{i_r, j_r} and s_l in Σ_{i_l, j_l}. Note that only the first coordinate of the pair (I_r, I_l) is considered.
$\omega(s_r) = 0$ and $\omega(s_l) = 0$. Obviously, in this case the loop formed by I_r and I_l remains an interior loop

$$((i_r - 1, j_l + 1), [i_r, j_r], (j_r + 1, i_l - 1), [i_l, j_l]),$$

whose energy is given by $I(i_r - 1, j_l + 1, j_r + 1, i_l - 1)$.
$\omega(s_r) \geq 1$ and $\omega(s_l) = 0$. In this case, $s_l = \{\varnothing\}$. I_r and I_l create a multi-loop, in which s_r and the substructure G_{j_r+1, i_l-1} are nested.
$\omega(s_r) = 0$ and $\omega(s_l) \geq 1$. Completely analogous to the previous case.
$\omega(s_r) \geq 1$ and $\omega(s_l) \geq 1$. In this case, I_r and I_l create a multi-loop, in which s_r, s_l and G_{j_r+1, i_l-1} are nested; see Fig. 6.29.
Accordingly, we select all pairs of structures (s_r, s_l) satisfying

$$\epsilon(s_r) + \epsilon(s_l) = \min \begin{cases} I(i_r - 1, j_l + 1, j_r + 1, i_l - 1) \\ M + 2P_1 + \epsilon(OS_{\mathrm{mul}}(i_r, j_r)) + (j_l - i_l + 1) \cdot Q_{\mathrm{mul}} \\ M + 2P_1 + \epsilon(OS_{\mathrm{mul}}(i_l, j_l)) + (j_r - i_r + 1) \cdot Q_{\mathrm{mul}} \\ M + 2P_1 + \epsilon(OS_{\mathrm{mul}}(i_r, j_r)) + \epsilon(OS_{\mathrm{mul}}(i_l, j_l)). \end{cases}$$

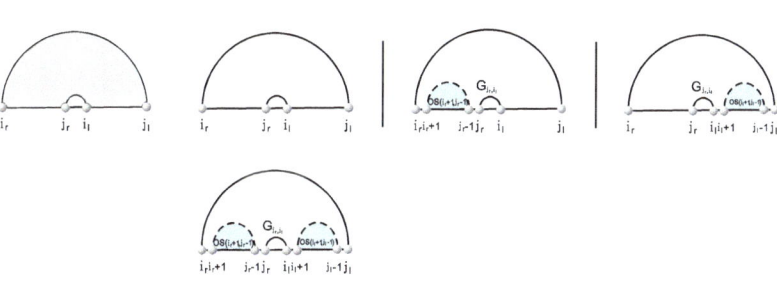

Fig. 6.29. Saturation of an interval contained in an interior loop, which is obtained by I_r and I_l, where $r < l$.

Accordingly, we inductively saturate all intervals and in case of interior loops interval pairs and thereby derive $OSM(S_{i,j})$. Then we select an energy-minimal $OSM(i, j)$ substructure from the set of all $OSM(S_{i,j})$ for any skeleton $S_{i,j}$.

As for the construction of $OS(i, j)$ via $OSM(i', j')$, we consider position i in $OS(i, j)$. If i is paired, then i is contained in some $OSM(i, s)$. Then $OS(i, j)$ induces a substructure S_2 over $\{s + 1, \ldots, j\}$. By construction $OS(i, j) = OSM(i, s) \dot\cup S_2$, whence $S_2 = OS(s + 1, j)$ and in particular we have

$$\epsilon(OS(i, j)) = \epsilon(OSM(i, s)) + \epsilon(OS(s + 1, j)).$$

Suppose next i is unpaired in $OS(i, j)$. Since ϵ is a loop-based energy, we can conclude $OS(i, j) = \{\varnothing\} \dot\cup OS(i + 1, j)$, i.e., we have

$$\epsilon(OS(i, j)) = \epsilon(OS(i + 1, j)) + Q,$$

where Q represents the energy contribution of a single, unpaired nucleotide. Accordingly, we can inductively construct $OS(i, j)$ via the criterion (Fig. 6.30)

$$\epsilon(OS(i, j)) = \min\{\epsilon(OS(i + 1, j)) + Q, \epsilon(OSM(i, s)) + \epsilon(OS(s + 1, j))\}$$

for $i < s \leq j$.

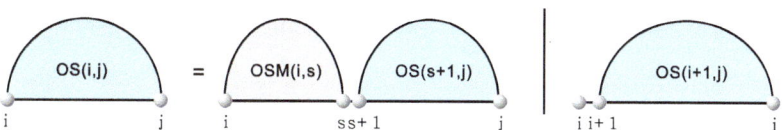

Fig. 6.30. Constructing $OS(i, j)$: inductive decomposition of the optimal structure, $OS(i, j)$, into saturated skeleta, $OSM(i, s)$, and unpaired nucleotides.

Now we can inductively construct the array of structures $OS(i,j)$ and $OSM(i,j)$ via OS and OSM structures over smaller intervals. As a result, we finally obtain the structure $OS(1,n)$, i.e., the mfe structure; see Fig. 6.31.

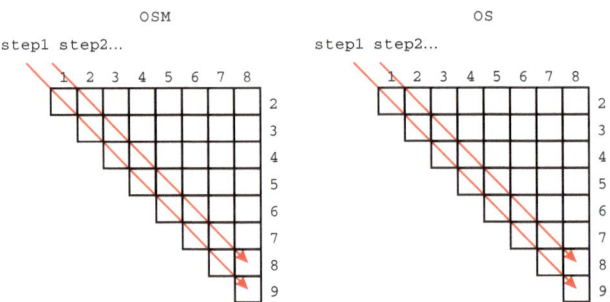

Fig. 6.31. Inductive construction of OS and OSM structures: in the sth step, we first construct $OSM(i, i+s)$, for any $0 < i < n-s+1$. We then construct $OS(i, i+s)$ recruiting OSM structures over intervals of lengths strictly smaller than s.

7

Neutral networks

In this chapter we study the structure of neutral networks as induced subgraphs of sequence space. Since exhaustive computation of sequence to structure maps using folding algorithms is at present time only feasible for sequences of length <40, we will study the structure of neutral networks using the language of random graphs. For data on sequence to structure maps into RNA secondary structures, obtained by computer folding algorithms, see [55, 56]. In [71] data on sequence to structure maps into RNA pseudoknot structures based on cross are being presented. The above papers allow to contrast the random graph model with biophysical folding maps. Our presentation is based on the papers [102, 103, 105, 106].

7.1 Neutral networks as random graphs

Let S be a fixed RNA pseudoknot structure. We recall that $C[S]$ denotes its set of compatible sequences, i.e., all sequences that have at any two paired positions one of the six nucleotide pairs $(\mathbf{A}, \mathbf{U})$, $(\mathbf{U}, \mathbf{A})$, $(\mathbf{G}, \mathbf{U})$, $(\mathbf{U}, \mathbf{G})$, $(\mathbf{G}, \mathbf{C})$, $(\mathbf{C}, \mathbf{G})$. In fact, the structure S gives rise to consider a new adjacency relation within $C[S]$. To this end we reorganize a sequence $(x_1, \ldots, x_n)$ into the tuple

$$((u_1, \ldots, u_{n_u}), (p_1, \ldots, p_{n_p})), \qquad (7.1)$$

where the u_j denote the unpaired nucleotides and the $p_j = (x_i, x_k)$ all base pairs of S, respectively; see Fig. 7.1. We proceed by considering $v_u = (u_1, \ldots, u_{n_u})$ and $v_p = (p_1, \ldots, p_{n_p})$ as elements of the cubes $Q_4^{n_u}$ and $Q_6^{n_p}$, implying the new adjacency relation for elements of $C[S]$, that is, the set of compatible sequences $C[S]$ can be endowed with a natural graph structure induced by $Q_4^{n_u} \times Q_6^{n_p}$, where "$\times$" denotes the direct product of graphs. Clearly, this decomposition is valid whether or not we have crossing arcs. The set of compatible sequences has an additional property [106]:

C. Reidys, *Combinatorial Computational Biology of RNA*,
DOI 10.1007/978-0-387-76731-4_7,
© Springer Science+Business Media, LLC 2011

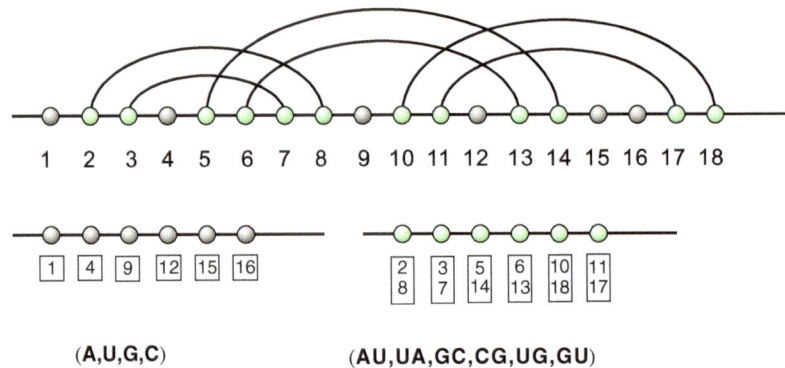

Fig. 7.1. Deriving the two subcubes, $Q_4^{n_u}$ and $Q_6^{n_p}$: A structure gives rise to re-arrange a compatible sequence into an unpaired and a paired segment. The former is a sequence over the original alphabet **A**, **U**, **G**, **C** and for the latter we derive a sequence over the alphabet of base pairs $(\mathbf{A}, \mathbf{U})$, $(\mathbf{U}, \mathbf{A})$, $(\mathbf{G}, \mathbf{U})$, $(\mathbf{U}, \mathbf{G})$, $(\mathbf{G}, \mathbf{C})$, $(\mathbf{C}, \mathbf{G})$.

Theorem 7.1. (Intersection theorem) *Suppose S_1, S_2 are two arbitrary structures in which each nucleotide is either unpaired or establishes a Watson–Crick base pair. Then we have*

$$C[S_1] \cap C[S_2] \neq \varnothing.$$

Proof. We identify S_1 and S_2 with the involutions I_1 and I_2, obtained by considering unpaired bases as fixed points and paired bases as transpositions. Two involutions generate a dihedral group acting upon the set $[n]$. Any cycle derived from this action $(i_1, \dots, i_m)$, where $i_h \in [n]$, has even length, $2s$, and assigning s **G**-**C** pairs, we observe that there exists at least one sequence compatible to both S_1 and S_2, whence the theorem.

Accordingly, the intersection theorem, originally formulated for RNA secondary structures, holds also for RNA pseudoknot structures; see Fig. 7.2.

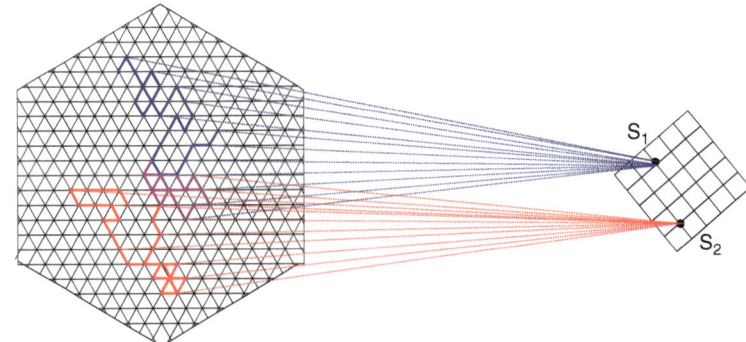

Fig. 7.2. The intersection theorem: two neutral networks come close in sequence space.

It has led to several remarkable discoveries [117] and is of relevance in the context of multistability of RNA molecules [44], that is, the existence of non-native conformations, exhibiting energies comparable to the molecule's ground state. The latter are separated from the native state by high-energy barriers [37, 45, 64].

We proceed by discussing the relation between the graph $Q_4^{n_u} \times Q_6^{n_p}$ and sequence space in Fig. 7.3.

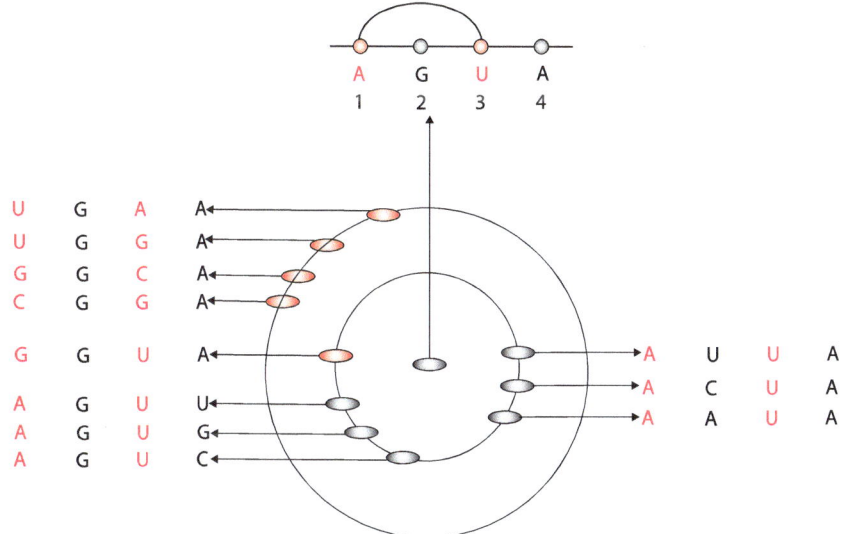

Fig. 7.3. Compatible neighbors in sequence space: diagram representation of an RNA structure (*top*) and its induced compatible neighbors in sequence space. Note that each base pair gives rise to five compatible neighbors exactly one of which being in Hamming distance 1.

Accordingly, there are two types of compatible neighbors in sequence space: u- and p-neighbors: a u-neighbor has Hamming distance 1 and differs exactly by a point mutation at an unpaired position. Analogously a p-neighbor differs by a compatible base pair mutation, see Fig. 7.3. A p-neighbor has either Hamming distance 1 $((\mathbf{G}, \mathbf{C}) \mapsto (\mathbf{G}, \mathbf{U}))$ or Hamming distance 2 $((\mathbf{G}, \mathbf{C}) \mapsto (\mathbf{C}, \mathbf{G}))$. We call a u- or a p-neighbor, y, a compatible neighbor. If y is contained in the neutral network we refer to y as a neutral neighbor. This suggests to consider the compatible and neutral distance, denoted by $C(v, v')$ and $N(v, v')$, denoting the minimum length of a $C[S]$-path and path in the neutral network between v and v', respectively.

The next step is to decide whether or not some compatible sequence is contained in the neutral network. For this purpose we employ the ansatz of [106] and select the vertices v_u and v_p with the independent probability λ_u and λ_p, respectively. The probability λ_u and λ_p is easily measured *locally* via computer folding maps: it coincides with the average fraction of neutral neighbors within the compatible neighbors. Explicitly, λ_u is the percentage of sequences that differ by a neutral mutation in an unpaired position, while λ_p corresponds to the percentage of neutral sequences that are compatible via a base pair mutation (for instance, $(\mathbf{A}, \mathbf{U}) \mapsto (\mathbf{G}, \mathbf{C})$); see Fig. 7.3.

The above construction reduces the random graph analysis of neutral networks to random subgraphs of the subcubes $Q_4^{n_u}$ and $Q_6^{n_p}$; see eq. (7.1) and Fig. 7.1. From a conceptual point of view these two cubes "only" differ by the respective alphabet length. This allows us to analyze the structure of neutral networks via random subgraphs of n-cubes.

In Section 7.2 we investigate the evolution of the largest component. Our main result here is

Suppose $\lambda_n = \frac{1+\chi_n}{n}$, where $\epsilon \geq \chi_n \geq n^{-\frac{1}{3}+\delta}$, $\delta > 0$. Then we have

$$\lim_{n \to \infty} \mathbb{P}\left(|C_n^{(1)}| \sim \pi(\chi_n) \frac{1+\chi_n}{n} 2^n \text{ and } C_n^{(1)} \text{ is unique} \right) = 1.$$

The result establishes the existence of a giant component at vertex selection probabilities "just" above $\lambda_n = 1/n$.

In Section 7.3 we investigate paths in random-induced subgraphs of n-cubes, once the giant has emerged. We will observe that there exists a threshold probability beyond which random-induced subgraphs can be viewed as "scaled" sequence spaces. The particular scaling factor, Δ, is explicitly computed and our main result reads

Suppose v, v' are contained in Γ_n, then
(a) for $\lambda_n < n^{\delta - \frac{1}{2}}$, for any $\delta > 0$, there exists a.s. no $\Delta > 0$ satisfying

$$d_{\Gamma_n}(v, v') \leq \Delta \, d_{Q_2^n}(v, v');$$

(b) for $\lambda_n \geq n^{\delta - \frac{1}{2}}$, for some $\delta > 0$, there exists a.s. some finite $\Delta = \Delta(\delta) > 0$ such that

$$d_{\Gamma_n}(v, v') \leq \Delta \, d_{Q_2^n}(v, v').$$

Finally, in Section 7.4 we localize the threshold value for connectivity of generalized n-cubes, Q_α^n [102]. Our proof is constructive and confirms our findings in Section 7.3: There exist many vertex disjoint paths between two vertices in the random graph. The particular construction has led to several computational studies on the connectivity of neutral networks [51, 55]:

$$\lim_{n \to \infty} \mathbb{P}(\Gamma_n \text{ is connected}) = \begin{cases} 0 & \text{for } \lambda < 1 - \sqrt[\alpha-1]{\alpha-1}, \\ 1 & \text{for } \lambda > 1 - \sqrt[\alpha-1]{\alpha-1}. \end{cases}$$

7.2 The giant

Maybe the most prominent feature in the evolution of random subgraphs of n-cubes is the sudden emergence of the giant, i.e., a unique largest component.

Burtin was the first [19] to study the connectedness of random subgraphs of n-cubes, Q_2^n, obtained by selecting all Q_2^n-edges independently (with probability p_n). He proved that a.s. all such subgraphs are connected for $p > 1/2$ and are disconnected for $p < 1/2$. Erdős and Spencer [38] refined Burtin's result and, more importantly in our context, they conjectured that there exists a.s. a giant component for $p_n = \frac{1+\epsilon}{n}$ and $\epsilon > 0$. Their conjecture was proved by Ajtai et al. [2] who established the existence of a unique giant component for $p_n = \frac{1+\epsilon}{n}$. Key ingredients in their proof are Harper's isoperimetric inequality [61] and a two round randomization, used for showing the non-existence of certain splits. Several variations including the analysis of the giant component in random graphs with given average degree sequence have been studied [90, 93]. Bollobás et al. [14] analyzed the behavior for ϵ tending to 0 and showed in particular that the constant for the giant component for fixed $\epsilon > 0$ coincides with the probability of infinite survival of the associated Poisson branching process. Borg et al. [16] refined their results, using the isoperimetric inequality [61] and Ajtai et al.'s two round randomization idea.

Considerably less is known for random-induced subgraphs of the n-cube obtained by independently selecting each Q_2^n-vertex with probability λ_n. Bollobás et al. have shown in [15] for $\lambda_n = (1 + \chi)/n$, where $\chi > 0$ is constant, that

$$|C_n^{(1)}| = (1 + o(1))\alpha(\chi)\frac{1 + \chi}{n}2^n,$$

where $0 < \alpha(\chi) < 1$ is the unique solution of the equation $x + e^{(1+\chi)x} = 1$. We will show, following [105], that this giant emerges for even smaller vertex selection probabilities. In the following we will work in binary n-cubes, Q_2^n. All results and proofs easily extend to the case of arbitrary alphabets.

We remark that the existence of a giant alone does not imply that the random graphs are well suited for neutral evolution. The relevant property will be identified in Section 7.3. Intuitively the largest component in its "early" stage is locally "tree-like." This structure is not suited for preserving sequence-specific information.

In the following let $k \in \mathbb{N}$ be a sufficiently large but fixed natural number, set $u_n = n^{-\frac{1}{3}}$, and

$$\pi(\chi_n) = \begin{cases} \alpha(\epsilon) & \text{for} \quad \chi_n = \epsilon \\ 2(1 + o(1))\chi_n & \text{for} \quad o(1) = \chi_n \geq n^{-\frac{1}{3}+\delta}. \end{cases}$$

Furthermore let

$$\nu_n = \lfloor \frac{1}{2k(k+1)}u_n n \rfloor,$$

$$\iota_n = \lfloor \frac{k}{2(k+1)}u_n n \rfloor,$$

$$z_n = k\nu_n + \iota_n,$$

$$\varphi_n = \pi(\chi_n)\nu_n(1 - e^{-(1+\chi_n)u_n/4}).$$

A k-cell (cell) is a Γ_n-subcomponent of size at least $c_k (u_n n) \varphi_n^k$, where $c_k > 0$.

One important observation in the context of the following two lemmas is the particular organization of a sequence $(x_1, \ldots, x_n)$. It facilitates the continuous switching between considering Q_2^n as a combinatorial graph and as a Cayley graph over the vector space $\mathbb{F}_2^n$, which allows us to use the notion of linear independence. We write a Q_2^n-vertex $v = (x_1, \ldots, x_n)$ as

$$
(\underbrace{x_1^{(1)}, \ldots, x_{\nu_n}^{(1)}}_{\nu_n \text{ coordinates}}, \underbrace{x_1^{(2)}, \ldots, x_{\nu_n}^{(2)}}_{\nu_n \text{ coordinates}}, \ldots, \underbrace{x_1^{(k+1)}, \ldots, x_{\iota_n}^{(k+1)}}_{\iota_n \text{ coordinates}}, \underbrace{x_{z_n+1}, \ldots, x_n}_{\substack{n - z_n \geq \\ n - \lfloor \frac{1}{2} u_n n \rfloor \text{ coordinates}}}). \quad (7.2)
$$

For any $1 \leq s \leq \nu_n$, $r = 1, \ldots, k$ we set $e_s^{(r)}$ to be the $(s + (r-1)\nu_n)$th unit vector, i.e., $e_s^{(r)}$ has exactly one 1 at its $(s + (r-1)\nu_n)$th coordinate. Similarly let $e_s^{(k+1)}$ $(1 \leq s \leq \iota_n)$ denote the $(s + k\nu_n)$th unit vector. We use the standard notation for the $z_n + 1 \leq t \leq n$ unit vectors, i.e., e_t is the vector where $x_t = 1$ and $x_j = 0$, otherwise.

Let us outline the strategy for proving the existence of the giant component:

- In Section 7.2.1 we prove the cell lemma, Lemma 7.3. It guarantees that many vertices are contained in cells, with probability only slightly smaller than $\pi(\chi_n)$. Its proof is based on Lemma 7.2, which supplies certain trees that serve as the "building blocks" for these cells.

- In Section 7.2.2 we study vertices in small components. The main result here is Lemma 7.6. It shows that the number of vertices contained in cells is sharply concentrated at $\pi(\chi_n) |\Gamma_n|$. Technicalities aside, of importance here is Lemma 7.5 which establishes a lower bound on the probability of a vertex being contained in a component of size at most $n^{\frac{1}{2}}$.

- In Section 7.2.3 we prove the main theorem. For this purpose we prove the split lemma, Lemma 7.10, which guarantees the existence of many vertex-independent paths between certain splits of Γ_n. These paths will eventually connect the cells and merging them into the giant component.

7.2.1 Cells

Lemma 7.2. *Suppose $\lambda_n = \frac{1+\chi_n}{n}$ and $1 > \chi_n \geq n^{-\frac{1}{3}+\delta}$, where $\delta > 0$. Then each Γ_n-vertex is contained in a Γ_n-subcomponent of size $\lfloor \frac{1}{4} u_n n \rfloor$ with probability at least $\pi(\chi_n)$ (Fig. 7.4).*

Proof. We consider the following branching process in the subcube $Q_2^{n-z_n}$, using the notation of eq. (7.2). Without loss of generality we initialize the process at $v = (0, \ldots, 0)$ (abusing notation we shall denote $(0, \ldots, 0)$ by 0) and set $E_0 = \{e_{z_n+1}, \ldots, e_n\}$ and $L_*[0] = \{(0, \ldots, 0)\}$. We consider the $n - \lfloor \frac{3}{4} u_n n \rfloor$ smallest neighbors of v. Starting with the smallest, we select each of them with independent probability $\lambda_n = \frac{1+\chi_n}{n}$. Suppose $v + e_j$ is the first being selected.

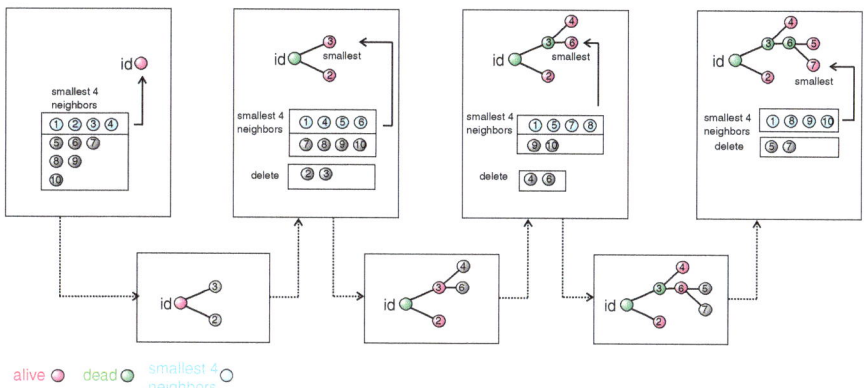

Fig. 7.4. Illustration of Lemma 7.2: constructing an acyclic, connected Q_2^n-subgraph H via a branching process embedded in $Q_2^{n-z_n}$.

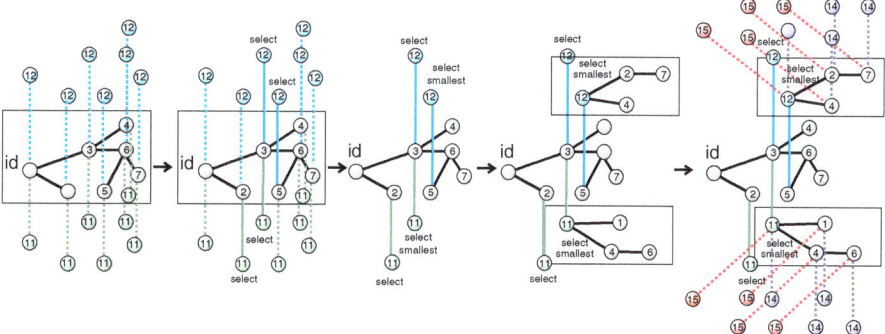

Fig. 7.5. Enlarging the acyclic, connected subgraph H displayed in Fig. 7.4 by successive translations and applications of Lemma 7.2 at smallest vertices.

Then we set $E_1 = E_0 \setminus \{e_j\}$, $N_1[0] = \{v + e_j\}$ and proceed inductively setting $E_t = E_{t-1} \setminus \{e_w\}$ and $N_t[0] = N_{t-1}[0] \cup \{v + e_w\}$ for each neighbor $v + e_w$ being selected, subject to the condition $|E_t| > n - (\lfloor \frac{3}{4} u_n n \rfloor - 1)$. This procedure generates the set containing all selected 0-neighbors, which we denote by $N_*[0]$. We consider $L_*[1] = N_*[0] \cup L_*[0] \setminus \{0\}$. If $\varnothing \neq L_*[1]$ we proceed by choosing its smallest element, v_1^*. By construction, v_1^* has at least $n - \lfloor \frac{3}{4} u_n n \rfloor$ neighbors of the form $v_1^* + e_r$, where $e_r \in E_t$. We iterate the process selecting from the smallest $n - \lfloor \frac{3}{4} u_n n \rfloor$ neighbors of v_1^* and set $L_*[2] = (N_*[1] \cup L_*[1]) \setminus \{v_1^*\}$. We then proceed inductively, setting $L_*[r + 1] = (N_*[r] \cup L_*[r]) \setminus \{v_r^*\}$. By construction, this process generates an induced acyclic, connected subgraph of $Q_2^{n-z_n}$. It stops in case of $L_*[r] = \varnothing$ for some $r \geq 1$ or

$$|E_s| = n - \left(\lfloor \frac{3}{4} u_n n \rfloor - 1 \right),$$

in which case $\lfloor \frac{1}{4} u_n n \rfloor - 1$ vertices have been connected. Corollary 2.30 guarantees that this $Q_2^{n-z_n}$-tree has size $\lfloor \frac{1}{4} u_n n \rfloor$ with probability at least $\pi(\chi_n)$.

We refer to the particular branching process used in Lemma 7.2 as γ-process. The γ-process produces a subcomponent of size $\lfloor \frac{1}{4} u_n n \rfloor$, which we refer to as γ-subcomponent. Note that in this process we did not use the first z_n coordinates of a vertex. In the following lemma we will use the first $k \nu_n$ of them in order to build cells; see Fig. 7.5.

Lemma 7.3. (Cell lemma) *Let* $k \in \mathbb{N}$ *be arbitrary but fixed,* $\lambda_n = \frac{1+\chi_n}{n}$, *and* $\varphi_n = \pi(\chi_n)\nu_n(1 - e^{-(1+\chi_n)u_n/4})$. *Then there exists some* $\rho_k > 0$ *such that each* Γ_n-*vertex is with probability at least*

$$\pi_k(\chi_n) = \pi(\chi_n)\left(1 - e^{-\rho_k \varphi_n}\right) \tag{7.3}$$

contained in a k-cell.

Proof. Since all translations are Q_2^n-automorphisms we can, without loss of generality, assume that $v = (0, \ldots, 0)$ (abusing notation we shall denote $(0, \ldots, 0)$ by 0). Using the notation of eq. (7.2) we recruit the $n - z_n$-unit vectors e_t for a γ-process. The γ-process of Lemma 7.2 yields a γ-subcomponent, $C_0^{(0)}$, of size $\lfloor \frac{1}{4} u_n n \rfloor$ with probability $\geq \pi(\chi_n)$. We consider for $1 \leq i \leq k$ the sets of ν_n elements $B_i = \{e_1^{(i)}, \ldots, e_{\nu_n}^{(i)}\}$ and set $H = \langle e_{z_n+1}, \ldots, e_n \rangle$. By construction we have

$$\left\langle B_i \cup \left\langle \bigcup_{1 \leq j \leq i-1} B_j \right\rangle \oplus H \right\rangle = \langle B_i \rangle \oplus \left\langle \bigcup_{1 \leq j \leq i-1} B_j \right\rangle \oplus H.$$

In particular, for any $1 \leq s \neq j \leq \nu_n$: $e_s^{(1)} - e_j^{(1)} \in H$ is equivalent to $e_s^{(1)} = e_j^{(1)}$. Since all vertices are selected independently and $|C_0^{(0)}| = \lfloor \frac{1}{4} u_n n \rfloor$, for fixed $e_s^{(1)} \in B_1$ the probability of not selecting a vertex $v' \in e_s^{(1)} + C_0^{(0)}$ is given by

$$\mathbb{P}\left(\{e_s^{(1)} + \xi \mid \xi \in C_0^{(0)}\} \cap \Gamma_n = \varnothing\right) = \left(1 - \frac{1+\chi_n}{n}\right)^{\lfloor \frac{1}{4} u_n n \rfloor} \sim e^{-(1+\chi_n)\frac{1}{4}u_n}. \tag{7.4}$$

We set $\mu_n = (1 - e^{-(1+\chi_n)\frac{1}{4}u_n})$, i.e., $\mu_n = \mathbb{P}\left((e_s^{(1)} + C_0^{(0)}) \cap \Gamma_n \neq \varnothing\right)$ and introduce the r.v.

$$X_1 = \left|\left\{e_s^{(1)} \in B_1 \mid \exists \xi \in C_0^{(0)}; \, e_s^{(1)} + \xi \in \Gamma_n\right\}\right|.$$

Obviously, $\mathbb{E}(X_1) = \mu_n \nu_n$ and using the large deviation result of eq. (2.39) we can conclude that

$$\exists \rho > 0; \quad \mathbb{P}\left(X_1 < \frac{1}{2}\mu_n \nu_n\right) \leq e^{-\rho \mu_n \nu_n}.$$

Suppose for $e_s^{(1)}$ there exists some $\xi \in C_0^{(0)}$ such that $e_s^{(1)} + \xi \in \Gamma_n$ (that is, $e_s^{(1)}$ is counted by X_1). We then select the smallest element of the set $\{e_s^{(1)} + \xi \mid \xi \in C_0^{(0)}, e_s^{(1)} + \xi \in \Gamma_n\}$, say $e_s^{(1)} + \xi_{0,e_s^{(1)}}$, and initiate a γ-process using the $n - z_n$ elements $\{e_{z_n+1}, \ldots, e_n\}$ at $e_s^{(1)} + \xi_{0,e_s^{(1)}}$. The process yields a γ-subcomponent, $C_{e_s^{(1)} + \xi_{0,e_s^{(1)}}}^{(1)}$, of size $\lfloor \frac{1}{4} u_n n \rfloor$ with probability at least $\pi(\chi_n)$. For any two elements $e_s^{(1)}, e_j^{(1)}$ with $e_s^{(1)} + \xi_{0,e_s^{(1)}}, e_j^{(1)} + \xi_{0,e_j^{(1)}} \in \Gamma_n$ the respective γ-subcomponents, $C_{e_s^{(1)} + \xi_{0,e_s^{(1)}}}^{(1)}$ and $C_{e_j^{(1)} + \xi_{0,e_j^{(1)}}}^{(1)}$, are vertex disjoint since $\langle B_1 \cup H \rangle = \langle B_1 \rangle \oplus H$. Let $\tilde{X}_1$ be the r.v. counting the number of these new, pairwise vertex disjoint sets of γ-subcomponents of size $\lfloor \frac{1}{4} u_n n \rfloor$. By construction each of them is connected to 0. We immediately observe $\mathbb{E}(\tilde{X}_1) \geq \pi(\chi_n)\mu_n\nu_n$ and set $\varphi_n = \pi(\chi_n)\mu_n\nu_n$. Using the large deviation result in eq. (2.39) we derive

$$\exists \rho_1 > 0; \quad \mathbb{P}\left(\tilde{X}_1 < \frac{1}{2}\varphi_n\right) \leq e^{-\rho_1\varphi_n}.$$

We proceed by proving that for each $1 \leq i \leq k$ there exists a sequence of r.v.s $(\tilde{X}_1, \tilde{X}_2, \ldots, \tilde{X}_i)$ where $\tilde{X}_j$ counts the number of pairwise disjoint sets of γ-subcomponents added at step j, where $1 \leq j \leq i$, such that
(a) all sets, $C_\alpha^{(j)}$, $1 \leq j \leq i$, are pairwise vertex disjoint and have size $\lfloor \frac{1}{4} u_n n \rfloor$ and
(b) all $C_\alpha^{(j)}$ are connected to 0 and

$$\exists \rho_i > 0; \quad \mathbb{P}\left(\tilde{X}_i < \frac{1}{2^i}\varphi_n^i\right) \leq e^{-\rho_i\varphi_n}, \text{ where } \varphi_n = \pi(\chi_n)\mu_n\nu_n.$$

We prove the assertion by induction on i. Without loss of generality we may assume $i < k$. Indeed, in our construction of $\tilde{X}_1$, we established the induction basis. In order to define $\tilde{X}_{i+1}$ we use the set $B_{i+1} = \{e_1^{(i+1)}, \ldots, e_{\nu_n}^{(i+1)}\}$. For each $C_\alpha^{(i)}$ counted by $\tilde{X}_i$ (i.e., the subcomponents that were connected in step i) we form the set $e_s^{(i+1)} + C_\alpha^{(i)}$. By induction hypothesis two different $C_\alpha^{(i)}, C_{\alpha'}^{(i)}$, counted by $\tilde{X}_i$, are vertex disjoint and connected to 0. In view of

$$\langle B_{i+1} \rangle \bigoplus \left\langle \bigcup_{1 \leq j \leq i} B_j \right\rangle \bigoplus H$$

we can conclude

$$(s \neq s' \lor \alpha \neq \alpha') \implies (e_s^{(i+1)} + C_\alpha^{(i)}) \cap (e_{s'}^{(i+1)} + C_{\alpha'}^{(i)}) = \varnothing.$$

Furthermore, the probability that we have for fixed $C_\alpha^{(i)}$: $(e_s^{(i+1)} + C_\alpha^{(i)}) \cap \Gamma_n = \varnothing$, for some $e_s^{(i+1)} \in B_{i+1}$, is exactly as in eq. (7.4):

$$\mathbb{P}\left(\left(e_s^{(i+1)} + C_\alpha^{(i)}\right) \cap \Gamma_n = \varnothing\right) = \left(1 - \frac{1+\chi_n}{n}\right)^{\lfloor \frac{1}{4} u_n n \rfloor} \sim e^{-(1+\chi_n)\frac{1}{4} u_n}.$$

As it is the case for the induction basis, $\mu_n = (1 - e^{-(1+\chi_n)\frac{1}{4} u_n})$ is the probability that $(e_s^{(i+1)} + C_\alpha^{(i)}) \cap \Gamma_n \neq \varnothing$. We proceed by defining the r.v.

$$X_{i+1} = \sum_{C_\alpha^{(i)}} \left| \left\{ e_s^{(i+1)} \in B_{i+1} \mid \exists \xi \in C_\alpha^{(i)}; e_s^{(i+1)} + \xi \in \Gamma_n \right\} \right|.$$

The r.v. X_{i+1} counts the number of events where $(e_s^{(i+1)} + C_\alpha^{(i)}) \cap \Gamma_n \neq \varnothing$ for each $C_\alpha^{(i)}$, respectively. Equivalently, for fixed $C_\alpha^{(i)}$ and $e_s^{(i+1)} \in B_{i+1}$ let

$$e_s^{(i+1)} + \xi_{\alpha, e_s^{(i+1)}} = \min \left\{ e_s^{(i+1)} + \xi_\alpha \mid \xi_\alpha \in C_\alpha^{(i)}, e_s^{(i+1)} + \xi_\alpha \in \Gamma_n \right\}.$$

Then X_{i+1} counts exactly the minimal elements

$$e_s^{(i+1)} + \xi_{\alpha, e_s^{(i+1)}}, e_{s'}^{(i+1)} + \xi_{\alpha', e_{s'}^{(i+1)}}, \dots$$

for all $C_\alpha^{(i)}, C_{\alpha'}^{(i)}, \dots$ and any two can be used to construct pairwise vertex disjoint γ-subcomponents of size $\lfloor \frac{1}{4} u_n n \rfloor$. We next define $\tilde{X}_{i+1}$ to be the r.v. counting the number of events that the γ-process in H initiated at the $e_s^{(i+1)} + \xi_{\alpha, e_s^{(i+1)}} \in \Gamma_n$ yields a γ-subcomponent of size $\lfloor \frac{1}{4} u_n n \rfloor$. By construction each of these is connected to a unique $C_\alpha^{(i)}$. Since $\langle B_{i+1} \rangle \bigoplus \langle \bigcup_{1 \leq j \leq i} B_j \rangle \bigoplus H$ all newly added sets are pairwise vertex disjoint to all previously added subcomponents. We derive

$$\mathbb{P}\left(\tilde{X}_{i+1} < \frac{1}{2^{i+1}} \varphi_n^{i+1}\right) \leq \underbrace{\mathbb{P}\left(\tilde{X}_i < \frac{1}{2^i} \varphi_n^i\right)}_{\text{failure at step } i} +$$

$$\underbrace{\mathbb{P}\left(\tilde{X}_{i+1} < \frac{1}{2^{i+1}} \varphi_n^{i+1} \wedge \tilde{X}_i \geq \frac{1}{2^i} \varphi_n^i\right)}_{\text{failure at step } i+1 \text{ conditional to } \tilde{X}_i \geq \frac{1}{2^i} \varphi_n^i}$$

$$\leq e^{-\rho_i \varphi_n} + e^{-\rho \varphi_n^{i+1}} (1 - e^{-\rho_i \varphi_n}), \quad \rho > 0$$

$$\leq e^{-\rho_{i+1} \varphi_n}.$$

Therefore each Γ_n-vertex is, with probability at least $\pi(\chi_n)(1 - e^{-\rho_k \varphi_n})$, contained in a subcomponent of size at least $c_k (u_n n) \varphi_n^k$, for $c_k > 0$ and the proof of the lemma is complete.

Lemma 7.3 gives rise to introduce the induced subgraph $\Gamma_{n,k} = Q_2^n[A]$ where

$$A = \{v \mid v \text{ is contained in a } \Gamma_n\text{-subcomponent of size } \geq c_k (u_n n) \varphi_n^k, c_k > 0\}.$$

In case of $\epsilon \geq \chi_n \geq n^{-\frac{1}{3}+\delta}$ we have $1 - e^{-\frac{1}{4}(1+\chi_n)u_n} \geq u_n/4$ and consequently $\varphi_n \geq c'\,(1+o(1))\chi_n u_n^2\, n \geq c_0\, n^\delta$ for some $c', c_0 > 0$. Furthermore

$$\left\lfloor \frac{1}{4}u_n n \right\rfloor \varphi_n^k \geq c_k\, n^{\frac{2}{3}} n^{k\delta}, \quad c_k > 0.$$

Accordingly, choosing k sufficiently large, each Γ_n-vertex is contained in a cell with probability at least

$$\pi(\chi_n)\left(1 - e^{-\rho_k n^\delta}\right), \quad 0 < \delta,\ 0 < \rho_k.$$

7.2.2 The number of vertices contained in cells

Let us begin with a technical lemma, which shows that the number of vertices contained in components of size $\leq n^a$, where $a > 0$ is sharply concentrated [14]. This result holds since sufficiently small components are "almost" independent. Let $U_n = U_n(a)$ denote the set of vertices contained in such small components.

Lemma 7.4. *Let $a > 0$ be a fixed constant and $\lambda_n = \frac{1+\chi_n}{n}$, where $1 > \chi_n \geq n^{-\frac{1}{3}+\delta}$. Then*

$$\mathbb{P}\left(\,\big|\,|U_n| - \mathbb{E}[|U_n|]\,\big| \geq \frac{1}{n}\mathbb{E}[|U_n|]\right) = o(1).$$

Proof. Let C be a Q_2^n-component of size strictly smaller than $\tau = n^a$ and let v be a fixed C-vertex. We shall denote the ordered pair (C, v) by C_v and the indicator variable of the pair C_v by X_{C_v}. Clearly, we have

$$|U_n| = \sum_{C_v} X_{C_v},$$

where the summation is taken over all ordered pairs (C, v) with $|C| < \tau$. Considering isolated points, we immediately obtain $\mathbb{E}[|U_n|] \geq c|\Gamma_n|$ for some $1 \geq c > 0$.

Claim. The random variable $|U_n|$ is sharply concentrated.

We prove the claim by estimating $\mathbb{V}[|U_n|]$ via computing the correlation terms $\mathbb{E}[X_{C_v} X_{D_{v'}}]$ and applying Chebyshev's inequality. Suppose $C_v \neq D_{v'}$. There are two ways by which $X_{C_v}, X_{D_{v'}}$ viewed as r.v. over Q_{2,λ_n}^n, can be correlated. First, v, v' can belong to the same component, i.e., $C = D$, in which case we write $C_v \sim_1 D_{v'}$. Clearly,

$$\sum_{C_v \sim_1 D_{v'}} \mathbb{E}[X_{C_v} X_{D_{v'}}] \leq \tau\, \mathbb{E}[|U_n|].$$

Second, correlation arises when v, v' belong to two different components C_v, $D_{v'}$ having minimal distance 2 in Q_2^n. In this case we write $C_v \sim_2 D_{v'}$. Then there exists some Q_2^n-vertex, w, such that $w \in \mathsf{d}(C_v) \cap \mathsf{d}(D_{v'})$ and we derive

$$\mathbb{P}(d(C_v, D_{v'}) = 2) = \frac{1 - \lambda_n}{\lambda_n} \, \mathbb{P}(C_v \cup D_{v'} \cup \{w\} \text{ is a } \Gamma_n\text{-component})$$
$$\leq n \, \mathbb{P}(C_v \cup D_{v'} \cup \{w\} \text{ is a } \Gamma_n\text{-component}).$$

We can now immediately give the upper bound

$$\sum_{C_v \sim_2 D_{v'}} \mathbb{E}[X_{C_v} X_{D_{v'}}] \leq n \, (2\tau + 1)^3 \, |\Gamma_n|.$$

The uncorrelated pairs $(X_{C_v}, X_{D_{v'}})$, writing $C_v \not\sim D_{v'}$, can easily be estimated by

$$\sum_{C_v \not\sim D_{v'}} \mathbb{E}[X_{C_v} X_{D_{v'}}] = \sum_{C_v \not\sim D_{v'}} \mathbb{E}[X_{C_v}]\mathbb{E}[X_{D_{v'}}] \leq \mathbb{E}[|U_n|]^2.$$

Consequently we arrive at

$$\mathbb{E}[|U_n|(|U_n| - 1)]$$
$$= \sum_{C_v \sim_1 D_{v'}} \mathbb{E}[X_{C_v} X_{D_{v'}}] + \sum_{C_v \sim_2 D_{v'}} \mathbb{E}[X_{C_v} X_{D_{v'}}] + \sum_{C_v \not\sim D_{v'}} \mathbb{E}[X_{C_v} X_{D_{v'}}]$$
$$\leq \tau \, \mathbb{E}[|U_n|] + n \, (2\tau + 1)^3 |\Gamma_n| + \mathbb{E}[|U_n|]^2.$$

Using

$$\mathbb{V}[|U_n|] = \mathbb{E}[|U_n|(|U_n| - 1)] + \mathbb{E}[|U_n|] - \mathbb{E}[|U_n|]^2$$

and $\mathbb{E}[U_n] \geq c \, |\Gamma_n|$ we obtain

$$\frac{\mathbb{V}[|U_n|]}{\mathbb{E}[|U_n|]^2} \leq \frac{O(n^a) + O(n^{3a+1})}{\mathbb{E}[|U_n|]} = o\left(\frac{1}{n^2}\right).$$

Chebyshev's inequality guarantees

$$\mathbb{P}\left(||U_n| - \mathbb{E}[|U_n|]| \geq \frac{1}{n} \mathbb{E}[|U_n|]\right) \leq n^2 \, \frac{\mathbb{V}[|U_n|]}{\mathbb{E}[|U_n|]^2},$$

whence the claim and the lemma follows.

By linearity of expectation, Lemma 7.3 implies a lower bound on the expected number of Γ_n-vertices contained in cells. Lemma 7.4 lifts this observation to an a.s. statement about the number of vertices contained in cells. We will formalize this conclusion in Lemma 7.6.

However, so far we only have a lower bound on the number of vertices contained in cells. It thus remains to prove that the so-derived lower bound is sharp. To this end we show that there are many Γ_n-vertices contained in components of size $\leq n^{1/2}$. The idea here will be to show that the probability that these small components are not trees is small compared to the probability of simply "dying" out due to not selecting neighboring vertices. The key observation is the following lower bound on the probability of small Q_2^n-components that contain the fixed vertex v, denoted by C_v:

Lemma 7.5. *For any vertex* $v \in Q_2^n$ *holds*

$$\mathbb{P}\left(|C_v| < n^{1/2}\right) \geq 1 - (1 + o(1))\pi(\chi_n).$$

We postpone the proof of Lemma 7.5 to the end of this section and proceed by proving the concentration result on the number of vertices contained in cells.

Lemma 7.6. *Let* $\lambda_n = \frac{1+\chi_n}{n}$ *where* $1 > \chi_n \geq n^{-\frac{1}{3}+\delta}$. *Then, for sufficiently large* $k \in \mathbb{N}$

$$|\Gamma_{n,k}| \sim \pi(\chi_n)|\Gamma_n| \quad a.s..$$

Proof. *Claim 1.* $|\Gamma_{n,k}| \geq ((1 - o(1))\,\pi(\chi_n))\,|\Gamma_n|$ a.s..
According to Lemma 7.3 we have $\mathbb{E}[|U_n|] < (1 - \pi_k(\chi_n))\,|\Gamma_n|$ and we can conclude using Lemma 7.4 and $\mathbb{E}[|U_n|] = O(|\Gamma_n|)$

$$|U_n| < \left(1 + O(n^{-1})\right)\mathbb{E}[|U_n|] < \left(1 - (\pi_k(\chi_n) - O(n^{-1}))\right)|\Gamma_n| \quad a.s..$$

In view of eq. (7.3) and $\chi_n \geq n^{-\frac{1}{3}+\delta}$ we have for arbitrary but fixed k,

$$\pi_k(\chi_n) - O(n^{-1}) = (1 - o(1))\,\pi(\chi_n).$$

Therefore we derive

$$|\Gamma_{n,k}| \geq (1 - o(1))\,\pi(\chi_n)\,|\Gamma_n| \quad a.s., \tag{7.5}$$

and Claim 1 follows.
Claim 2. For sufficiently large k, $|\Gamma_{n,k}| \leq ((1 + o(1))\,\pi(\chi_n))\,|\Gamma_n|$ a.s. holds.
According to Lemma 7.5 we have

$$\mathbb{P}\left(|C_v| < n^{1/2}\right) \geq 1 - (1 + o(1))\pi(\chi_n).$$

By linearity of expectation we derive $(1 - (1 + o(1))\pi(\chi_n))|\Gamma_n| \leq \mathbb{E}[|U_n|]$ and according to Lemma 7.4 $(1 - O(n^{-1}))\mathbb{E}[|U_n|] < |U_n|$ a.s. In view of $n^{-1} = o(\pi(\chi_n))$ we consequently arrive at

$$(1 - (1 + o(1))\,\pi(\chi_n))\,|\Gamma_n| \leq |U_n| \quad a.s..$$

Since $|U_n| \leq |\Gamma_n| - |\Gamma_{n,k}|$ we obtain

$$(1 + o(1))\,\pi(\chi_n)\,|\Gamma_n| \geq |\Gamma_{n,k}| \quad a.s.. \tag{7.6}$$

Combining eqs. (7.5) and (7.6) we derive

$$(1 - o(1))\,\pi(\chi_n)|\Gamma_n| \leq |\Gamma_{n,k}| \leq (1 + o(1))\,\pi(\chi_n)|\Gamma_n| \quad a.s.,$$

whence the lemma.

It thus remains to prove Lemma 7.5. As mentioned above, the intuition here is that these small components are "typically" acyclic. Let us therefore first have a look at the situation for trees.

We consider the rooted tree T_n with root v^*. Then v^* has n and all other T_n-vertices have $n-1$ descendants. Selecting the T_n-vertices with independent probability λ_n, we obtain the probability space T_{n,λ_n}, whose elements, A_n, are random-induced subtrees. We shall be interested in the A_n-component which contains the root, denoted by C_{v^*}. Let ξ_{v^*} and ξ_v, for $v \neq v^*$ be two r.v. such that $\mathsf{Prob}(\xi_{v^*} = \ell) = B_n(\ell, \lambda_n)$ and $\mathsf{Prob}(\xi_v = \ell) = B_{n-1}(\ell, \lambda_n)$, respectively. We assume that ξ_{v^*} and ξ_v count the offspring produced at v^* and $v \neq v^*$. Then the induced branching process initialized at v^*, $(Z_i)_{i \in \mathbb{N}_0}$ constructs C_{v^*}. Let $\pi_0(\chi_n)$ denote its survival probability, then we have in view of assertions (1) and (2) of Corollary 2.29:

$$\pi_0(\chi_n) = (1 + o(1))\, \pi(\chi_n). \tag{7.7}$$

Lemma 7.7. (Bollobás et al. [14]) For any $a > 0$, there exists some $\kappa_a > 0$ such that

$$\mathbb{P}(|C_{v^*}| < n^a) \geq 1 - \pi_0(\chi_n) - O(e^{-\kappa_a\, n^a}).$$

Proof. We begin by expressing $\mathbb{P}(|C_{v^*}| < n^a)$ as follows:

$$\mathbb{P}(|C_{v^*}| < n^a) = \underbrace{\mathbb{P}(|C_{v^*}| < \infty)}_{=1 - \pi_0(\chi_n)} - \mathbb{P}(n^a \leq |C_{v^*}| < \infty).$$

According to [14] we have

$$\mathbb{P}(|C_{v^*}| = i) = (1 + o(1)) \cdot \frac{(\lambda_n \cdot (n-1))^{i-1}}{i\sqrt{2\pi i}} \left[\frac{(n-1)(1-\lambda_n)}{(n-2)}\right]^{i(n-2)+2}, \tag{7.8}$$

where $i = i(n) \to \infty$ as $n \to \infty$. We express

$$\mathbb{P}(n^a \leq |C_{v^*}| < \infty) = \sum_{i \geq n^a} \mathbb{P}(|C_{v^*}| = i)$$

and observe that eq. (7.8) implies the upper bound

$$\leq \sum_{i \geq n^a} (1 + o(1)) \frac{(\lambda_n \cdot (n-1))^{i-1}}{i\sqrt{2\pi i}} \left[\frac{(n-1)(1-\lambda_n)}{(n-2)}\right]^{i(n-2)+2}$$

$$\leq \sum_{i \geq n^a} \left[(1 + \epsilon_n)e^{-\epsilon_n}\right]^i \leq \sum_{i \geq n^a} c(\epsilon)^i = O(e^{-\kappa_a\, n^a}),$$

where $0 < c(\epsilon) < 1$ and $0 < \kappa_a$. Consequently, we arrive at

$$\mathbb{P}(|C_{v^*}| < n^a) \geq 1 - \pi_0(\chi_n) - O(e^{-\kappa_a\, n^a}). \tag{7.9}$$

In the following we shall present a process by which a small Q_2^n-component can generically be generated. We then show that the probability to stay acyclic *and* forming a component of size $\leq n^{\frac{1}{2}}$ is much larger than the probability of forming a cycle. This will allow us to establish a lower bound on the probability $\mathbb{P}(|C_v| \leq n^{\frac{1}{2}})$ in terms of the probability $\mathbb{P}(|C_{v^*}| \leq n^{\frac{1}{2}})$, i.e., the probability of forming such a component in the rooted tree T_n.

We next present the particular process by which we generate a random, connected, induced subgraph $H_v^\dagger$ of Q_2^n that contains v [14]. Let $n_0 = \lfloor n^{1/2} \rfloor$ and let S be a stack. We initialize the generation by setting $H_v^\dagger = \{v\}$. We select the v-neighbors, one by one, in increasing order, with probability λ_n. For each selected neighbor v_i, we

- put the corresponding edge (v, v_i) into S,
- add v_i to $H_v^\dagger$, and
- check condition (h1) "$|H_v^\dagger| = n_0$."

If (h1) holds we stop, otherwise we proceed examining the next v-neighbor. Suppose (h1) does not hold and all v-neighbors have been examined. If S is empty, we stop. Otherwise we proceed inductively as follows: we remove the first element, (u, w), from S and consider the w-neighbors, except u, one by one, in increasing order. For each selected neighbor, r, we

- insert the edge (w, r) into the back of S,
- add r to $H_v^\dagger$, and
- check condition (h1) "$|H_v^\dagger| = n_0$" and (h2) "$H_v^\dagger$ contains a cycle."

In case (h1) or (h2) holds we stop. Otherwise, we continue examining w-neighbors in increasing order until all w-neighbors are considered. If S is empty we stop and otherwise we consider the next element from S and iterate the process.

By construction, $H_v^\dagger$ can contain only cycles that contain the vertex that was added in the last step of the process.

Lemma 7.8. *Suppose* $\frac{1+\chi_n}{n} = \lambda_n$, *where* $1 > \chi_n \geq n^{-\frac{1}{3}+\delta}$. *Then*

$$\mathbb{P}\left(|H_v^\dagger| < n_0 \wedge H_v^\dagger \text{ is a acyclic}\right) \geq 1 - (1 + o(1))\pi(\chi_n).$$

Proof. We first prove

$$\mathbb{P}\left(H_v^\dagger \text{ contains a cycle}\right) \leq O(n^{-\frac{1}{2}}). \tag{7.10}$$

Let $C_{2\ell}$ be such a cycle of length 2ℓ, then we have $2 \leq \ell \leq n_0/2$. To prove eq. (7.10) we observe that, according to Lemma 2.25, there are at most $\binom{2\ell}{\ell} n^\ell \ell! = O\left(\frac{4\ell n}{e}\right)^\ell$ Q_2^n-cycles of length 2ℓ that contain a fixed vertex v_0. Let w be the last vertex added to $H_v^\dagger$.

Suppose that the $H_v^\dagger$-cycle C_ℓ of length 2ℓ does not contain v. We consider the vertices contained in C_ℓ. By construction, each vertex $v_0 \neq w$ has been examined only once and w has been examined at most $n^{\frac{1}{2}} - 1 \leq n^{\frac{1}{2}}$ times. Therefore, the probability for such a $C_{2\ell}$ is bounded from above by

$$n_0 2\ell \underbrace{\binom{2\ell}{\ell} n^\ell \ell!}_{O\left(\frac{4\ell n}{e}\right)^\ell} \left(\frac{2}{n}\right)^{2\ell-1} \frac{2}{n^{\frac{1}{2}}} = O\left(\ell n^1 \left(\frac{16\ell}{e\,n}\right)^\ell\right),$$

where the terms are interpreted as follows:

- n_0 represents to the number of ways to select w within $H_v^\dagger$,
- 2ℓ represents the number of possible positions for w in $C_{2\ell}$,
- $\left(\frac{2}{n}\right)^{2\ell-1}$ is the upper bound probability of selecting $2\ell - 1$ vertices that were examined exactly once,
- $n^{\frac{1}{2}} \cdot 2/n$ is the upper bound probability of $C_{2\ell}$ to contain w.

Taking the sum over all possible length (note that $\ell \geq 2$, i.e., we cannot have $\ell = 1$ corresponding to a cycle of length 2) we conclude that the probability of this event is bounded by $O(n^{-1})$.

Suppose next that the $H_v^\dagger$-cycle $C_{2\ell}$ of length 2ℓ does contain v. By construction, each vertex, except of w and v, has been examined exactly once. w has been examined at most $n^{\frac{1}{2}}$ times and v has not been considered at all. Thus the probability for such a $C_{2\ell}$ is bounded by

$$2\ell \binom{2\ell}{\ell} n^\ell \ell! \left(\frac{2}{n}\right)^{2\ell-2} \frac{2}{n^{\frac{1}{2}}} = O\left(\ell n^{\frac{3}{2}} \left(\frac{16\ell}{e\,n}\right)^\ell\right).$$

Note here that

- since v is fixed, there appears no factor n_0 and
- the term $\left(\frac{2}{n}\right)^{2\ell-2}$ reflects the fact that $2\ell - 2$ vertices occur that have only been examined once.

Again, taking the sum over all possible length $\ell \geq 2$, we conclude that the probability of this event is bounded by $O(n^{-\frac{1}{2}})$ and eq. (7.10) follows.

To prove Lemma 7.8 we note that if $|H_v^\dagger| < n_0$ and $H_v^\dagger$ is acyclic, then the generation of $H_v^\dagger$ represents a particular way to simulate the branching process Z^0 in the n-cube Q_2^n. Consequently,

$$\mathbb{P}\left(|H_v^\dagger| < n_0 \wedge H_v^\dagger \text{ is acyclic}\right) \leq \mathbb{P}\left(|C_{v^*}| < n_0\right)$$

and the discrepancy lies in the probability of exactly those events for which a covering map from $T_n(v^*)$ into Q_2^n (mapping v^* into v) produces a cycle in Q_2^n. The latter are bounded from above by $\mathbb{P}(H_v^\dagger \text{ contains a cycle})$, whence

$$\mathbb{P}\left(|H_v^\dagger| < n_0 \wedge H_v^\dagger \text{ is acyclic}\right) + \mathbb{P}\left(H_v^\dagger \text{ contains a cycle}\right) \geq \mathbb{P}\left(|C_{v^*}| < n_0\right).$$

In view of $\mathbb{P}(|C_{v^*}| < n^{\frac{1}{2}}) \geq 1 - \pi_0(\chi_n) - O(e^{-\kappa n^{\frac{1}{2}}})$, where $\kappa = \kappa_{1/2} > 0$ and $\mathbb{P}\left(H_v^\dagger \text{ contains a cycle}\right) \leq O(n^{-\frac{1}{2}})$ we have

$$\mathbb{P}\left(|H_v^\dagger| < n_0 \wedge H_v^\dagger \text{ is acyclic}\right) \geq 1 - \pi_0(\chi_n) - O(e^{-\kappa n^{\frac{1}{2}}}) - O\left(n^{-\frac{1}{2}}\right).$$

In view of $\pi_0(\chi_n) = (1 + o(1))\pi(\chi_n)$ and $\pi(\chi_n) \geq n^{-\frac{1}{3}+\delta}$ the lemma follows.

Proof of Lemma 7.5. Let D_v be a tree containing v of size $< n_0$ in Q_2^n. Since there is only one way by which the procedure $H_v^\dagger$ can generate D_v we have

$$\mathbb{P}(C_v = D_v) \geq \mathbb{P}\left(H_v^\dagger = D_v\right)$$

and consequently, taking the sum over all such trees we obtain

$$\mathbb{P}\left(|C_v| < n_0 \wedge C_v \text{ is a tree}\right) \geq \mathbb{P}\left(|H_v^\dagger| < n_0 \wedge H_v^\dagger \text{ is acyclic}\right).$$

According to Lemma 7.8 we have

$$\mathbb{P}\left(|H_v^\dagger| < n_0 \wedge H_v^\dagger \text{ is acyclic}\right) \geq 1 - (1 + o(1))\pi(\chi_n).$$

Consequently we arrive at

$$\begin{aligned}
\mathbb{P}\left(|C_v| < n_0\right) &\geq \mathbb{P}\left(|C_v| < n_0 \wedge C_v \text{ is a tree}\right) \\
&\geq \mathbb{P}\left(|H_v^\dagger| < n_0 \wedge H_v^\dagger \text{ is a acyclic}\right) \\
&\geq 1 - (1 + o(1))\pi(\chi_n)
\end{aligned}$$

and Lemma 7.5 is proved.

7.2.3 The largest component

The first objective of this section is to prove Lemma 7.10, where we establish the existence of many vertex disjoint, short paths between certain splits of the $\Gamma_{n,k}$. For this purpose we observe

Lemma 7.9. *Let $k \in \mathbb{N}$ and $\lambda_n = \frac{1+\chi_n}{n}$ and $1 > \chi_n \geq n^{-\frac{1}{3}+\delta}$. Then we have*

$$\exists \Delta > 0; \forall v \in \mathbb{F}_2^n, \quad \mathbb{P}\left(|\mathsf{S}(v,2) \cap \Gamma_{n,k}| < \frac{1}{2}\left(\frac{k}{2(k+1)}\right)^2 n^\delta\right) \leq e^{-\Delta n^\delta}.$$

$$(7.11)$$

Let $D_\delta = \left\{v \mid |\mathsf{S}(v,2) \cap \Gamma_{n,k}| < \frac{1}{2}\left(\frac{k}{2(k+1)}\right)^2 n^\delta\right\}$, then there exists some $\Delta > \tilde{\Delta} > 0$ such that

$$|D_\delta| \leq 2^n e^{-\tilde{\Delta} n^\delta} \quad a.s.$$

Proof. To prove the lemma, we use the last (see eq. (7.2)) $\iota_n = \lfloor \frac{k}{2(k+1)} u_n n \rfloor$ elements $e_1^{(k+1)}, \ldots, e_{\iota_n}^{(k+1)}$. We consider for arbitrary $v \in Q_2^n$

$$\mathsf{S}^{(k+1)}(v,2) = \left\{v + e_i^{(k+1)} + e_j^{(k+1)} \mid 1 \leq i < j \leq \iota_n, \right\}.$$

Clearly, $|S^{(k+1)}(v,2)| = \binom{\ell_n}{2}$ holds. By construction, for any two $S^{(k+1)}(v,2) \cap \Gamma_n$-vertices, the Γ_n-subcomponents of size $\geq c_k(u_n n)\varphi_n^k$ constructed via Lemma 7.3 are vertex disjoint. Furthermore, each Γ_n-vertex belongs to $\Gamma_{n,k}$ with probability $\geq \pi_k(\chi_n)$. Let Z be the r.v. counting the number of vertices in $S^{(k+1)}(v,2) \cap \Gamma_{n,k}$. Then we have

$$\mathbb{E}[Z] \geq \left(\frac{k}{2(k+1)}\right)^2 \frac{u_n^2}{2} n\, \pi(\chi_n) \quad a.s..$$

Equation (7.11) follows from eq. (2.39), $u_n^2 n\chi_n \geq n^\delta$ and

$$\mathbb{P}(|S(v,2) \cap \Gamma_{n,k}| < \eta) \leq \mathbb{P}(|S^{(k+1)}(v,2) \cap \Gamma_{n,k}| < \eta).$$

Let

$$D_\delta = \left\{ v \mid |S(v,2) \cap \Gamma_{n,k}| < \frac{1}{2}\left(\frac{k}{2(k+1)}\right)^2 n^\delta \right\}.$$

By linearity of expectation $\mathbb{E}(|D_\delta|) \leq 2^n e^{-\Delta n^\delta}$ holds and using Markov's inequality,

$$\forall t > 0; \quad \mathbb{P}(X > t\mathbb{E}(X)) \leq 1/t,$$

we derive $|D_\delta| \leq 2^n e^{-\tilde{\Delta} n^\delta}$ a.s. for some $0 < \tilde{\Delta} < \Delta$.

Now we are in position to prove the split lemma:

Lemma 7.10. (Split lemma) *Suppose* $\lambda_n = \frac{1+\chi_n}{n}$ *where* $1 > \chi_n \geq n^{-\frac{1}{3}+\delta}$. *Let* (A,B) *be a split of the* $\Gamma_{n,k}$-*vertex set with the properties*

$$\exists\, 0 < \sigma_0 \leq \sigma_1 < 1; \quad \frac{1}{n^2} 2^n \leq |A| = \sigma_0|\Gamma_{n,k}| \quad \text{and} \quad \frac{1}{n^2} 2^n \leq |B| = \sigma_1|\Gamma_{n,k}|.$$
$$\tag{7.12}$$

Then there exists some $t > 0$ *such that a.s.* $\mathsf{d}(A)$ *is connected to* $\mathsf{d}(B)$ *in* Q_2^n *via at least*

$$\frac{t}{n^4} 2^n / \binom{n}{7}$$

vertex disjoint (independent) paths of length ≤ 3.

Proof. We consider $\mathsf{B}(A,2)$ and distinguish the cases

$$|\mathsf{B}(A,2)| \leq \frac{2}{3} 2^n \quad \text{and} \quad |\mathsf{B}(A,2)| > \frac{2}{3} 2^n.$$

Suppose first $|\mathsf{B}(A,2)| \leq \frac{2}{3} 2^n$ holds. According to Theorem 2.27 and eq. (7.12), we have

$$\exists\, d_1 > 0; \quad |\mathsf{d}(\mathsf{B}(A,2))| \geq \frac{d_1}{n^3} 2^n.$$

Lemma 7.9 guarantees that a.s. all except of at most $2^n e^{-\tilde{\Delta} n^\delta}$ Q_2^n-vertices are within distance 2 to some $\Gamma_{n,k}$-vertex. Hence there exist at least $\frac{d}{n^3} 2^n$ vertices of $\mathsf{d}(\mathsf{B}(A,2))$ that are contained in $\mathsf{B}(B,2)$, i.e.,

$$|\mathsf{d}(\mathsf{B}(A,2)) \cap \mathsf{B}(B,2)| \geq \frac{d}{n^3}\, 2^n \quad \text{a.s.}.$$

For each $\beta_2 \in \mathsf{d}(\mathsf{B}(A,2)) \cap \mathsf{B}(B,2)$ there exists a path $(\alpha_1, \alpha_2, \beta_2)$, starting in $\mathsf{d}(A)$ with terminus β_2. In view of $\mathsf{B}(B,2) = \mathsf{d}(\mathsf{B}(B,1)) \dot\cup \mathsf{B}(B,1)$, we distinguish the following cases:

$$|\mathsf{d}(\mathsf{B}(A,2)) \cap \mathsf{d}(\mathsf{B}(B,1))| \geq \frac{1}{n^3} d_{2,1}\, 2^n \quad \text{and} \quad |\mathsf{d}(\mathsf{B}(A,2)) \cap \mathsf{B}(B,1)| \geq \frac{1}{n^3}\, d_{2,2}\, 2^n.$$

Suppose we have $|\mathsf{d}(\mathsf{B}(A,2)) \cap \mathsf{d}(\mathsf{B}(B,1))| \geq \frac{1}{n^3} d_{2,1}\, 2^n$. For each $\beta_2 \in \mathsf{d}(\mathsf{B}(B,1))$, we select some element $\beta_1(\beta_2) \in \mathsf{d}(B)$ and set $B^* \subset \mathsf{d}(B)$ to be the set of these endpoints. Clearly at most n elements in $\mathsf{B}(B,2)$ can produce the same endpoint, whence

$$|B^*| \geq \frac{1}{n^4} d_{2,1}\, 2^n.$$

Let $B_1 \subset B^*$ be maximal subject to the condition that for any pair of B_1-vertices (β_1, β_1') we have $d(\beta_1, \beta_1') > 6$. Then we have $|B_1| \geq |B^*|/\binom{n}{7}$ since $|\mathsf{B}(v,6)| = \sum_{i=0}^{6} \binom{n}{i} \leq \binom{n}{7}$. Any two of the paths from $\mathsf{d}(A)$ to $B_1 \subset \mathsf{d}(B)$ are of the form $(\alpha_1, \alpha_2, \beta_2, \beta_1)$ and vertex disjoint since each of them is contained in $\mathsf{B}(\beta_1, 3)$. Therefore, there are a.s. at least

$$\frac{1}{n^4} d_{2,1}\, 2^n \,\Big/\, \binom{n}{7}$$

vertex disjoint paths connecting $\mathsf{d}(A)$ and $\mathsf{d}(B)$. Suppose next $|\mathsf{d}(\mathsf{B}(A,2)) \cap \mathsf{B}(B,1)| \geq \frac{1}{n^3} d_{2,2}\, 2^n$. We conclude in complete analogy that there exist a.s. at least

$$\frac{1}{n^3} d_{2,2}\, 2^n \,\Big/\, \binom{n}{5}$$

vertex disjoint paths of the form $(\alpha_1, \alpha_2, \beta_2)$ connecting $\mathsf{d}(A)$ and $\mathsf{d}(B)$. It remains to consider the case $|\mathsf{B}(A,2)| > \frac{2}{3} 2^n$. By construction both A and B satisfy eq. (7.12), respectively, whence we can without loss of generality assume that also $|\mathsf{B}(B,2)| > \frac{2}{3} 2^n$ holds. In this case we have

$$|\mathsf{B}(A,2) \cap \mathsf{B}(B,2)| > \frac{1}{3}\, 2^n$$

and for each $\alpha_2 \in \mathsf{B}(A,2) \cap \mathsf{B}(B,2)$ we select $\beta_1 \in \mathsf{d}(B)$. We derive in analogy to the previous arguments that there exist a.s. at least

$$\frac{1}{n} d_2\, 2^n \,\Big/\, \binom{n}{5}$$

pairwise vertex disjoint paths of the form $(\alpha_1, \alpha_2, \beta_1)$ and the proof of the lemma is complete.

Theorem 7.11. *Let Q^n_{2,λ_n} be the random graph consisting of Q^n_2-subgraphs, Γ_n, induced by selecting each Q^n_2-vertex with independent probability λ_n. Suppose $\lambda_n = \frac{1+\chi_n}{n}$, where $1 > \chi_n \geq n^{-\frac{1}{3}+\delta}$, $\delta > 0$. Then we have*

$$\lim_{n\to\infty} \mathbb{P}\left(|C^{(1)}_n| \sim \pi(\chi_n)\frac{1+\chi_n}{n}2^n \text{ and } C^{(1)}_n \text{ is unique}\right) = 1.$$

In Fig. 7.6 we illustrate the result for random-induced subgraphs of Q^{15}_2.

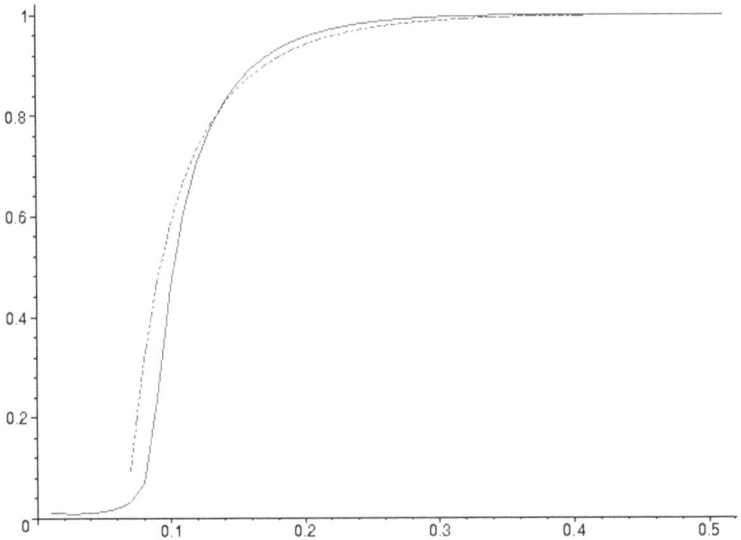

Fig. 7.6. The evolution of the giant as a function of λ_n: we display the theoretical growth implied by Theorem 7.11, $\pi(\chi_n)\frac{1+\chi_n}{n}2^n$ (*dashed curve*) versus the average size of the giant component obtained from a 100 random-induced subgraphs of Q^{15}_2 (*solid curve*).

Proof. Claim. We have $|C^{(1)}_n| \sim |\Gamma_{n,k}|$ a.s..

Let $\mu_2 > \mu_1 > 1$ be constants satisfying $\frac{1}{\mu_1} + \frac{1}{\mu_2} = 1$. To prove the claim we use an idea introduced by Ajtai et al. [2] and select Q^n_2-vertices in two rounds. First, we select Q^n_2-vertices with independent probability $\frac{1+\chi_n/\mu_1}{n}$ and subsequently with $\frac{\chi_n}{\mu_2 n}$. The probability for some vertex not to be chosen in both randomizations is

$$\left(1 - \frac{1+\chi_n/\mu_1}{n}\right)\left(1 - \frac{\chi_n/\mu_2}{n}\right) = 1 - \frac{1+\chi_n}{n} + \frac{(1+\chi_n/\mu_1)\chi_n/\mu_2}{n^2}$$

$$\geq 1 - \frac{1+\chi_n}{n}.$$

Hence selecting first with probability $\frac{1+\chi_n/\mu_1}{n}$ (first round) and then with $\frac{\chi_n/\mu_2}{n}$ (second round) a vertex is selected with probability less than $\frac{1+\chi_n}{n}$ (all preceding lemmas hold for the first randomization $\frac{1+\chi_n/\mu_1}{n}$). We now select in our first round each Q_2^n-vertex with probability $\frac{1+\chi_n/\mu_1}{n}$. According to Lemma 7.6, we have

$$|\Gamma_{n,k}^{\mu_1}| \sim \pi(\chi_n/\mu_1)\,|\Gamma_n| \quad \text{a.s..}$$

Suppose $\Gamma_{n,k}^{\mu_1}$ contains a component, A, such that

$$\frac{1}{n^2}\,2^n \le |A| \le (1-b)\,|\Gamma_{n,k}^{\mu_1}|, \quad b > 0.$$

Then there exists a split of $\Gamma_{n,k}^{\mu_1}$, (A,B), satisfying the assumptions of Lemma 7.10. We observe that Lemma 7.3 limits the number of ways these splits can be constructed. In view of

$$\left\lfloor \frac{1}{4}u_n n \right\rfloor \varphi_n^k \ge c_k\,n^{\frac{2}{3}}n^{k\delta}, \quad c_k > 0,$$

each A-vertex is contained in a component of size at least $c_k\,n^{\frac{2}{3}}n^{k\delta}$. Therefore there are at most

$$2^{\left(2^n/(c_k\,n^{\frac{2}{3}}n^{k\delta})\right)}$$

ways to choose A in such a split. According to Lemma 7.10 there exists $t > 0$ such that a.s. $\mathsf{d}(A)$ is connected to $\mathsf{d}(B)$ in Q_2^n via at least $\frac{t}{n^4}2^n/\binom{n}{7}$ vertex disjoint paths of length ≤ 3. We now select Q_2^n-vertices with probability $\frac{\chi_n/\mu_2}{n}$. None of the above $\ge \frac{t}{n^4}2^n/\binom{n}{7}$ paths can be selected during this process. Since any two paths are vertex disjoint the expected number of such splits is less than

$$2^{\left(2^n/(c_k\,n^{\frac{2}{3}}n^{k\delta})\right)}\left(1-\left(\frac{\chi_n/\mu_2}{n}\right)^4\right)^{\frac{t}{n^4}2^n/\binom{n}{7}} \sim$$
$$2^{\left(2^n/(c_k\,n^{\frac{2}{3}}n^{k\delta})\right)}e^{-\frac{t(\chi_n/\mu_2)^4}{n^8}2^n/\binom{n}{7}}.$$

Hence choosing k sufficiently large, we can conclude that a.s. there cannot exist such a split of $\Gamma_{n,k}^{\mu_1}$. Therefore $C_n^{(1)}$ has a.s. at least $\sim \pi(\chi_n/\mu_1)|\Gamma_n|$ elements. Since $\pi(\chi_n/\mu_1)$ is continuous and monotonously decreasing in the parameter μ_1, for any $0 < q < 1$ there exists a $\mu_1 > 1$ such that

$$\pi(\chi_n/\mu_1)|\Gamma_n| \sim \frac{q+1}{2}\pi(\chi_n)|\Gamma_n|$$

which implies

$$|C_n^{(1)}| \sim \pi(\chi_n)\,|\Gamma_n|$$

and the claim is proved. In particular, for $\chi_n = \epsilon$, Lemma 2.28 $(0 < \alpha(\epsilon) < 1)$ implies that there exists a giant Γ_n-component. It remains to prove that $C_n^{(1)}$ is unique. By construction any largest component, C_n', is necessarily contained in $\Gamma_{n,k}$. In the proof of the claim we have shown that a.s. there cannot exist another component C_n' in Γ_n with the property $|C_n'| \sim c_0 |\Gamma_{n,k}|$, $0 < c_0 < 1$. Therefore $C_n^{(1)}$ is unique and the proof of the theorem is complete.

Theorem 7.12 is the analogue of Ajtai et al.'s result [2] (for random subgraphs of n-cubes obtained by selecting Q_2^n-edges independently).

Theorem 7.12. *Let Q_{2,λ_n}^n be the random graph consisting of Q_2^n-subgraphs, Γ_n, induced by selecting each Q_2^n-vertex with independent probability λ_n and suppose $\epsilon > 0$. Then*

$$\lim_{n \to \infty} \mathbb{P}(\Gamma_n \text{ has an unique giant component}) = \begin{cases} 1 & \text{for } \lambda_n \geq \frac{1+\epsilon}{n} , \\ 0 & \text{for } \lambda_n \leq \frac{1-\epsilon}{n} . \end{cases}$$

7.3 Neutral paths

In view of the fact that the connectivity of random graphs does not imply that they are well suited for evolutionary optimization, we study random-induced subgraphs beyond the emergence of the giant. We ask whether there are any structural changes within the giant component aside from its growth. In this section we follow the ideas in [104] and ask what happens in the random graph if we increase the vertex selection probability λ_n. One key property in this context is the particular path connectivity within the giant, in particular the emergence of "short" paths. To be precise we ask for which λ_n does there exist some constant $\Delta > 0$ such that

(†) $\exists \Delta > 0; \quad d_{\Gamma_n}(v, v') \leq \Delta d_{Q_2^n}(v, v')$ a.s. provided v, v' are in Γ_n.

The following theorem [104] establishes the threshold value for the existence of the above constant Δ. The result is of relevance in the context of local connectivity of neutral networks, a structural property which allows populations of RNA strings to preserve sequence-specific information [104].

Theorem 7.13. *Let v, v' be arbitrary but fixed Q_2^n-vertices, having distance $d_{Q_2^n}(v, v') = d$, $d \geq 2$, $d \in \mathbb{N}$. Let Γ_n denote the random subgraph of Q_2^n, obtained by independently selecting Q_2^n-vertices with probability λ_n. Suppose v, v' are contained in Γ_n, then the following assertions hold:*
(a) Suppose $\lambda_n < n^{\delta - \frac{1}{2}}$, for any $\delta > 0$. Then there exists a.s. no $\Delta > 0$ satisfying

$$d_{\Gamma_n}(v, v') \leq \Delta d_{Q_2^n}(v, v').$$

(b) Suppose $\lambda_n \geq n^{\delta - \frac{1}{2}}$, for some $\delta > 0$. Then there exists a.s. some finite $\Delta = \Delta(\delta) > 0$ such that

$$d_{\Gamma_n}(v, v') \leq \Delta d_{Q_2^n}(v, v').$$

Proof. Suppose $d_{Q-2^n}(v, v')$ and $\Delta > 0$ are fixed. Let $Z = Z(d, \Delta)$ be the r.v. counting the paths of length $\leq \Delta d$ from v to v'. According to Lemma 2.25 we have

$$\mathbb{E}[Z] \leq \sum_{2\ell+d \leq \Delta d} \binom{2\ell+d}{\ell+d} \binom{\ell+d}{\ell} n^\ell \, \ell! \, d! \, \lambda_n^{2\ell+d-1}.$$

Since $\lambda_n < n^{\delta-\frac{1}{2}}$ for any $\delta > 0$, we obtain

$$\sum_{2\ell+d \leq \Delta d} \binom{2\ell+d}{\ell+d} \binom{\ell+d}{\ell} n^\ell \, \ell! \, d! \, \lambda_n^{2\ell+d-1}$$

$$\leq \sum_{2\ell+d \leq \Delta d} \binom{2\ell+d}{\ell+d} \binom{\ell+d}{\ell} \ell! \, d! \, n^{\delta \, 2\ell} \left[\frac{1}{n^{\frac{1}{2}-\delta}} \right]^{d-1}.$$

For given $d \geq 2$ and Δ, the quantity ℓ is bounded and choosing δ sufficiently small we derive the upper bound

$$\mathbb{E}[Z] \leq O(n^{-\mu}) \quad \text{for some} \quad \mu > 0, \tag{7.13}$$

proving assertion (a).

To prove (b) we consider a specific subset of paths, $\mathbb{A}_\sigma$, where σ is some permutation of d elements. The $\mathbb{A}_\sigma$-elements are called α-paths and given by the following data:

- Some family $\mathcal{F}^+ = (e_{j_1}, \ldots, e_{j_\ell})$, where $d < j_i \leq n$ and $|\{j_i \mid 1 \leq i \leq \ell\}| = \ell$
- The fixed family $\mathcal{G} = (e_{\sigma(1)}, \ldots, e_{\sigma(d)})$
- The family $\mathcal{F}^- = (e_{j_\ell}, \ldots, e_{j_1})$, i.e. $\mathcal{F}^-$ is the "mirror-image" of $\mathcal{F}^+$

Let X_α be the indicator r.v. for the event "α is a path in Γ_n." Clearly, $A = \sum_{\alpha \in \mathbb{A}_\sigma} X_\alpha$ is the r.v. counting the number of α-paths contained in Γ_n. Let $n' = n - d$. By construction of α-paths and linearity of expectation we observe

$$\mathbb{E}[A] = \ell! \binom{n'}{\ell} \lambda_n^{2\ell+d-1} = (n')_\ell \, \lambda_n^{2\ell+d-1},$$

where $(n)_\ell = n(n-1) \cdots (n-(\ell-1))$. Since $\lambda_n \geq n^{-\frac{1}{2}+\delta}$ for some $0 < \delta$

$$\mathbb{E}[A] \geq \left[\frac{(n'-\ell)}{n} \right]^\ell n^{2\ell\delta} \left[n^{-\frac{1}{2}+\delta} \right]^{d-1}.$$

The idea is now to use Janson's inequality (Theorem 2.31) in order to show that a.s. at least one α-path is contained in Γ_n. For this purpose we estimate the correlation between the indicator r.v. X_α and $X_{\alpha'}$. The key term we have to analyze is

$$\Omega = \sum_{\alpha \in \mathbb{A}_\sigma} \sum_{\substack{\alpha' \in \mathbb{A}_\sigma; \\ \alpha' \cap \alpha \neq \varnothing}} \mathbb{E}[X_\alpha X_{\alpha'}].$$

Let $u_s = v + (\sum_{i=1}^{s} e_{j_i})$, where $s \leq \ell$. Since $\mathcal{F}^-$ is the mirror image of the sequence $(e_{j_1}, \ldots, e_{j_\ell})$ we inspect

$$|\alpha \cap \alpha'| = 2|\{u_s \in \alpha \cap \alpha' \mid 1 \leq s \leq \ell\}| + \begin{cases} d-1 & \text{if } u_\ell \in \alpha \cap \alpha', \\ 0 & \text{otherwise.} \end{cases} \quad (7.14)$$

Indeed, only if α and α' intersect at u_ℓ, the subsequent d steps of $\mathcal{G}$ coincide. In view of eq. (7.14), we distinguish the cases

$$\text{(i)} \ u_\ell \notin \alpha \cap \alpha' \quad \text{and} \quad \text{(ii)} \ u_\ell \in \alpha \cap \alpha'.$$

Case (i): In this case we have $|\alpha \cap \alpha'| = 2h$, where $1 \leq h \leq \ell - 1$. For fixed h, there are exactly $\binom{\ell-1}{h}$ ways to select the h vertices where α and α' intersect. For each such selection, there at most $h! \, (n' - h)_{\ell-h}$ paths α', whence

$$|\{\alpha' \mid |\alpha' \cap \alpha| = 2h\}| \leq \binom{\ell-1}{h} h! \, (n' - h)_{\ell-h}.$$

The probability for choosing a correlated α'-path is given by $\lambda_n^{2[2\ell+d-1]-2h}$ and we compute

$$\sum_{\alpha \in \mathbb{A}_\sigma} \sum_{\substack{\alpha' \in \mathbb{A}_\sigma; \\ u_\ell \notin \alpha' \cap \alpha \neq \varnothing}} \mathbb{E}[X_\alpha X_{\alpha'}] = \mathbb{E}[A] \sum_{h=1}^{\ell-1} |\{\alpha' \mid |\alpha' \cap \alpha| = 2h\}| \lambda_n^{[2\ell+d-1]-2h}$$

$$\leq \mathbb{E}[A] \sum_{h=1}^{\ell-1} h! \binom{\ell-1}{h} (n' - h)_{\ell-h} \lambda_n^{[2\ell+d-1]-2h}$$

$$= \mathbb{E}[A]^2 \sum_{h=1}^{\ell-1} h! \binom{\ell-1}{h} (n')_h^{-1} \lambda_n^{-2h}$$

$$\leq \mathbb{E}[A]^2 \sum_{h=1}^{\ell-1} h! \binom{\ell-1}{h} \frac{n^h}{(n')_h} n^{-2h\delta},$$

where the last inequality is implied by $\lambda_n \geq n^{-\frac{1}{2}+\delta}$. We have for sufficiently large n

$$\sum_{h=1}^{\ell-1} h! \binom{\ell-1}{h} \frac{n^h}{(n')_h} n^{-2h\delta} = \underbrace{(\ell-1) \frac{n}{n'} n^{-2\delta}}_{h=1} + \underbrace{O\left(n^{-4\delta}\right)}_{h>1}.$$

Consequently, in case of (i), we can give the following upper bound:

$$\sum_{\alpha \in \mathbb{A}_\sigma} \sum_{\substack{\alpha' \in \mathbb{A}_\sigma; \\ u_\ell \notin \alpha' \cap \alpha \neq \varnothing}} \mathbb{E}[X_\alpha X_{\alpha'}] \leq \left[(\ell-1) \frac{n}{n'} n^{-2\delta} + O(n^{-4\delta}) \right] \mathbb{E}[A]^2. \quad (7.15)$$

Case (ii): The key observation is that for fixed α, there are at most $\ell!$ paths α' that intersect α at least in u_ℓ. Each of these appears with probability at most 1, whence

$$\sum_{\alpha \in \mathbb{A}_\sigma} \sum_{\substack{\alpha' \in \mathbb{A}_\sigma; \\ u_\ell \in \alpha' \cap \alpha \neq \varnothing}} \mathbb{E}[X_\alpha X_{\alpha'}] \leq \ell! \, \mathbb{E}[A]. \tag{7.16}$$

Using eqs. (7.15) and (7.16), we arrive at

$$\Omega \leq \left(\underbrace{(\ell-1)\frac{n}{n'} \, n^{-2\delta} + O(n^{-4\delta})}_{(i)} + \underbrace{\frac{\ell!}{\mathbb{E}[A]}}_{(ii)} \right) \mathbb{E}[A]^2.$$

According to Theorem 2.31, we have $\mathbb{P}(A \leq (1-\gamma)\mathbb{E}[A]) \leq e^{-\frac{\gamma^2 \mathbb{E}[A]}{2+2\Omega/\mathbb{E}[A]}}$, i.e.

$$\mathbb{P}(A \leq (1-\gamma)\mathbb{E}[A]) \leq \exp\left[-\frac{\gamma^2}{2/\mathbb{E}[A] + 2\left((\ell-1)\frac{n}{n'} \, n^{-2\delta} + O(n^{-4\delta}) + \frac{\ell!}{\mathbb{E}[A]}\right)} \right]. \tag{7.17}$$

In view of $\mathbb{E}[A] \geq \left[\frac{(n'-\ell)}{n}\right]^\ell n^{2\ell\delta} \left[n^{-\frac{1}{2}+\delta}\right]^{d-1}$, we observe, for sufficiently large ℓ,

$$\left[\frac{\gamma^2}{2/\mathbb{E}[A] + 2\left((\ell-1)\frac{n}{n'} \, n^{-2\delta} + O(n^{-4\delta}) + \frac{\ell!}{\mathbb{E}[A]}\right)} \right] = O(n^{2\delta}).$$

Setting $\gamma = 1$, eq. (7.17) becomes

$$\mathbb{P}(A = 0) \leq e^{-c'n^{2\delta}} \quad \text{for some } c' > 0 . \tag{7.18}$$

Since an α-path has length $2\ell + d$, eq. (7.18) proves **(b)** and the proof of the theorem is complete.

As a result of the constructive proof of Theorem 7.13 we are now in position to compute the probabilities of short paths connecting two vertices of fixed distance d; see Problem 7.6.

7.4 Connectivity

In this section we localize the threshold value for connectivity of random-induced subgraphs of n-cubes.

Lemma 7.14. *Let Q_α^n be a generalized n-cube, $\lambda > 1 - \sqrt[\alpha-1]{\alpha^{-1}}$, and Γ_n an induced Q_α^n-subgraph obtained by selecting each Q_α^n-vertex with independent probability λ. Then we have*

$$\lim_{n\to\infty} \mathbb{P}(\forall\, v, v' \in \Gamma_n, \, d_{Q_\alpha^n}(v, v') = k; \, v \text{ is connected to } v') = 1.$$

Proof. Claim 1. Suppose $\lambda > 1 - \sqrt[\alpha-1]{\alpha-1}$. Then for arbitrary $\ell \in \mathbb{N}$, Γ_n contains a.s. exclusively vertices of degree $\geq \ell$.

To prove the claim we first observe that $\lambda > 1 - \sqrt[\alpha-1]{\alpha-1}$ is equivalent to $(1 - \lambda)^{(\alpha-1)}\alpha < 1$. We fix $\ell \in \mathbb{N}$. Using linearity of expectation, the expected number of vertices of degree $\leq \ell$ is given by

$$\alpha^n \sum_{i=0}^{\ell} \binom{(\alpha-1)n}{i} \lambda^i (1-\lambda)^{(\alpha-1)n-i} \leq \ell \left((\alpha-1)n\right)^\ell \alpha^n (1-\lambda)^{(\alpha-1)n-\ell}$$

$$= c'n^\ell \left[\alpha(1-\lambda)^{(\alpha-1)}\right]^n, \quad c' > 0$$

$$\sim e^{-cn}, \quad c > 0.$$

Since we have for any r.v. X with positive integer values: $\mathbb{E}(X) \geq \mathbb{P}(X > 0)$, Claim 1 follows.

According to Claim 1 we can now choose for $v, v' \in \Gamma_n$ with $d(v, v') = k$ and $\ell \in \mathbb{N}$ the two sets of neighbors $\{v^{(j_h)} \mid 1 \leq h \leq \ell\}$ and $\{v'^{(i_h)} \mid 1 \leq h \leq \ell\}$. W.l.o.g. we may assume that $\{j_h\} = \{1, \ldots, \ell\}$ and $\{i_h\} = \{\ell+1, \ldots, 2\ell\}$ and that v, v' differ exactly in the positions $2\ell+1, \ldots, 2\ell+k$. Furthermore we may assume that v, v' and $v^{(i)}, v'^{(\ell+i)}$ differ by 0 and 1 entries, i.e., are of the form

$$v = (\underbrace{0, \ldots, 0}_{\ell}, \underbrace{0, \ldots, 0}_{\ell}, \underbrace{0, \ldots, 0}_{k}, x_{2\ell+k+1}, \ldots, x_n),$$

$$v' = (\underbrace{0, \ldots, 0}_{\ell}, \underbrace{0, \ldots, 0}_{\ell}, \underbrace{1, \ldots, 1}_{k}, x_{2\ell+k+1}, \ldots, x_n),$$

$$v^{(i)} = (\underbrace{0, \ldots, 1, 0, \ldots, 0}_{\text{1 in } i\text{th-position}}, \underbrace{0, \ldots, 0}_{\ell}, \underbrace{0, \ldots, 0}_{k}, x_{2\ell+k+1}, \ldots, x_n),$$

$$v'^{(\ell+i)} = (\underbrace{0, \ldots, 0}_{\ell}, \underbrace{0, \ldots, 1, 0, \ldots, 0}_{\text{1 in } (\ell+i)\text{th-position}}, \underbrace{1, \ldots, 1}_{k}, x_{2\ell+k+1}, \ldots, x_n).$$

For each pair of elements $(v^{(i)}, v'^{(\ell+i)})$ with $1 \leq i \leq \ell$ we consider the sets $\mathsf{B}^{n-(2\ell+k)}(v^{(i)}, 1)$ and $\mathsf{B}^{n-(2\ell+k)}(v'^{(\ell+i)}, 1)$ where

$$\mathsf{B}^{n-(2\ell+k)}(w, 1) = \{e_h + w \mid 2\ell + k < h \leq n\}.$$

$(v^{(i)}, v'^{(\ell+i)})$ is connected by the Q_α^n-path

$$\gamma_i = (v^{(i)}, e_{\ell+i}, \underbrace{e_{2\ell+1}, \ldots, e_{2\ell+k}}_{k}, e_i, v'^{(\ell+i)}), \quad 1 \leq i \leq \ell. \quad (7.19)$$

γ_i is contained in Γ_n with probability at least λ^{k+3}. Since all neighbors of v and v' are of the form $v^{(i)}$, for $1 \leq i \leq \ell$ and $v'^{(\ell+i)}$ for $\ell+1 \leq \ell+i \leq 2\ell$, for $i \neq j$ any two paths

$$\gamma_i = (v^{(i)}, e_{\ell+i}, e_{2\ell+1}, \ldots, e_{2\ell+k}, e_i, v'^{(\ell+i)}), \quad (7.20)$$

$$\gamma_j = (v^{(j)}, e_{\ell+j}, e_{2\ell+1}, \ldots, e_{2\ell+k}, e_j, v'^{(\ell+j)}) \quad (7.21)$$

are vertex disjoint. The probability of selecting a pair of vertices $(v^{(i)} + e_h, v'^{(\ell+i)} + e_h)$ is λ^2. Any two pairs

$$(v^{(i)} + e_h, v'^{(\ell+i)} + e_h), \quad (v^{(i)} + e_{h'}, v'^{(\ell+i)} + e_{h'}), \quad 1 \le i \le \ell, \ h \ne h'$$

have the vertex disjoint paths $\gamma_i + e_h, \gamma_i + e_{h'}$ since $h, h' > 2\ell + k$. Two paths $\gamma_i + e_h$ and $\gamma_j + e_h$ of two pairs

$$(v^{(i)} + e_h, v'^{(\ell+i)} + e_h) \quad \text{and} \quad (v^{(j)} + e_h, v'^{(\ell+j)} + e_h)$$

are in view of eqs. (7.20) and (7.21) also disjoint.

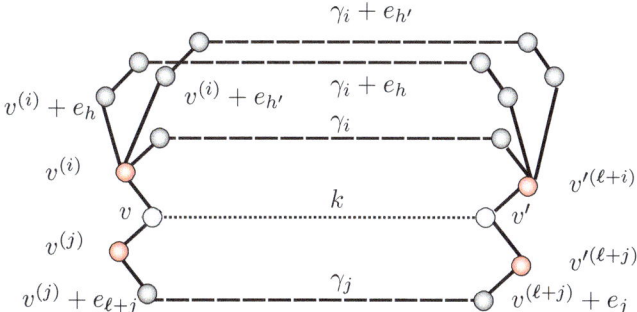

Fig. 7.7. The paths between v and v' in Q_α^n ($d(v, v') = k$): γ_j connects $(v^{(j)}, v'^{(\ell+j)})$ in the form $\gamma_j = (v^{(j)}, e_{\ell+j}, e_{2\ell+1}, \ldots, e_{2\ell+k}, e_j, v'^{(\ell+j)})$; γ_i connects $(v^{(i)}, v'^{(\ell+i)})$ in the same way; $\gamma_i + e_{h'}$ and $\gamma_i + e_h$ are obtained by shifting γ_i.

The expected number of pairs $(v^{(i)} + e_h, v'^{(\ell+i)} + e_h)$, where $1 \le i \le \ell$, $2\ell + k < h$ such that no path $\gamma_i + e_h$ is selected, see Fig. 7.7, is less than

$$\alpha^n \beta_n^{k+3}(1 - \lambda^2 \lambda^{k+3})^{\ell(n-(2\ell+k))} = \alpha^n \beta_n^{k+3}(1 - \lambda^{k+5})^{-\ell(2\ell+k)} \left(1 - \lambda^{k+5}\right)^{\ell n},$$

where $\beta_n = (\alpha - 1)n$. By choosing ℓ large enough we can satisfy

$$(1 - \lambda^{k+5})^\ell < (1 - \lambda)^{(\alpha-1)},$$

which implies

$$((\alpha - 1)n)^{k+3}(1 - \lambda^{k+5})^{-\ell(2\ell+k)} \left[\alpha(1 - \lambda)^{(\alpha-1)}\right]^n$$

which obviously tends to zero. Accordingly, there exists a.s. at least one path of the form $\gamma_i + e_h$ (eq. (7.19)) which connects v and v' in Γ_n and the proof of the lemma is complete.

Theorem 7.15. *Let Q_α^n be a generalized n-cube and $\mathbb{P}$ be the probability*

$$\mathbb{P}(\Gamma_n) = \lambda^{|\Gamma_n|}(1 - \lambda)^{\alpha^n - |\Gamma_n|}.$$

Then the following assertions hold:

$$\lim_{n\to\infty} \mathbb{P}(\Gamma_n \text{ is connected}) = \begin{cases} 0 & for \ \lambda < 1 - \sqrt[\alpha-1]{\alpha^{-1}}, \\ 1 & for \ \lambda > 1 - \sqrt[\alpha-1]{\alpha^{-1}}. \end{cases}$$

An illustration of this result for random-induced subgraphs of Q_2^{15} is given in Fig. 7.8.

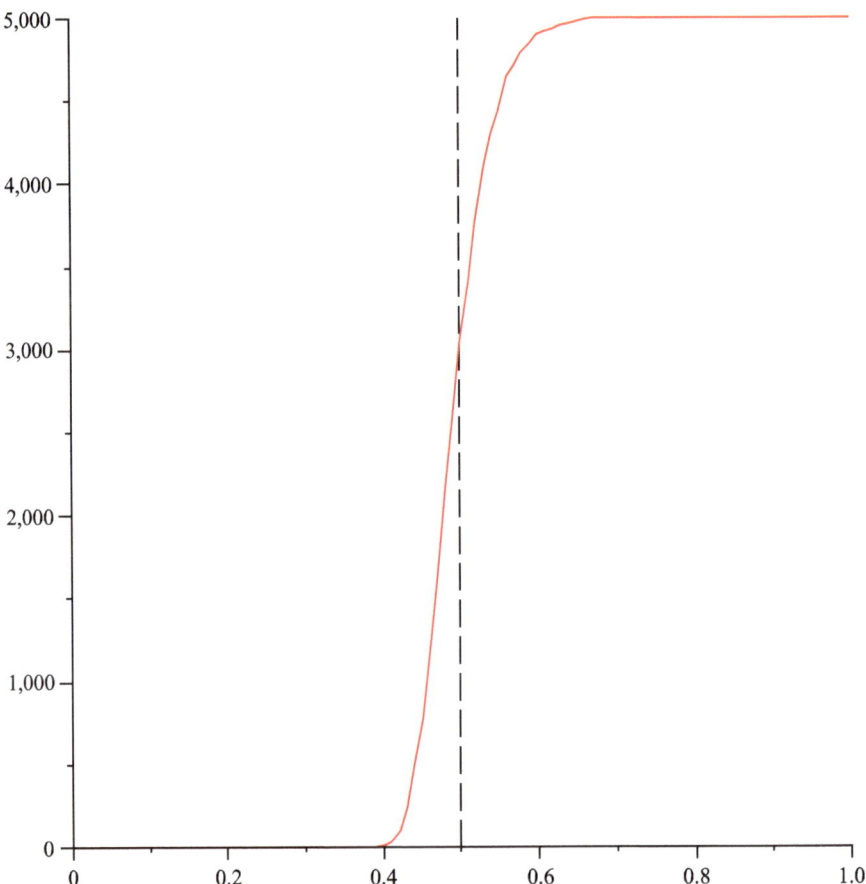

Fig. 7.8. The fraction of connected subgraphs as a function of λ. The data points are based on 5000 randomly generated induced subgraphs of Q_2^{15}.

Proof. Suppose first we have $\lambda > 1 - \sqrt[\alpha-1]{\alpha^{-1}}$. For any two vertices $w, w' \in \Gamma_n$ where, $s = d(w, w')$, we fix a shortest Q_α^n-path, $\gamma_{w,w'}$ connecting them. Let W be the r.v. counting the vertices in Q_α^n that have no Γ_n-neighbor and let w_{i_j} denote the vertex of the jth step of $\gamma_{w,w'}$. Since

$$\mathbb{P}(\mathsf{B}(w_{i_j},1)\cap \Gamma_n=\varnothing,1\le j\le s)\le \mathbb{E}(W)=\alpha^n(1-\lambda)^{(\alpha-1)n},$$

we observe that a.s. for all $1\le j\le s$, $\mathsf{B}(w_{i_j},1)\cap \Gamma_n\ne \varnothing$ holds. Let $a_j\in$ $\mathsf{B}(w_{i_j},1)\cap \Gamma_n$. All pairs (a_j,a_{j+1}) have distance $d(a_j,a_{j+1})\le 3$ and are by Lemma 7.14 a.s. connected. We can therefore select a Γ_n-path, γ_j connecting a_j and a_{j+1}. Concatenating all paths γ_j produces a Γ_n-path connecting w and w' whence for $\lambda>1-\sqrt[\alpha-1]{\alpha-1}$ Γ_n is a.s. connected.

Claim. For $\lambda<1-\sqrt[\alpha-1]{\alpha-1}$ the random graph Γ_n contains a.s. isolated points. We consider $\mathsf{B}(v,1)\subset Q_\alpha^n$ and define I_v to be the indicator r.v. of the event

$$\{\Gamma_n\mid v\in \Gamma_n\ \wedge\ \mathsf{S}(v,1)\cap \Gamma_n=\varnothing\}.$$

Clearly, $\mathbb{P}(I_v=1)=\mathbb{E}(I_v)=\lambda(1-\lambda)^{(\alpha-1)n}$ and we set

$$\Omega=\sum_{\{(v,v')\mid v\ne v',\,\mathsf{B}(v,1)\cap \mathsf{B}(v',1)\ne \varnothing\}}\mathbb{P}(I_v\cdot I_{v'}=1).$$

Suppose for $v\ne v'$, $\mathsf{B}(v,1)\cap \mathsf{B}(v',1)\ne \varnothing$. Then either $d(v,v')=1$ and $|\mathsf{B}(v,1)\cap \mathsf{B}(v',1)|=\alpha$, in which case $I_v\cdot I_{v'}=0$, or $d(v,v')=2$ and $|\mathsf{B}(v,1)\cap \mathsf{B}(v',1)|=2$. Therefore

$$\mathsf{B}(v,1)\cap \mathsf{B}(v',1)\ne \varnothing\quad\Longrightarrow\quad \mathbb{P}(I_v\cdot I_{v'}=1)=\lambda^2(1-\lambda)^{2(\alpha-1)n-2}.$$

Set $Z=\sum_{v\in Q_\alpha^n}I_v$ then $\mathbb{E}(Z)=\alpha^n\lambda(1-\lambda)^{(\alpha-1)n}$. Since $\lambda<1-\sqrt[\alpha-1]{\alpha-1}$ we have $\mathbb{E}(Z)\sim e^{cn}$, for $c>0$. We next compute

$$\Omega=\alpha^n(\alpha-1)^2\binom{n}{2}\lambda^2(1-\lambda)^{-2}\left[(1-\lambda)^{(\alpha-1)n}\right]^2$$

$$=(\alpha-1)^2\binom{n}{2}(1-\lambda)^{-2}\lambda(1-\lambda)^{(\alpha-1)n}\,\mathbb{E}(Z)$$

$$\sim e^{-c'n}\mathbb{E}(Z),\quad c'>0.$$

Janson's inequality (Theorem 2.31) guarantees

$$\mathbb{P}(Z\le (1-\gamma)\mathbb{E}[Z])\le e^{-\frac{\gamma^2\mathbb{E}[Z]}{2+2\Omega/\mathbb{E}[Z]}}. \tag{7.22}$$

Equation (7.22) shows that a.s. the r.v. Z cannot be smaller than $(1-\gamma)\mathbb{E}[Z]$ for any $\gamma>0$, which implies that Γ_n contains a.s. isolated points. Therefore, we have proved that for $\lambda<1-\sqrt[\alpha-1]{\alpha-1}$ Γ_n is not connected.

7.5 Exercises

7.1. (Intersection theorem) Given a set of $\ell>2$ different k-noncrossing, σ-canonical structures, $M=\{S_1,\dots,S_\ell\}$:

- Does there always exist some sequence that is compatible to all of them?
- Given an arbitrary tangle and assume that Watson–Crick as well as **G-U** base pairs are the base pairing rules for all arcs. Is there always a sequence compatible to a tangle?
- Consider the graph over $[n]$ obtained by taking the union of all arcs contained in the structures $S_1, \ldots, S_\ell$. Examine which graph properties guarantee that the intersection theorem holds.

7.2. Prove:

Lemma 7.16. *Let $v \in Q_2^n$ be a fixed vertex. Let $C_v(s, m)$ be the set of connected induced subgraphs, C_v, that contain v, have size s and a boundary of size $|d(C_v)| = m$. Let $c_v(s, m)$ denote the cardinality of $C_v(s, m)$. Then we have*

$$c_v(s, m) \leq \frac{(s + m)^{s+m}}{s^s m^m}.$$

7.3. Prove:

Proposition 7.17. *Suppose $\lambda_n = \frac{1+\epsilon}{n}$ and ω_n tends to zero arbitrarily slowly as n tends to infinity. Then for arbitrary but fixed $k \in \mathbb{N}$, $\nu_n = \lfloor \frac{\omega_n n}{2k(k+1)} \rfloor$ and $\varphi_n = \pi(\epsilon)\nu_n(1 - e^{-(1+\epsilon)\omega_n/4})$, where $\pi(\epsilon) > 0$ there exists $\rho_k > 0$ such that each Γ_n-vertex is with probability at least*

$$\pi_k(\epsilon) = \pi(\epsilon)\left(1 - e^{-\rho_k \varphi_n}\right)$$

contained in a Γ_n-subcomponent of size at least $c_k (\omega_n n)\varphi_n^k$, where $c_k > 0$ and

$$|\Gamma_{n,k}| \sim \pi(\epsilon)|\Gamma_n| \qquad a.s.$$

7.4. Prove: For $o(1) = \chi_n \geq n^{-\frac{1}{3}+\delta}$, $\Gamma_{n,k}$ is a.s. 4-dense in Q_2^n.

7.5. Prove: for $\chi_n \geq n^{-\frac{1}{3}+\delta}$, the largest component, $C_n^{(1)}$, is a.s. 4-dense in Q_2^n.

7.6. (Reidys [104]) Let $b, d \in \mathbb{N}$, $b, d \geq 2$, v, v' be arbitrary but fixed Q_b^n-vertices, having distance $d_{Q_b^n}(v, v') = d$ and $n' = n - d$. Suppose we select Q_b^n-vertices with the probability $0 < \lambda < 1$. Then there exists a Γ_n-path connecting v and v' of length exactly $2 + d$ with probability at least

$$\sigma_{\lambda, d}^{[b]}(n) = 1 - \exp\left(-\frac{(b-1)n'\lambda^{2+(d-1)}}{4}\right),$$

provided v, v' are contained in Γ_n.

Note that this problem "almost" implies the connectivity theorem for random subgraphs of n-cubes. In order to recover the connectivity theorem we only need to observe that at the threshold any Γ_n-vertex has arbitrarily large *finite* degree. This allows us to employ the above statement "in parallel" for each of those vertices; see [106] for details.

7.7. (Reidys [104]) Let v, v' be arbitrary but fixed Q_2^n-vertices, having distance $d_{Q_2^n}(v, v') = d$, $d \geq 2$, and $n' = n - (d-1)$. Suppose we select Q_2^n-vertices with the probability $0 < \lambda < 1$. Then there exists a Γ_n-path connecting v and v' of length exactly $4 + d$ with probability at least

$$\tau_{\lambda, d}(n) = 1 - \exp\left(-\left[\frac{2}{\lambda^2 \left[\frac{n'-2}{n'-1}\right] n'} + \frac{2(2 + \lambda^2)}{n'(n'-1)\lambda^{4+(d-1)}}\right]^{-1}\right),$$

provided v, v' are contained in Γ_n.

References

1. M. Abramowitz and I.A. Stegun, editors. *Handbook of Mathematical Functions with Formulas, Graphs, and Mathematical Tables.* 55. NBS Applied Mathematics, Dover, NY, 1964.
2. M. Ajtai, J. Komlós, and E. Szemerédi. Largest random component of a k-cube. *Combinatorica*, 2:1–7, 1982.
3. T. Akutsu. Dynamic programming algorithms for RNA secondary structure prediction with pseudoknots. *Discr. Appl. Math.*, 104:45–62, 2000.
4. D. Aldous and P. Diaconis. Strong uniform times and finite random walks. *Adv. Appl. Math.*, 2:69–97, 1987.
5. D. André. Solution directe du problème résolu par M. Bertrand,. *C R d Acad Sci*, 105:436–437, 1887.
6. L. Babai. Local expansion of vertex transitive graphs and random generation in finite groups. *Proc 23 ACM Symp Theory Comput (ACM New York)*, 1: 164–174, 1991.
7. L. Babai and V.T. Sos. Sidon sets in groups and induced subgraphs of Cayley graphs. *Eur. J. Combinator.*, 1:1–11, 1985.
8. P. Babitzke and C. Yanofsky. Reconstitution of *bacillus subtilis* trp attenuation in vitro with trap, the trp RNA-binding attenuation protein. *Proc. Natl. Acad. Sci. USA*, 90:133–137, 1990.
9. R.T. Batey, R.P. Rambo, and J.A. Doudna. Tertiary motifs in RNA structure and folding. *Angew. Chem.*, 38:2326–2343, 1999.
10. T. Baumstark, A.R. Schroder, and D. Riesner. Viroid processing: Switch from cleavage to ligation is driven by a change from a tetraloop to a loop e conformation. *EMBO. J.*, 16:599–610, 1997.
11. E.A. Bender. Central and local limit theorem applied to asymptotic enumeration. *J. Comb. Theory, Ser. A*, 15:91–111, 1973.
12. C.K. Biebricher, S. Diekmann, and R. Luce. Structural analysis of self-replicating RNA synthesized by qb replicase. *J. Mol. Biol.*, 154:629–648, 1982.
13. C.K. Biebricher and R. Luce. In vitro recombination and terminal elongation of RNA by qb replicase. *EMBO. J.*, 11:5129–5135, 1992.
14. B. Bollobás, Y. Kohayakawa, and T. Luczak. The evolution of random subgraphs of the cube. *Random Struct. Algorithms*, 3:55–90, 1992.

C. Reidys, *Combinatorial Computational Biology of RNA*,
DOI 10.1007/978-0-387-76731-4,
© Springer Science+Business Media, LLC 2011

15. B. Bollobás, Y. Kohayakawa, and T. Luczak. On the evolution of random boolean functions. *in: P. Frankl, Z. Füredi, G. Katona, D. Miklós (eds), External Probls for Finite Sets (Visegrád), 3 of Bolyai Soc mathematical Studies*, pages 137–156, János Bolyai Mathematical Society, Budapest, 1994.

16. C. Borgs, J.T. Chayes, H. Remco, G. Slade, and J. Spencer. Random subgraphs of finite graphs: III. the phase transition for the n-cube. *Combinatorica*, 26: 359–410, 2006.

17. M. Bousquet-Mélou and G. Xin. On partitions avoiding 3-crossings. *Séminaire Lotharingien de Combinatoire*, 54, 2006.

18. P. Brion and E. Westhof. Hierarchy and dynamics of RNA folding. *Annu. Rev. Biophys. Biomol. Struct.*, 26:113–137, 1997.

19. J.D. Burtin. The probability of connectedness of a random subgraph of an n-dimensional cube. *Probl Infom Transm*, 13:147–152, 1977.

20. N.T. Cameron and L. Shapiro. Random walks, trees and extensions of RIORDAN group techniques. *Talk, in: Annual Joint Mathematics Meetings, Baltimore, MD,* 2003.

21. S. Cao and S.J. Chen. Predicting RNA pseudoknot folding thermodynamics. *Nucl. Acids. Res.*, 34(9):2634–2652, 2006.

22. R. Cary and G. Stormo. Graph-theoretic approach to RNA modeling using comparative data. *Proc. Int. Conf. Intell. Syst. Mol. Biol.*, 3:75–80, 1995.

23. M. Chamorro, N. Parkin, and H.E. Varmus. An RNA pseudoknots and an optimal heptameric shift site are required for highly efficient ribosomal frameshifting on a retroviral messenger RNA. *Proc. Natl. Acad. Sci. USA*, 89, 1992.

24. M. Chastain and I. Tinoco. A base-triple structural domain in RNA. *Biochemistry*, 31:12733–12741, 1992.

25. W.Y.C. Chen, E.Y.P. Deng, R.R.X. Du, R.P. Stanley, and C.H. Yan. Crossing and nesting of matchings and partitions. *Trans. Amer. Math. Soc.*, 359: 1555–1575, 2007.

26. W.Y.C. Chen, H.S.W. Han, and C.M. Reidys. Random k-noncrossing RNA structures. *Proc. Natl. Acad. Sci. USA*, 106(52):22061–22066, 2009.

27. W.Y.C. Chen, J. Qin, and C.M. Reidys. Crossing and nesting in tangled-diagrams. *Elec. J. Comb.*, 15, 86, 2008.

28. W.Y.C. Chen, J. Qin, C.M. Reidys, and D. Zeilberger. Efficient counting and asymptotics of k-noncrossing tangled-diagrams. *Elec. J. Comb.*, 16(1), 37, 2008.

29. C. DeLisi and D.M. Crothers. Prediction of RNA secondary structures. *Proc. Natl. Acad. Sci, USA*, 68:2682–2685, 1971.

30. E. Deutsch and L. Shapiro. A survey of the fine numbers. *Discrete Math.*, 241:241–265, 2001.

31. R.M. Dirks and N.A. Pierce. An algorithm for computing nucleic acid base-pairing probabilities including pseudoknots. *J. Comput. Chem.*, 25(10): 1295–1304, 2004.

32. J.A. Doudna and T.R. Cech. The chemical repertoire of natural ribozymes. *Nature*, 418(11), 2002.

33. P. Duchon, P. Flajolet, G. Louchard, and G. Schaeffer. Boltzmann samplers for the random generation of combinatorial structures. *Combin. Probab. Comput.*, 13:577–625, 2004.

34. S.R. Eddy. How do RNA folding algorithms work? *Nature Biotechnology*, 22:1457–1458, 2004.

35. J. Edmonds. Maximum matching and polyhedron with 0, 1-vertices. *J. Res. Nat. Bur. Stand.*, 69B:125–130, 1965.

36. G. P. Egorychev. *Integral Representation and the computation of combinatorial sums*, volume 59. American Mathematical Society, NY, 1984.

37. V.L. Emerick and S.A. Woodson. Self-splicing of the *tetrahymena* pre-rrna is decreased by misfolding during transcription. *Biochemistry*, 32:14062–14067, 1993.

38. P. Erdős and J. Spencer. The evolution of the *n*-cube. *Comput. Math. Appl.*, 5:33–39, 1979.

39. G. Fayat, F.J. Mayaux, C. Sacerdot, M. Fromant, M. Springer, M. Grunberg-Manago, and S. Blanquet. *Escherichia coli* phenylalanyl-trna synthetase operon region: Evidence for an attenuation mechanism and identification of the gene for the ribosomal protein l20. *J. Mol. Biol.*, 171:239–261, 1983.

40. W. Feller. *An introduction to probability theory and its application*. Addison-Wesley Publishing Company Inc., NY, 1991.

41. P. Flajolet, J.A. Fill, and N. Kapur. Singularity analysis, hadamard products, and tree recurrences. *J. Comp. Appl. Math.*, 174:271–313, 2005.

42. P. Flajolet and R. Sedgewick. *Analytic Combinatorics*. Cambridge University Press, Cambridge, England, 2009.

43. C. Flamm, W. Fontana, I.L. Hofacker, and P. Schuster. RNA folding kinetics at elementary step resolution. *RNA*, 6:325–338, 2000.

44. C. Flamm, I.L. Hofacker, S. Maurer-Stroh, P.F. Stadler, and M. Zehl. Design of multistable RNA molecules. *RNA*, 7:254–265, 2001.

45. J.R. Fresco, A. Adains, R. Ascione, D. Henley, and T. Lindahl. Tertiary structure in transfer ribonucleic acids. *Cold Spring Harbor Symp. Quant. Biol.*, 31:527–539, 1966.

46. J.R. Fresco, B.M. Alberts, and P. Doty. Some molecular details of the secondary structure of ribonucleic acid. *Nature*, 188:98–101, 1960.

47. H.N. Gabow. An efficient implementation of Edmonds' algorithm for maximum matching on graphs. *J. Asc. Com. Mach.*, 23:221–234, 1976.

48. I.M. Gessel and X.G. Viennot. Determinants, paths, and plane partitions. *preprint*, 1989.

49. I.M. Gessel and D. Zeilberger. Random walk in a Weyl chamber. *Proc. Am. Math. Soc.*, 115:27–31, 1992.

50. U. Göbel. *Neutral Networks of Minimum Free Energy RNA Secondary Structures*. PhD thesis, University of Vienna, 2000.

51. U. Goebel and C.V. Forst. RNA pathfinder–global properties of neutral networks. *Zeitschrift fuer physikalische Chemie*, 216, 2002.

52. D. Gouyou-Beauschamps. Standard young tableaux of height 4 and 5. *Europ. J. Combin.*, 10:69–82, 1989.

53. D.J. Grabiner and P. Magyar. Random walks in Weyl chambers and the decomposition of tensor powers. *J. Algebr. Combinator.*, 2:239–260, 1993.

54. L.C. Grove and C.T. Benson. *Finite reflection groups*. Springer, New York, 1985.

55. W. Grüner, R. Giegerich, D. Strothmann, C.M. Reidys, J. Weber, I.L. Hofacker, P.F. Stadler, and P. Schuster. Analysis of RNA sequence structure maps by exhaustive enumeration I. structures of neutral networks and shape space covering. *Chem. Mon.*, 127:355–374, 1996.

56. W. Grüner, R. Giegerich, D. Strothmann, C.M. Reidys, J. Weber, I.L. Hofacker, P.F. Stadler, and P. Schuster. Analysis of RNA sequence structure maps by exhaustive enumeration II. structures of neutral networks and shape space covering. *Chem. Mon.*, 127:375–389, 1996.

57. A.P. Gultyaev, F.H. Batenburg, and C.W. Pleij. Dynamic competition between alternative structures in viroid rnas simulated by an RNA folding algorithm. *J. Mol. Biol.*, 276:43–55, 1998.

58. Chernoff. H. A measure of the asymptotic efficiency for tests of a hypothesis based on the sum of observations. *Ann. Math. Stat.*, 23:493–509, 1952.

59. H.S.W. Han and C.M. Reidys. Pseudoknot RNA structures with arc-length ≥ 4. *J. Comp. bio.*, 9(15):1195–1208, 2008.

60. H.S.W. Han and C.M. Reidys. Stacks in canonical RNA pseudoknot structures. *Math. Bioscience*, 219, Issue 1:7–14, 2009.

61. L.H. Harper. Minimal numberings and isoperimetric problems on cubes. *Theory of Graphs, International Symposium, Rome*, 1966.

62. T.E. Harris. *The Theory of Branching Processes*. Springer, 1963.

63. C. Haslinger and P.F. Stadler. RNA structures with pseudo-knots. *Bull. Math. Biol.*, 61:437–467, 1999.

64. E.R. Hawkins, Chang S.H., and W.L. Mattice. Kinetics of the renaturation of yeast trnaleu3. *Biopolymers*, 16:1557–1566, 1977.

65. R. Hecker, Z.M. Wang, G. Steger, and D. Riesner. Analysis of RNA structures by temperature-gradient gel electrophoresis: Viroid replication and processing. *Gene*, 72:59–74, 1988.

66. P. Henrici. *Applied and Computational Complex Analysis*, volume 2. John Wiley, 1974.

67. I.L. Hofacker. Vienna RNA secondary structure server. *Nucl. Acids. Res.*, 31(13):3429–3431, 2003.

68. I.L. Hofacker, W. Fontana, P.F. Stadler, L.S. Bonhoeffer, M. Tacker, and P. Schuster. Fast folding and comparison of RNA secondary structures. *Monatsh. Chem.*, 125:167–188, 1994.

69. I.L. Hofacker, P. Schuster, and P.F. Stadler. Combinatorics of RNA secondary structures. *Discr. Appl. Math.*, 88:207–237, 1998.

70. J.A. Howell, T.F. Smith, and M.S. Waterman. Computation of generating functions for biological molecules. *SIAM: SIAM J Appl Math.*, 39:119–133, 1980.

71. F.W.D. Huang, L.Y.M. Li, and C.M. Reidys. Sequence-structure relations of pseudoknot RNA. *BMC Bioinformatics*, 10, Suppl 1, S39, 2009.

72. F.W.D. Huang, W.W.J. Peng, and C.M. Reidys. Folding 3-noncrossing RNA pseudoknot structures. *J. Comp. Biol.*, 16(11):1549–1575, 2009.

73. F.W.D. Huang and C.M. Reidys. Statistics of canonical RNA pseudoknot structures. *J. Theor. Biol.*, 253:570–578, 2008.

74. N. Iwahori. On the structure of a hecke ring of a chevalley group over a finite field. *J. Fac. Sci. Univ. Tokyo*, 10:215–236, 1964.

75. S. Janson. Poisson approximation for large deviations. *Random Struct. Algorithms*, 1:221–229, 1990.

76. E.Y. Jin, J. Qin, and C.M. Reidys. Combinatorics of RNA structures with pseudoknots. *Bull. Math. Biol.*, 70:45–67, 2008.

77. E.Y. Jin and C.M. Reidys. Asymptotic enumeration of RNA structures with pseudoknots. *Bull. Math. Biol.*, 70:951–970, 2008.

78. E.Y. Jin and C.M. Reidys. Combinatorial design of pseudoknot RNA. *Adv. Appl. Math.*, 42:135–151, 2009.

79. E.Y. Jin and C.M. Reidys. RNA pseudoknots structures with arc length-length ≥ 3 and stack-length-length ≥ σ. *Discr. Appl. Math.*, 158:25–36, 2010.

80. E.Y. Jin, C.M. Reidys, and R.R. Wang. Asymptotic analysis of *k*-noncrossing matchings. arXiv:0803.0848, 2008.

81. I.T. Jun, O.C. Uhlenbeck, and M.D. Levine. Estimation of secondary structure in ribonucleic acids. *Nature*, 230:362 – 367, 1971.

82. D. Kleitman. Proportions of irreducible diagrams. *Studies in Appl. Math.*, 49:297–299, 1970.

83. V.F. Kolchin. *Random Mappings.* Number 14 in Translations Series. Optimization Software, New York, 1986.

84. D.A.M. Konings and R.R. Gutell. A comparison of thermodynamic foldings with comparatively derived structures of 16s and 16s-like rRNAs. *RNA*, 1: 559–574, 1995.

85. J.S. Lodmell and Dahlberg A.E. A conformational switch in Escherichia coli 16s ribosomal RNA during decoding of messenger RNA. *Science*, 277: 1262–1267, 1997.

86. A. Loria and T. Pan. Domain structure of the ribozyme from eubacterial ribonuclease. *RNA*, 2:551–563, 1996.

87. R.B. Lyngsø and C.N.S. Pedersen. RNA pseudoknot prediction in energy-based models. *J. Comput. Biol.*, 7:409–427, 2000.

88. G. Ma and C.M. Reidys. Canonical RNA pseudoknot structures. *J. Comput. Biol.*, 15:1257–1273, 2008.

89. D.H. Mathews, J. Sabina, M. Zuker, and D.H. Turner. Expanded sequence dependence of thermodynamic parameters improves prediction of RNA secondary structure. *J. Mol. Biol.*, 288:911–940, 1999.

90. M.V. Meshikov. Coincidence of critical points in percolation problems. *Soviet Mathematics, Doklady*, 33:856–859, 1986.

91. D. Metzler and M. E. Nebel. Predicting RNA secondary structures with pseudoknots by mcmc sampling. *J. Math. Biol.*, 56(1–2):161–181, 2008.

92. S.G. Mohanty. *Lattice path counting and Applications.* Academic Press, NY, 1979.

93. M. Molloy and B. Reed. The size of the giant component of a random graph with given degree sequence. *Combin. Probab. Comput.*, 7:295–305, 1998.

94. M. E. Nebel. Combinatorial properties of RNA secondary structures. *J. Comp. Biol.*, 9(3):541–574, 2003.

95. M.E. Nebel, C.M. Reidys, and R.R. Wang. Loops in canonical RNA pseudoknot structures. *arXiv:0912.0429*, 2009.

96. R. Nussinov and A.B. Jacobson. Fast algorithm for predicting the secondary structure of single-stranded RNA. *Proc. Natl. Acad. Sci., USA*, 77:6309–6313, 1980.

97. R. Nussinov, G. Pieczenik, J.R. Griggs, and D. Kleitman. Algorithms for loop matchings. *SIAM J. of Appl. Math.*, 35:68–82, 1978.

98. A.M. Odlyzko. *Handbook of combinatorics.* Elsevier, Amsterdam, 2005.

99. T. Pan, D. Thirumalai, and S.A. Woodson. Folding of RNA involves parallel pathways. *J. Mol. Biol.*, 273:7–13, 1997.

100. A.T. Perrotta and M.D. Been. A toggle duplex in hepatitis delta virus self-cleaving RNA that stabilizes an inactive and a salt-dependent proactive ribozyme conformation. *J Mol Biol*, 279:361–373, 1998.

101. J. Reeder and R. Giegerich. Design, implementation and evaluation of a practical pseudoknot folding algorithm based on thermodynamics. *BMC Bioinformatics*, 5(104):1–12, 2004.

102. C.M. Reidys. Random induced subgraphs of generalized n-cubes. *Adv. Appl. Math.*, 19:360–377, 1997.

103. C.M. Reidys. Distance in random induced subgraphs of generalized n-cubes. *Combinator. Probab. Comput.*, 11:599–605, 2002.

104. C.M. Reidys. Local connectivity of neutral networks. *Bull. Math. Biol.*, 71: 265–290, 2008. in press.

105. C.M. Reidys. The largest component in random induced subgraphs of n-cubes. *Discr. Math.*, 309, Issue 10:3113–3124, 2009.

106. C.M. Reidys, P.F. Stadler, and P.K. Schuster. Generic properties of combinatory maps and neutral networks of RNA secondary structures. *Bull. Math. Biol.*, 59(2):339–397, 1997.

107. C.M. Reidys and R.R. Wang. Shapes of RNA pseudoknot structures. *J. Comp. Biol.*, 2009. to appear.

108. C.M. Reidys, R.R. Wang, and A.Y.Y. Zhao. Modular, k-noncrossing diagrams. *Electron. J. Combin.*, 17:76, 2010.

109. J. Ren, B. Rastegari, A. Condon, and H. Hoos. Hotknots: Heuristic prediction of RNA secondary structures including pseudoknots. *RNA*, 11:1494–1504, 2005.

110. M. Renault. Lost (and found) in translation: André's actual method and its application to the generalized ballot problem. *Amer. Math. Monthly.*, 115: 358–363, 2008.

111. E. Rivas and S.R. Eddy. A dynamic programming algorithm for RNA structure prediction including pseudoknots. *J. Mol. Biol.*, 285:2053–2068, 1999.

112. E. Rivas and S.R. Eddy. The language of RNA: A formal grammar that includes pseudoknots. *Bioinformatics*, 16:326–333, 2000.

113. J. Ruan, G. Stormo, and W. Zhang. An iterated loop matching approach to the prediction. *Bioinformatics*, 20:58–66, 2004.

114. B. Salvy and P. Zimmerman. Gfun: a maple package for the manipulation of generating and holonomic functions in one variable. *ACM TOMS*, 20:163–177, 1994.

115. C.E. Schensted. Longest increasing and decreasing subsequences. *Canad. J. Math.*, 13:179–191, 1961.

116. W.R. Schmitt and M.S. Waterman. Linear trees and RNA secondary structure. *Discr. Appl. Math.*, 51:317–323, 1994.

117. E.A. Schultes and P.B. Bartels. One Sequence, Two Ribozymes: Implications for the Emergence of New Ribozyme Folds. *Science*, 289:448–452, 2000.

118. P. Schuster, W. Fontana, P. F. Stadler, and I.L. Hofacker. From sequences to shapes and back: A case study in RNA secondary structures. *Proc. Roy. Soc. Lond. B*, 255:279–284, 1994.

119. Mapping RNA form and function. *Science*, 2, 2005.

120. D.B. Searls. The language of genes. *Nature*, 420:211–217, 2002.

121. L. Shapiro, S. Getu, W. Woan, and L. Woodson. The RIORDAN group. *Discr. Appl. Math.*, 34:229–239, 1991.

122. L.X. Shen, Z. Cai, and I. Tinoco. RNA structure at high resolution. *FASEB*, 9:1023–1033, 1995.

123. G.A. Soukup and R.R. Breaker. Engineering precision RNA molecular switches. *Proc. Natl. Acad. Sci. USA*, 96:3584–3589, 1999.

124. P. Stadler. private communication. unpublished.

125. R.P. Stanley. Differentiably finite power series. *Eur. J. Combinator.*, 1: 175–188, 1980.

126. R.P. Stanley. *Enumerative Combinatorics*, volume 1. Cambridge University Press, Cambridge, England, 2000.

127. R.P. Stanley. *Enumerative Combinatorics*, volume 2. Cambridge University Press, Cambridge, England, 2000.

128. D.W. Staple and S.E. Butcher. Pseudoknots: RNA structures with diverse functions. *PLoS Biol. 3*, 6:956–959, 2005.

129. S. Sundaram. The Cauchy identity for Sp(2n). *J. Comb. Theory, Ser. A*, 53:209–238, 1990.

130. J. Tabaska, R. Cary, H. Gabow, and G. Stormo. An RNA folding method capable of identifying pseudoknots and base triples. *Bioinformatics*, 14: 691–699, 1998.

131. E. ten Dam, I. Brierly, S. Inglis, and C. Pleij. Identification and analysis of the pseudoknot containing *gag-pro* ribosomal frameshift signal of simian retrovirus-1. *Nucl. Acids Res.*, 22:2304–2310, 1994.

132. I. Tinoco, P.N. Borer, B. Dengler, M.D. Levine, O.C. Uhlenbeck, D.M. Crothers, and J. Gralla. Improved estimation of secondary structure in ribonucleic acids. *Nat. New Bio.*, 246:40–41, 1973.

133. I.Jr. Tinoco and C. Bustamante. How RNA folds. *J. Mol. Biol.*, 293:271–281, 1999.

134. E.C. Titchmarsh. *The theory of functions*. Oxford University Press, NY, 1939.

135. D.K. Treiber, M.S. Rook, P.P. Zarrinkar, and J.R. Williamson. Kinetic intermediate strapped by native interactions in RNA folding. *Science*, 279:1943–1946, 1998.

136. C. Tuerk, S. MacDougal, and L. Gold. RNA pseudoknots that inhibit human immunodeficiency virus type 1 reverse transcriptase. *Proc. Natl. Acad. Sci. USA*, 89, 1992.

137. Y. Uemura, A. Hasegawa, S. Kobayashi, and T. Yokomori. Tree adjoining grammars for RNA structure prediction. *Theor. Comput. Sci.*, 210:277–303, 1999.

138. U. von Ahsen. Translational fidelity: Error-prone versus hyperaccurate ribosomes. *Chem. Biol.*, 5:R3–R6, 1998.

139. B. Voß, R. Giegerich, and M. Rehmsmeier. Complete probabilistic analysis of RNA shapes. *BMC Biology*, 5(4):1–23, 2006.

140. W. Wasow. *Asymptotic Expansions for Ordinary Differential Equations*. Dover, 1987.

141. M.S. Waterman. Combinatorics of RNA hairpins and cloverleafs. *Stud. Appl. Math.*, 60:91–96, 1978.

142. M.S. Waterman. Secondary structure of single - stranded nucleic acids. *Adv. Math.I (suppl.)*, 1:167–212, 1978.

143. M.S. Waterman. Combinatorics of RNA hairpins and cloverleafs. *Stud. Appl. Math.*, 60:91–96, 19790.

144. M.S. Waterman and T.F. Smith. Rapid dynamic programming algorithms for RNA secondary structure. *Adv. Appl. Math.*, 7:455–464, 1986.

145. E. Westhof and L. Jaeger. RNA pseudoknots. *Curr. Opin. Chem. Biol.*, 2: 327–333, 1992.

146. H.S. Wilf. A unified setting for sequencing, ranking, and selection algorithms for combinatorial objects. *Adv. Math.*, 24:281–291, 1977.

147. H.S. Wilf. Combinatorial algorithms. *Academic Press*, NY, 1978.

148. R. Wong and M. Wyman. The method of Darboux. *J. Appr. Theor.*, 10: 159–171, 1974.

149. D. Zeilberger. A holonomic systems approach to special functions identities. *J. Comput. Appl. Math.*, 32:321–368, 1990.

150. M. Zuker and D. Sankoff. RNA secondary structures and their prediction. *Bull. Math. Biol.*, 46(4):591–621, 1984.

Index